D1428892

Crustacean Farming

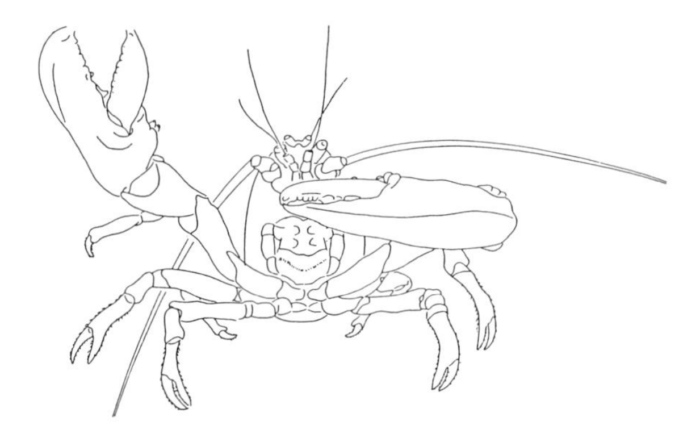

D.O'C. LEE and J.F. WICKINS

OXFORD
BLACKWELL SCIENTIFIC PUBLICATIONS
LONDON EDINBURGH BOSTON
MELBOURNE PARIS BERLIN VIENNA

Blackwell Scientific Publications
Editorial offices:
Osney Mead, Oxford OX2 0EL
25 John Street, London WC1N 2BL
23 Ainslie Place, Edinburgh EH3 6AJ
3 Cambridge Center, Cambridge,
Massachusetts 02142, USA
54 University Street, Carlton
Victoria 3053, Australia

Other Editorial Offices:
Arnette SA
2, rue Casimir-Delavigne
75006 Paris
France

Blackwell Wissenschaft
Meinekestrasse 4
D-1000 Berlin 15
Germany

Blackwell MZV
Feldgasse 13
A-1238 Wien
Austria

First published 1992

Printed and bound in Great Britain by
The University Press, Cambridge

DISTRIBUTORS

Marston Book Services Ltd
PO Box 87
Oxford OX2 0DT
(*Orders*: Tel: 0865 791155
Fax: 0865 791927
Telex: 837515)

USA
Blackwell Scientific Publications, Inc.
3 Cambridge Center
Cambridge, MA 02142
(*Orders*: Tel: 800 759-6102)

Canada
Oxford University Press
70 Wynford Drive
Don Mills
Ontario M3C 1J9
(*Orders*: Tel: 416 441-2941)

Australia
Blackwell Scientific Publications
(Australia) Pty Ltd
54 University Street
Carlton, Victoria 3053
(*Orders*: Tel: 03 347-0300)

British Library
Cataloguing in Publication Data
Lee, D. O'C.
Crustacean farming.
I. Title II. Wickins, J.F.
639.541

ISBN 0-632-02974-9

Library of Congress
Cataloging in Publication Data
Lee, D.O'C. (Daniel O'C.)
Crustacean farming/D.O'C. Lee and J.F. Wickins.
p. cm.
Includes bibliographical references and index.
ISBN 0-632-02974-9
1. Shellfish culture. 2. Shellfish trade.
3. Crustacea—Economic aspects.
I. Wickins, J.F. (John F.) II. Title.
SH370.L44 1991
338.3'715—dc20 91-19458
CIP

Contents

Acknowledgements

We are greatly indebted to numerous specialists, researchers and colleagues who provided us with original information, pre-publication manuscripts and encouragement during the preparation of this book:

Dean M. Akiyama, American Soybean Assoc., Singapore;
Adnan al Hajj, Consultant, Guayaquil, Ecuador;
Geoff Allan, Department of Agriculture, New South Wales;
Chris Austin, Queensland Institute of Technology, Queensland;
Conner Bailey, College of Agriculture, Auburn University, Alabama;
Colin Bannister, Ministry of Agriculture, Fisheries and Food, Lowestoft;
Matt Briggs and Janet Brown, Institute of Aquaculture, University of Stirling, Scotland;
James Brock, Aquaculture Veterinarian, Kailua, Hawaii;
Craig Burton, Sea Fish Industry Authority, Argyll, Scotland;
Pedro Cañavate, PEMARES, Cadiz, Spain;
Chau-Ling Chan, Lancaster Polytechnic, Lancashire;
Bill Cook and Pete Oxford, North West and North Wales Sea Fisheries Committee, Lancaster University, Lancaster;
Imre Csavas, Regional Aquaculture Officer, FAO Regional Office for Asia and the Pacific, Bangkok;
David Currie, Thai-Hawaiian Hatcheries, Bangkok;
Patrick Franklin, Macallister Elliot and Partners, Lymington, Hampshire;
Peter Fuke, Consultant, Chelmsford, Essex;
Dennis Hedgecock, Bodega Bay Laboratory, University of California;
Yves Henocque, Station Marine d'Endoume, Marseille;
David Holditch, Dept. of Zoology, University of Nottingham, Nottingham;
D.S. Holker, Secretary, Marron Growers Association of Western Australia;
Jay Huner, Dept. of Agricultural Sciences, University of south-western Louisiana;
Ray Ingle, The Natural History Museum, London;
Clive Jones, Dept. of Primary Industries, Queensland;
Ilan Karplus, Aquaculture Research Organisation, Beit-Dagan, Israel;
Max Keith and Peter Wood, Frippak Feeds, Aberdeen;
Jean-François LeBitoux, Centre Aquacole, Leucate;
Hervé Lucien-Brun, Sepia International, Paris;
Greg Maguire, Tasmanian State Institute of Technology, Tasmania;
Gay Marsden, Dept. of Primary Industries, Queensland;
Ronald D. Mayo, The Mayo Associates, Seattle, Washington;
Corny Mock, Cornelius Mock and Associates, Galveston, Texas;
Noel Morrissy, Fisheries Department, Western Australian Marine Research Laboratories;
Colin Nash, FAO, Rome;

Michael New, ASEAN EEC Aquaculture Development and Coordination Programme, Bangkok;
Paul Niemeier, National Marine Fisheries Service, Silver Spring, Maryland;
David O'Sullivan, Editor, Austasia Aquaculture Magazine;
Ian H. Pike, International Association of Fish Meal Manufacturers, Potters Bar, Hertfordshire;
John Portmann, Ministry of Agriculture, Fisheries and Food, Burnham on Crouch, Essex;
M.A. Robinson, Senior Fishery Statistician, FAO, Rome;
R.P. Romaire, Louisiana State University, Agricultural Center, Louisiana;
Bob Rosenberry, Editor, Aquaculture Digest, San Diego;
Bill Rowntree, Photographer, School of Ocean Sciences, University College of North Wales, Gwynedd;
Nathan Sammy, Department of Industries and Development, Darwin;
Rosalie A. Schnick, US Fish and Wildlife Service, Wisconsin;
Ephraim Seidman, Kibbutz Ma'agan Michael, Israel;
Robert Shleser, Aquacultural Concepts, San Juan, Puerto Rico;
Alan Stewart, University of Stirling, Scotland;
Albert Tacon, FAO, Rome;
Len Tong, Ministry of Agriculture and Fisheries, Wellington, New Zealand;
Granvil Treece, Sea Grant Mariculture Specialist, Texas A&M Unversity;
Gro I. van der Meeren, Austevoll Aquaculture Research Station, Norway;
Susan Waddy, Biological Station, St Andrews, New Brunswick;
Andy C. Watkins, Consultant, Guayaquil, Ecuador;
Priscilla Weeks, School of Human Sciences and Humanities, University of Houston, Texas;
Dennis M. Weidner, National Marine Fisheries Service, Silver Spring, Maryland;
John F. Wood, Natural Resources Institute, London;
Pat J. Wood, Consultant, Mazatlan, Mexico.

We are also grateful to the following colleagues and friends who made time to provide constructive comments on early drafts of individual chapters:

D.J. Alderman, Ministry of Agriculture, Fisheries and Food, Weymouth, Dorset;
R.W. Beales, Overseas Development Administration, London;
T.W. Beard, S.J. Lockwood and B.E Spencer, Ministry of Agriculture, Fisheries and Food, Conwy, Gwynedd;
I. Chaston, School of Business Studies, Polytechnic Southwest, Plymouth, Devon;
M. Esseen, Gannet Fishing Co., Bangor, Gwynedd;
P. Franklin, Macallister Elliot and Partners, Lymington, Hampshire;
D.M. Holditch, Dept. of Zoology, University of Nottingham, Nottinghamshire;
A.N. Jolliffe, Overseas Development Administration, London;
C.M. Jones, Dept. of Primary Industries, Queensland;
D.A. Jones, School of Ocean Sciences, University College of North Wales, Gwynedd;
G. Parry-Jones, School of Accounting, Banking and Economics, University College of North Wales, Gwynedd.

It is a pleasure to thank Dr. S.J. Lockwood for providing one of us (DL) with an eminently suitable habitat at the Fisheries Laboratory, Conwy, for the preparation of this book. On a more personal note, I (JFW) wish to acknowledge the unreserved support and understanding of my wife and family who ensured that I too had a stable environment in which to work. I (DL) would like to register my debt of gratitute to my relation, the late Ursula, Lady Hicks, whose legacy has enabled me to devote time to this book. Our thanks go also to Philip Wickins and John Hardwick for help with the choice and preparation of the cover design.

Chapter 1
Introduction

1.1 History

There is no doubt that over the past 40 years the idea of farming shrimp, crayfish, crabs or lobsters has become endowed with considerable 'investor appeal'. Since the early 1950s increasing personal disposable income in Japan and the West has allowed more and more people to explore the delights of eating crustaceans. As a result, consumption has soared and a host of entrepreneurs, businessmen and governmental agencies has rushed to exploit the aquaculture traditions and technologies of the Far East.

For hundreds, perhaps thousands of years, a variety of shrimp, prawn and crab species had been raised as an incidental crop from wild-caught juveniles entering coastal fish ponds throughout the Indo-pacific region. The advent of refrigeration and improved transportation gave the artisan farmers access to high-priced city and international markets and encouraged many to set aside ponds specifically for shrimps and prawns. The hatchery technologies developed by pioneers such as M. Hudinaga of Japan and S.W. Ling of Malaysia allowed much greater control of juvenile supplies. Hatcheries and shrimp farms became widely disseminated during the 1950s and 1960s both throughout the Far East and in the southern USA and Hawaii. Most failed to emulate oriental farming practices successfully during those early days and much money was lost. Nevertheless, valuable lessons were learnt and today shrimp and prawn farms contribute some 20–25% of world supplies.

This level of productivity has not been achieved without considerable social and environmental costs for some countries. Extensive construction of ponds is cause for concern because it often involves widespread clearance of mangrove forest resources, consequent loss of fish and shrimp nursery areas, coastal erosion and salinization of coastal lands. The rising demand for feedstuffs for shrimp and fish farming in particular has put pressure on supplies of low-value fish often consumed directly in developing countries. Ironically, in Taiwan it was the shrimp farmers themselves who suffered the greatest setbacks when, in 1988, environmental degradation brought about by their own activities resulted in severe disease outbreaks and the near collapse of the industry (Lin 1989).

In western temperate regions there is no long tradition of crustacean farming although since the turn of the century various attempts have been made in Europe and North America to restock natural waters with young lobsters and crayfish (Wickins 1982). However, studies made in North America between 1965 and 1975 demonstrated that clawed lobsters could be grown to commercial size in only two years instead of the five to seven years taken in the wild, simply by raising the water temperature and daily feeding. A plethora of commercial culture proposals followed. Many were based on assumptions not fully validated by research and again much money was, and continues to be, lost (e.g. McCoy 1986; Campbell 1989). Perhaps the greatest setback to aquaculture was that reported by Aiken & Waddy (1989) who wrote: 'In both countries (Canada and USA) a productive university-government research effort was extinguished by excessive promotion and premature entrepreneurial interest'.

Apart from the restocking of European inland waters, the only significant freshwater crayfish farming has, until recently, been that practised in the southern USA since about 1950. The last ten years however have witnessed an increase in crayfish pond culture, particularly in Europe and Australia. As in the early days of shrimp culture, euphoric predictions of crayfish yields abound in the trade press. These predictions stimulate entrepreneurs to propose culture projects often based on stocking densities and survival rates that can neither be supported nor refuted because relevant research and pilot studies have not been made. However, animals are being produced and this in itself may be sufficient to justify continued research support in some countries.

1.2 Objectives

The recent history of crustacean farming therefore is beset with failures as well as successes and it is this that has stimulated the preparation of this book. In it we attempt to provide the technical information required, and to address some of the problems faced by those new to the industry. The information will be relevant not only to all students of aquaculture but also to those who have responsibility for advising or making policy decisions concerning feasibility, investment, financing or implementation of crustacean aquaculture projects. Academic scientific information has been kept to a minimum in favour of basic biological and technical descriptions that have direct bearing on the reliability and costs of the various culture options. Shrimp farming in particular stands out for its influence on the economies of developing countries, and for this reason attention is paid to infrastructure and institutional factors as well as social and environmental impact. Representatives of all species that are

Table 1.1 Estimates of crustacean production (mt) from fisheries and aquaculture.

Species/group	Fishery production (1988)	Aquaculture production (1988–1990)
Marine shrimps & prawns	2 028 236–2 484 005	511 454–663 000*
Freshwater prawns	48 477	10–19 387
Freshwater crayfish	52 801–100 000	32 263*
Clawed lobsters	64 509	0*
Spiny lobsters	78 633	49
Other lobsters (*Nephrops*, Scyllaridae, Galatheidae)	80 664	0
Crabs	1 048 297	3277–7000*
Artemia	4000	350

*No attempt has been made to identify yields from restocking programmes.
Sources O'Sullivan 1986; Huner 1989; Rosenberry 1990; FAO 1990a,b; New 1990. Values for unspecified species/groups given by FAO (1990a,b) are excluded.

cultured commercially for the table, for restocking or that are thought to have potential for culture are discussed. A summary of important factors relating to their culture potential is given in Appendix 1.

Certain other species deserve mention and include those which form a significant by-catch to the main species being farmed, those captured at a large size and fattened, matured or induced to moult to take advantage of specialist or seasonal markets, and those that are cultivated for bait, display or research purposes. No doubt other species exist which may, for one reason or another, be worth cultivating. Possible candidates may be found among the crabs, the larger, filter-feeding freshwater atyid shrimps (*Atya gabonensis* of West Africa (maximum size 92–124 mm total length), *A. innocous* of the West Indies (21–34 mm carapace length)), or even goose barnacles (*Lepas* spp.) (Goldberg and Zabradnik 1984). However, accounts of large-scale culture trials with these novel candidates are either scarce or nonexistent. Crustaceans cultured as food for the rearing of other organisms include *Artemia* (Léger *et al.* 1986), *Acartia* (Ohno *et al.* 1990) and *Moina* (Ventura & Enderez 1980). *Artemia* culture technology is not discussed at length here; instead the reader is referred to the comprehensive works of Persoone *et al.* (1980), Sorgeloos *et al.* (1983; 1986), and O'Sullivan (1986).

1.3 Current status

By far the greatest tonnages of farmed crustaceans are marine and brackish water shrimp produced in south-east Asia and Ecuador (Table 1.1). Although estimates vary considerably, it seems likely that in 1990 around 600 000 tonnes (mt) were harvested, representing over 20% of the world total supply of 2.8 million mt. Farmed freshwater prawns (mainly *Macrobrachium rosenbergii*) total around 10–20 000 mt per year, of which Thailand produces 44%, Vietnam 32%, Taiwan 16% and others 8%. The total annual harvest of freshwater crayfish is probably around 50 000–100 000 mt of which up to 60 000 mt can come from the southern USA in productive years. Reliable figures for the production of cultured crabs and spiny lobsters are not readily available although Csavas (1989) estimates aquaculture production of these groups amounts to about 1% (6–7000 mt) of total crustacean production. As far as is known, no commercially viable farms for clawed lobsters exist.

1.3.1 Marine and brackish water shrimps

Such is the scale of shrimp farming that in the 1980s it induced significant changes in market structure and prices (Section 3.3.1). An abundance of shrimp in the medium size ranges (20–35 g) caused a fall in prices in south-east Asia from US$8.50 to 4.50 kg^{-1}. Part of the fall was believed to have been the result of a temporary reduction in consumption and consequent increased cold storage of shrimp in Japan (Rosenberry 1989). The decline in prices emphasised the vulnerability of intensive farming methods which have narrower profit margins than many low-cost, extensive farms (Sections 5.1, 10.8 and 10.9.1.4). Production in some countries has been adversely affected by a number of factors in recent years, notably the outbreaks of disease in Taiwan and Hawaii and unseasonable fluctuations in the oceanic currents that govern the availability of seedstock in Ecuador. The latter led to an official moratorium on new pond construction from 1984–9 and a widespread shortage of broodstock and juveniles. Major consumers of shrimp are the USA and Japan but expansion into European markets and increased sales of value-added product are thought to be vital to maintain prices (Rosenberry 1989) (Section 3.2.3).

1.3.2 Freshwater prawns

Production of tropical freshwater prawns, though much less than shrimp, remains of interest in many countries despite generally poorer export marketing opportunities. The heterogeneous growth typical of pond populations is a major constraint, but to combat this the benefits of selective stocking and harvesting regimes and the development of mono-sex populations are being actively investigated (Section 12.3). *Macrobrachium rosenbergii* is the prawn preferred by farmers although other large species (Table 4.6f) and their hybrids have been studied. The growth, yield and marketability of *M. rosenbergii* are now being compared with that of the Australian red claw crayfish (*Cherax quadricarinatus*) which grows in a similar culture environment, has no need for brackish water during its life cycle and is less aggressive (Section 4.7).

1.3.3 Crayfish

The red swamp crayfish (*Procambarus clarkii*) is the single most important species of this group with most production (85%), coming from wild and managed stocks in the USA. Significant harvests are also achieved in Spain and China, although natural as well as farmed (extensive) production in the USA and Kenya is highly variable due to climatic fluctuations. There is increasing commercial interest in crayfish farming in Western Europe and Australia. The potential for culture is good in Europe but fear of crayfish plague, to which native species are susceptible, has led to the importation of more resistant North American species, notably signal crayfish (*Pacifastacus leniusculus*), which unfortunately can carry and spread the disease. Several hundred hatcheries in Europe produce crayfish for restocking while in Britain most production is for the table (Holdich 1990). Australian species are reported to be highly susceptible to plague fungus, and their culture outside their native region could be risky. Ecological and commercial disaster could arise if North American or European crayfish from any source were taken into Australia.

Considerable interest has been aroused by claims that 'new' Australian species or strains have particularly good aquaculture potential. The culture of one, red claw, has already been implemented both in Australia and abroad despite limited knowledge of their culture performance (Jones 1990; Morrissy *et al.* 1990; Rubino *et al.* 1990). The experiences gained over the next few years therefore will be critical. Interest in value-added and soft-shell crayfish products is at a lower level but is increasing (Sections 3.3.3.2 and 7.5.7).

1.3.4 Clawed lobsters

Catches of North American lobsters have increased from 48 638 to 61 936 mt between 1984 and 1988, largely due to increased Canadian landings. European lobster catches remained steady at around 2000 mt over the same period. While the culture of both species is technically feasible, the lack of a suitable diet and the need for individual confinement during on-growing have so far prevented commercial viability.

After many years of concerted research effort into the development of battery culture technology and compounded diets, attention has been diverted towards investigating prospects for ranching hatchery-reared juveniles on the seabed (Sections 5.6 and 11.6.3).

1.3.5 Spiny lobsters

Australia, New Zealand, South Africa, Cuba, Brazil, Mexico and the USA are the main producers of the 79 000 mt of spiny lobsters fished annually. Spiny lobsters are usually sold frozen but the Japanese may pay up to US$100 kg^{-1} for live animals. Recent advances in the culture of larval stages in Japan are encouraging but considerable development of mass culture techniques will be required before commercial culture can succeed. In some areas where juveniles or adults occur naturally, fishermen provide shelters (often known as 'casas Cubanas') which modify the seabed habitat and concentrate fished populations (Section 7.9.8). On-growing or fattening of spiny lobsters in ponds, cages or tanks is possible when juvenile supplies are adequate, although only a few published records of the operation were found for use in this account.

1.3.6 Crabs

Crab culture has not attracted much commercial interest in the West, although the culture potential of the king crab (*Mithrax spinosissimus*) is being evaluated in the Caribbean. Some interest in the culture of mud crab (*Scylla serrata*) exists in Australia but the majority of farmed crabs come from extensive fattening or polyculture operations in south-east Asia. It may be possible to ranch some species (e.g. Japanese swimming crabs *Portunus* spp.), but the ownership problems that could arise with nomadic species are likely to limit such exercises to public control (Section 11.6.3). There is commercial interest in value-added and soft-shell crab products, particularly in the USA (Sections 3.3.5 and 7.10.9).

1.4 Advances and constraints

During the past decade, worthwhile advances have been made in the formulation of specialised compounded diets for broodstock, larvae and juvenile stages of shrimp. Progress with lobster diets has slowed because of reduced research effort, while the nutritional needs of several cultivated crayfish species have yet to be defined (Section 8.8).

The identification and treatment of a number of crustacean diseases have been reported and a vaccine developed against an important lobster disease, Gaffkaemia. Transplantation of wild-caught or cultured species around the world continues to spread diseases despite international recommendations on control of movements (Turner 1988) (Sections 8.9 and 11.4.3). In addition to transplantations, some animals inevitably escape from farm ponds or tanks and there is a risk of non-endemic species becoming established in the wild. The record of North American crayfish is particularly onerous in this respect (Section 11.4.2).

The tolerances of the different life cycle stages of several species to toxic levels of metabolites and other substances dissolved in water have been elucidated in part, but considerable work remains to be done especially in the determination of sub-lethal effects on crustacean growth, reproduction and susceptibility to disease (Section 8.5). The lack of basic scientific information on pond bottom chemistry and ecology has rapidly become a key issue on many intensive shrimp farms following recent commercial crop failures (Sections 8.3 to 8.7).

Uncertainties exist regarding changes in global climate patterns, and increased risks of pond damage and stock loss by flooding are expected in some areas. It is also likely that growth rates and harvesting schedules will be influenced by unpredictable and unseasonable temperature and rainfall patterns (Section 11.5.4). The adverse effects on shrimp and fish nurseries resulting from the destruction of mangroves to make farm ponds are now more widely appreciated, but concern over the impact of farming practices and effluent discharges on the environment is increasing (Section 11.5). Other factors which may occasionally handicap progress include a shortage of skilled technicians and a surfeit of unqualified or inexperienced consultants (Section 11.6.2.2).

Despite a number of recent problems related to markets and production, crustacean aquaculture seems likely to retain its momentum and attractiveness to investors for the foreseeable future. A careful approach and thorough project appraisal prior to commitment of capital are essential if past mistakes are to be avoided.

1.5 References

Aiken D.E. & Waddy S.L. (1989) Culture of the American lobster, *Homarus americanus*. In *Cold-water aquaculture in Atlantic Canada* (Ed. by A.D. Boghen). The Canadian Institute for Research on Regional Development, Moncton pp. 79–122.

Campbell M. (1989) Prince Edward Island. Lobster culture technology on hold. *Atlantic fish farming*, 20 Aug. 1989.

Csavas I. (1989) Recent developments in coastal aquaculture in the Asia-Pacific. *Infofish International*, (4) 47–51.

FAO (1990a) *Fishery Statistics 1988*, catches and landings, **66**.

FAO (1990b) Aquaculture production (1985–1988). Statistical tables *FAO Fisheries circular No. 815*, revision 2, FIDI/C815 rev. 2.

Goldberg H. & Zabradnik J.W. (1984) The feasibility of the gooseneck barnacle *Lepas anatifera* as a candidate for mariculture. *J. Shellfish Res.* **4** (1) 110–11.

Holdich D.M. (1990) Freshwater crayfish farming: production, prospects and environmental implications. *Scottish Shellfish Grower*, **6**, 10–11, 19.

Huner J.V. (1989) Overview of international and domestic freshwater crawfish production. *J. Shellfish Res.* **8** (1) 259–65.

Jones C.M. (1990) *The biology and aquaculture potential of the tropical freshwater crayfish* Cherax quadricarinatus. Information series QI 90028, Queensland Department of Primary Industry, Australia.

Léger Ph., Bengtson D.A., Simpson K.L. & Sorgeloos P. (1986) The use and nutritional value of *Artemia* as a food source. In *Oceanogr. Mar. Biol. Ann. Rev.* (Ed. by M. Barnes), **24**, 521–623.

Lin C.K. (1989) Prawn culture in Taiwan. What went wrong? *World Aquaculture*, **20** (2) 19–20.

McCoy H.D. II. (1986) Intensive culture systems past, present and future. Pt 1, *Aquaculture Magazine*, **12** (6) 32–5.

Morrissy N.M., Evans L.E. & Huner J.V. (1990) Australian freshwater crayfish: aquaculture species. *World Aquaculture*, **21** (2) 113–22.

New M.B. (1990) Freshwater prawn culture: a review. *Aquaculture*, **88** (2) 1–44.

Ohno A., Takahashi T. & Taki Y. (1990) Dynamics of exploited populations of the calanoid copepod, *Acartia tsuensis*. *Aquaculture*, **84** (1) 27–39.

O'Sullivan D. (1986) Growing brine shrimp — Australian companies see big potential. *Australian Fisheries*, Nov. 1986, 43–6.

Persoone G., Sorgeloos P., Roels O. & Jaspers E. (Eds.) (1980) *The brine shrimp* Artemia. Vol. 1–3. Universa Press Wetteren, Belgium.

Rosenberry R. (1989) *World shrimp farming 1989*. Rosenberry, San Diego, USA.

Rosenberry R. (1990) *World shrimp farming 1990*. Rosenberry, San Diego, USA.

Rubino M., Alon N., Rouse W.D. & Armstrong J. (1990) Marron aquaculture research in the United States and the Caribbean. *Aquaculture Magazine*, **16** (3) 27–44.

Sorgeloos P., Bossuyt E., Lavens P., Léger P., Vanhaecke P. & Versichele D. (1983) The use of *Artemia* in crustacean hatcheries and nurseries. In *Handbook of Mariculture, Vol. 1 Crustacean aquaculture* (Ed. by J.P. McVey), pp. 71–96. CRC Press, Boca Raton, Florida.

Sorgeloos P., Lavens P., Léger Ph., Tackaert W. & Versichele D. (1986) *Manual for the culture and use of brine shrimp* Artemia *in aquaculture*. Artemia Reference Centre, State University of Ghent.

Turner G.E. (Ed.) (1988) Codes of practice and manual of procedures for consideration of introductions and transfers of marine and freshwater organisms. *Int. Counc. Explor. Sea, Cooperative Research Report No. 159*. Copenhagen.

Ventura R.F. & Enderez E.M. (1980) Preliminary studies on *Moina* sp. production in freshwater tanks. *Aquaculture*, **21** (1) 93–6.

Wickins J.F. (1982) Opportunities for farming crustaceans in western temperate regions. In *Recent advances in aquaculture* (Ed. by J.F. Muir & R.J. Roberts), pp. 87–177. Croom Helm, London.

Chapter 2
Biology

2.1 Terminology

The common names shrimp, prawn, lobster, spiny lobster and crayfish are traditionally applied to different species in different parts of the world. For example, in Britain and Australia crayfish are freshwater crustaceans, but in the USA they are often called crawfish. In Britain the term crawfish is restricted to members of the Palinuridae, but in this book we use the more widely used name of spiny lobster for this group. In Australia spiny lobsters are also known as rock lobsters. The marine and brackish water Penaeidae are called shrimp in the USA, prawns in Australia, India and South Africa while either term may be used in Japan and Taiwan. In Great Britain small specimens are called shrimp, large specimens, prawns, while in the USA prawn is the name given to the large freshwater carideans of the genus *Macrobrachium*. The FAO convention is to call marine and brackish water forms, shrimp, and freshwater forms, prawns.

To help clarify the situation, at least as far as readers of this book are concerned, the names and relationships of the main cultivable groups are given in Table 2.1. Scientific and common names of commercially important species are included in Appendix I. The majority are classified in the order Decapoda of the class Crustacea and are characterised by having five pairs of walking legs, the first often bearing substantial chelae as in the case of clawed lobsters and crayfish. Major features of crustacean anatomy relevant to the understanding of their culture biology are shown in Figure 2.1. A glossary of scientific and technical terms is given in Appendix III.

2.2 Life history

To appreciate the different culture conditions required for farmed crustaceans, knowledge of the different life cycles involved is important (Figures 2.2a–e). The short account that follows contains the minimum information needed to understand the concepts discussed in later chapters. More detailed treatment may be found in Wickins (1976); Cobb & Phillips (1980); Bliss (1980–1985); Holdich & Lowery (1988); Dall *et al.* (1990).

The sexes are separate in most cultivated decapods. A few species, for example the spot prawn, *Pandalus platyceros*, change sex at some time during their lives. Mating generally occurs when the female is in a soft shelled condition (i.e. newly moulted) and results in the deposition of one or more spermatophores (containing many sperm) in, on or close to the genital openings of the female. Mating with hard shelled females occurs in spiny lobsters and in penaeid shrimp that have 'open' thelyca (see Glossary), but has only rarely been observed in lobsters. These crustaceans may rely on bristles

Table 2.1 Classification of cultivable crustacea.

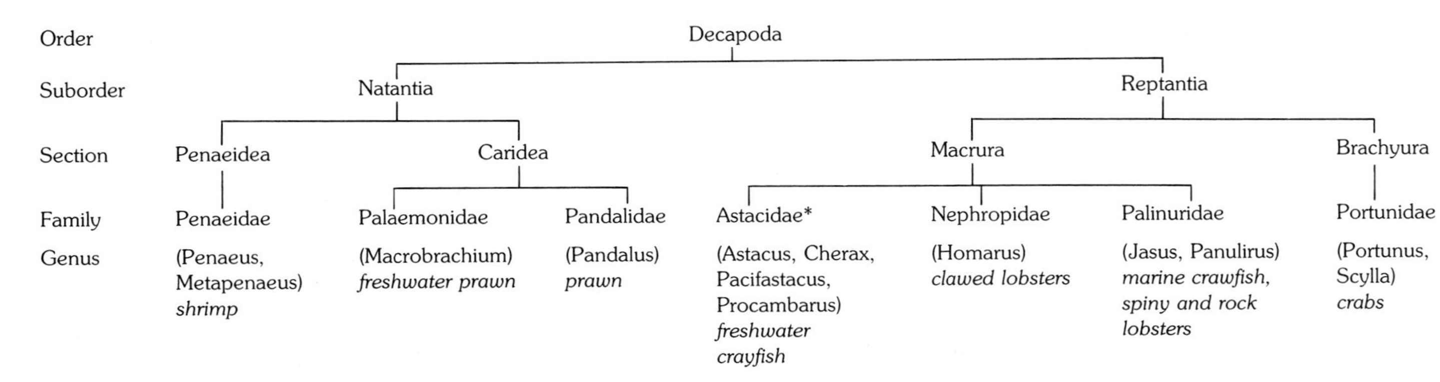

Order	Decapoda							
Suborder	Natantia			Reptantia			
Section	Penaeidea	Caridea		Macrura			Brachyura
Family	Penaeidae	Palaemonidae	Pandalidae	Astacidae*	Nephropidae	Palinuridae	Portunidae
Genus	(Penaeus, Metapenaeus) *shrimp*	(Macrobrachium) *freshwater prawn*	(Pandalus) *prawn*	(Astacus, Cherax, Pacifastacus, Procambarus) *freshwater crayfish*	(Homarus) *clawed lobsters*	(Jasus, Panulirus) *marine crawfish, spiny and rock lobsters*	(Portunus, Scylla) *crabs*

*Strictly, cultivable crayfish fall into three families: Astacidae, Cambaridae and Parastacidae.

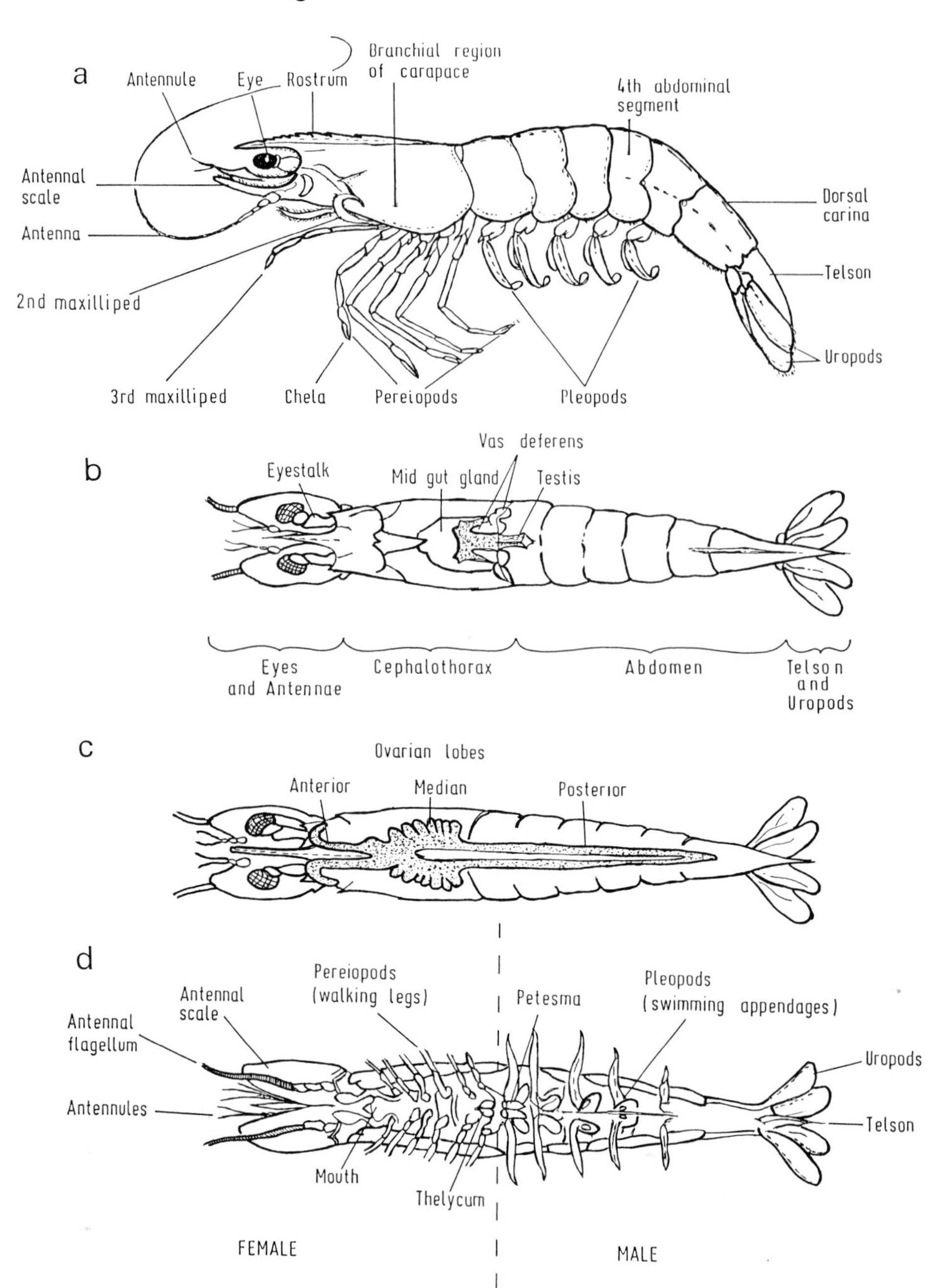

Fig. 2.1 The generalised anatomy of a penaeid shrimp: (a) lateral view; (b) dorsal view; (c) dorsal view of female with ripe ovary; (d) ventral view to show the position of the copulatory structures in both males and females (after Wickins 1976).

or cement to hold the spermatophores in place externally. Lobsters and certain species of penaeid shrimp have a 'closed' thelycum or pouch to retain the spermatophores until spawning occurs. Spawning is the release of eggs either directly into the sea in the case of penaeid shrimp, or to the brood chamber beneath the abdomen in all other cultured groups. The eggs are fertilized as they are spawned but in species with 'internal' sperm storage this may occur several hours or even months after mating according to species. Sperm from one mating are sufficient to fertilize more than one batch of eggs in lobsters, spiny lobsters and penaeid shrimp with 'closed' thelyca.

Penaeid eggs hatch a few hours after spawning and each larva is left to fend for itself as it develops through about twelve free-swimming planktonic stages of nauplius, protozoea and mysis before metamorphosing into a post-larva (Figure 2.2.a). The nauplii feed on internal stores of yolk while the protozoea stages filter unicellular algae from the water. The mysis stages feed voraciously on zooplankton (rotifers, *Artemia* nauplii) and in this respect they are like the larvae of caridean prawns and lobsters (Figure 2.2.b), the phyllosoma larvae of spiny lobsters (Figure 2.2.c), and crabs (Figure 2.2.d). Incubation in the non-penaeid decapods lasts from a few weeks in prawns to as long as 4 to 9.5 months in lobsters. Throughout this time the female tends and ventilates the clutch until hatching occurs. During incubation the early nauplius and protozoea stages are often by-passed in the egg so that when hatching occurs the larvae are almost immediately able to catch and feed on zooplankton. The most extreme cases of abbreviated development occur among the crayfish where there is no free-living larval phase, and post-larvae hatch directly (Figure 2.2.e). The post-larvae cling to their mother for the first one to three moults nourished by internal yolk until they are able to begin foraging for food. All farmed crustaceans are cannibalistic and unless the young can escape the mother they risk being eaten.

2.3 Nutrition

The juveniles of most species are omnivorous scavengers but can exist for considerable periods browsing on detritus. Compared to shrimp and lobsters, adult crayfish tend to take more detritus and vegetable matter but all tend to manipulate and fragment large pieces of food prior to ingestion, a habit which causes much fouling in intensive culture conditions. The formulation of compounded diets for crustaceans differs in several important details from mammalian and avian formulations. The ingredients must be finely ground and bound together to reduce disintegration in water. Since many key nutrients (for example vitamin C) are highly soluble, special insoluble forms may need to be used. The incorporation of chemical attractants can also be advantageous in reducing time available for nutrient leaching. Specific requirements include cholesterol and some long chain poly-unsaturated fatty acids (PUFAs) that many crustaceans are not able to synthesize rapidly enough for good growth and survival. The sources and component profiles of proteins and lipids used in diets are critical and in general, mammalian and avian offal are not good ingredients. Fish meals are widely used but are substantially improved by the addition of crustacean, molluscan and polychaete flesh, especially for broodstock diets. The frequency of feeding and the times of day the food is given are important aspects of husbandry that must be in sympathy with the animal's physiological needs if good growth is to be achieved (Section 8.8).

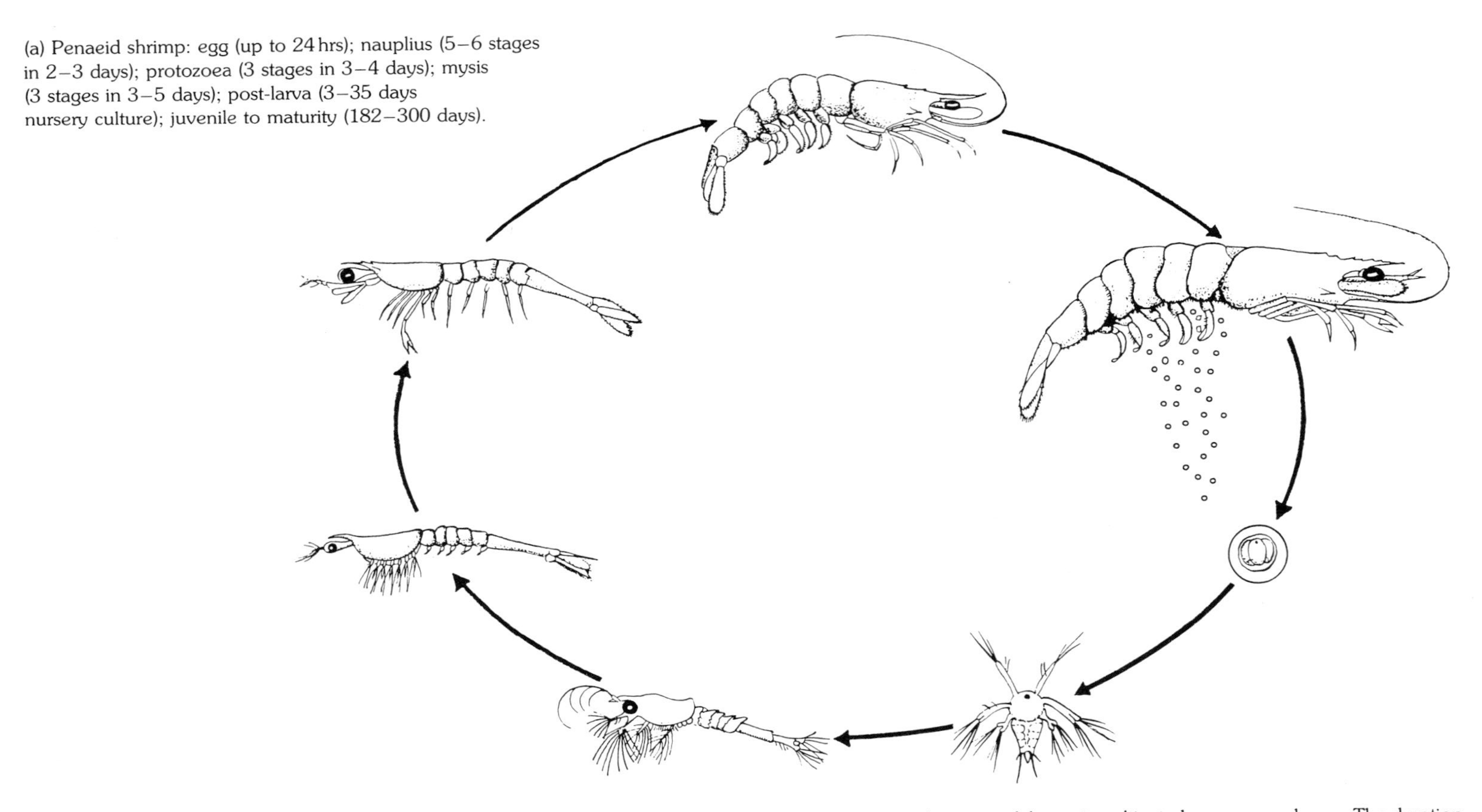

Fig. 2.2 The generalized life cycles of crustaceans. Typical changes in body form expressed during development of the main cultivated groups are shown. The duration and number of moult stages varies with species and temperature, but the ranges for the species listed in Tables 4.6.e,f,g and h (reading clockwise from the adult) are shown in these figures.

(b) Caridean prawns and lobsters (the illustration shows *Macrobrachium rosenbergii* and the adult is a male): egg (incubated by the female for 21–252 days); zoea (4–11 stages in 9–40 days); post-larva/juvenile to maturity (180–300 days).

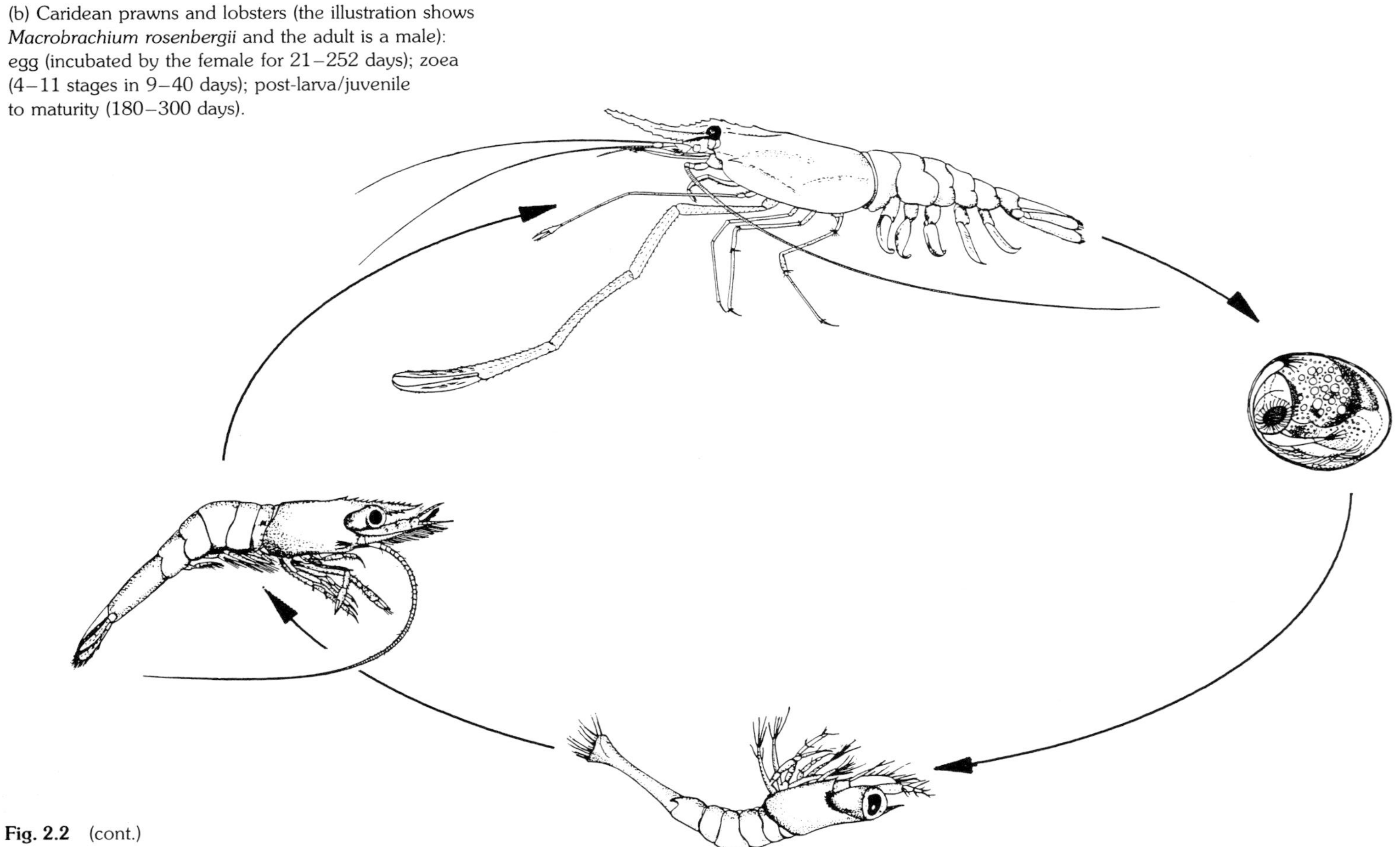

Fig. 2.2 (cont.)

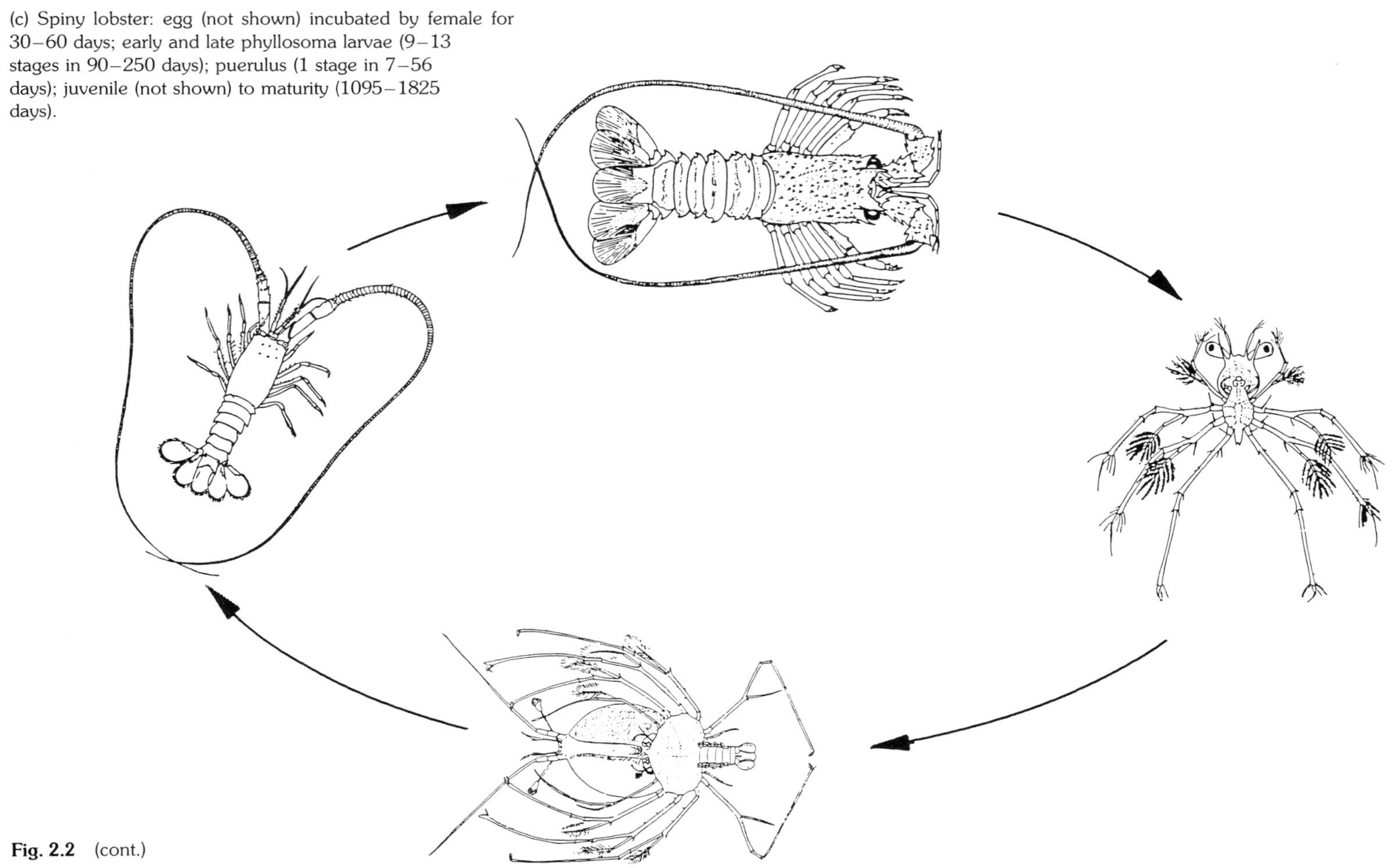

(c) Spiny lobster: egg (not shown) incubated by female for 30–60 days; early and late phyllosoma larvae (9–13 stages in 90–250 days); puerulus (1 stage in 7–56 days); juvenile (not shown) to maturity (1095–1825 days).

Fig. 2.2 (cont.)

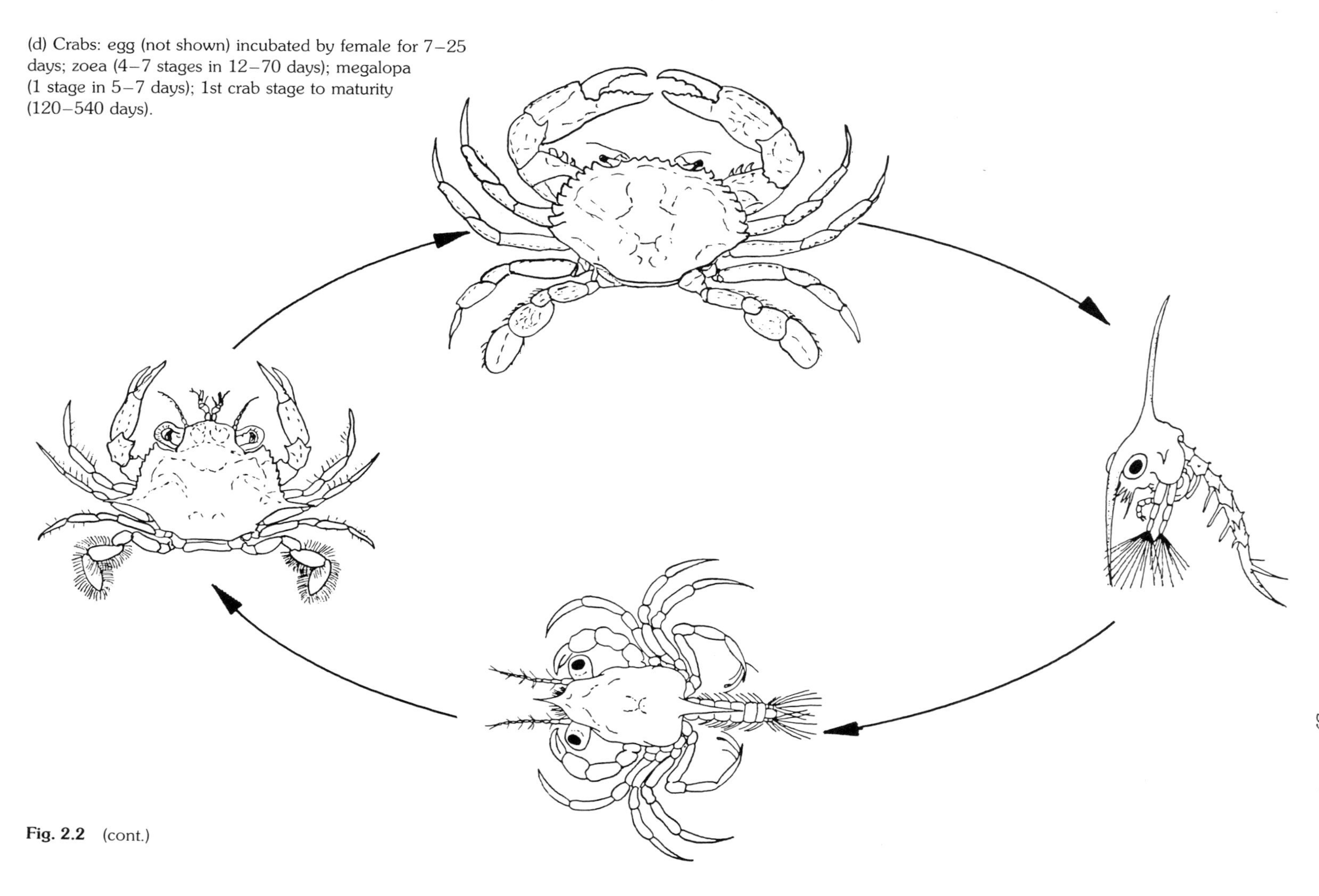

(d) Crabs: egg (not shown) incubated by female for 7–25 days; zoea (4–7 stages in 12–70 days); megalopa (1 stage in 5–7 days); 1st crab stage to maturity (120–540 days).

Fig. 2.2 (cont.)

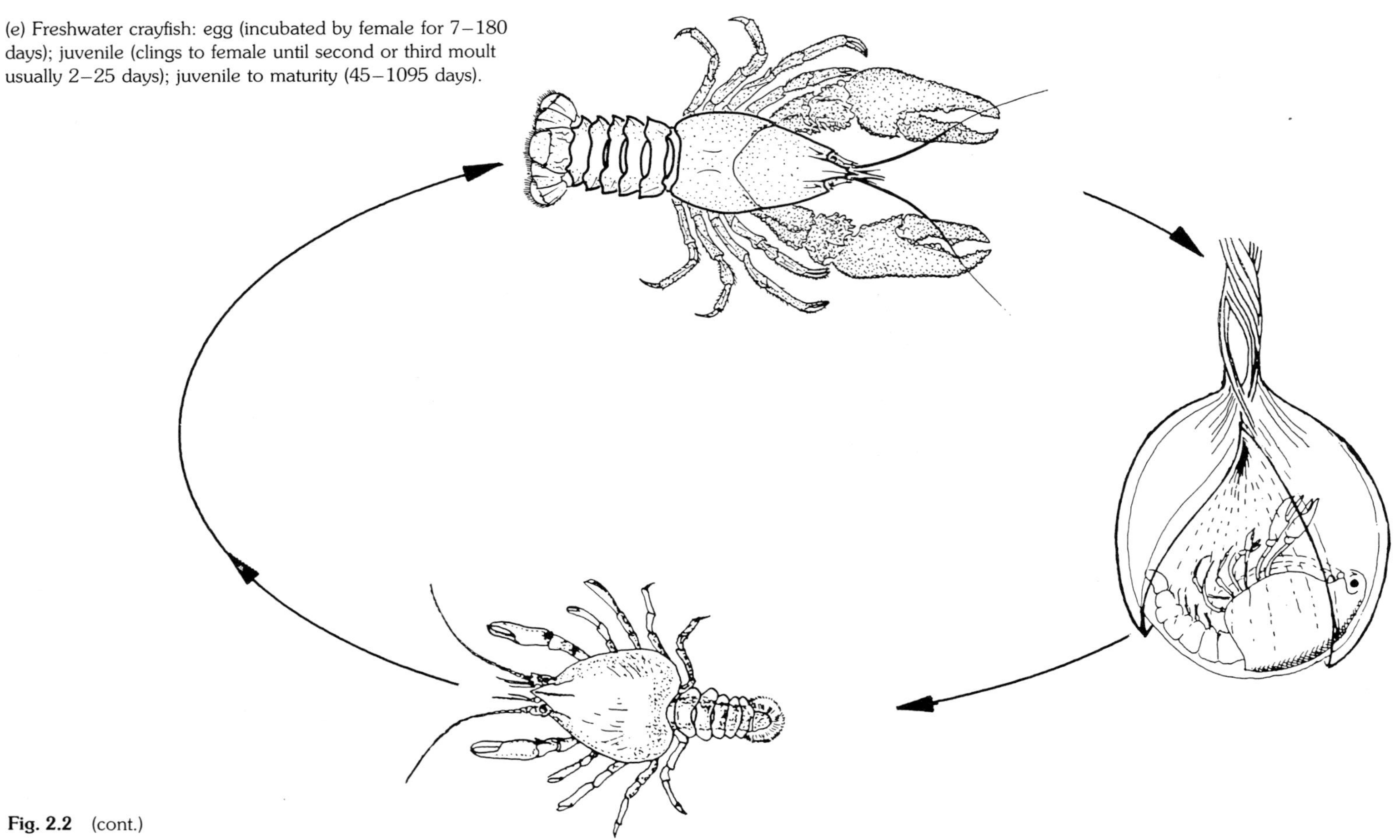

(*e*) Freshwater crayfish: egg (incubated by female for 7–180 days); juvenile (clings to female until second or third moult usually 2–25 days); juvenile to maturity (45–1095 days).

Fig. 2.2 (cont.)

2.4 Moulting, growth and maturation

The external shell (exoskeleton) of crustaceans is capable only of limited expansion. Growth occurs through moulting (shedding the exoskeleton or ecdysis) at intervals throughout life. The rate of growth is a function of the frequency of moulting and the increase in size at each moult. Adverse nutritional or environmental conditions can decrease both functions. The main sequence of events in the cycle is:

(1) Accumulation of mineral and organic reserves;
(2) Removal of material from the old shell and formation of the new exoskeleton;
(3) Ecdysis, accompanied by an uptake of water;
(4) Molecular strengthening of the exoskeleton by rearrangement of organic matrices and deposition of inorganic salts;
(5) Replacement of fluid by tissue growth.

The frequency of moulting varies naturally between species, with size and with age. Young shrimp larvae moult two or three times in a day, juveniles every three to 25 days depending on temperature and species, while adult lobsters and crayfish may only moult once every one or two years. Crustaceans often eat cast shells, a convenient source of minerals which would otherwise be lost. Mineralization of the new shell is affected by the availability of particular ions (calcium, bicarbonates and pH) in the surrounding waters, in the diet, and in freshwater animals from materials stored in the body prior to moulting. The changes that arise in water composition during intensive culture and particularly in recirculation systems can have a major effect on the mineralization process and on the animal's ability to control blood pH (Sections 8.4 and 8.5).

Newly-moulted individuals are particularly vulnerable to cannibalism especially under crowded culture conditions. Neither the presence of shelters nor the availability of adequate food eliminates cannibalism, although their absence may increase it. Attempts to reduce cannibalism by synchronizing moulting or ameliorating aggressive behaviour through claw removal or by giving drugs have met with little success.

Apart from the specialized markets for soft shelled crayfish and crabs, only hard shelled crustaceans fetch worthwhile prices. It is therefore necessary to minimise the proportion of the population moulting at the time of harvesting. Species grown in outdoor ponds may tend to moult in phase with the lunar cycle or in response to a change of water. Such effects are taken into account when the decision to harvest is made.

The moulting cycle and sexual maturation are two vital physiological processes influenced by a complex of glands situated in the crustacean eyestalks. Surgical removal of both sets of glands and the resulting decrease in the moult inhibiting hormone levels substantially increase growth in lobsters (Koshio *et al.* 1989) and spiny lobsters (Radhakrishnan & Vijayakumaran 1984), provided the diet is sufficient to support the extra growth. Removal of only one set of the glands from penaeid shrimp that do not readily mature in captivity (unilateral eyestalk ablation) is sufficient to reduce the circulating gonad-inhibiting hormone to levels that permit rapid maturation. These surgical techniques provide researchers with useful tools in the search for factors that might be useful for the less traumatic control of crustacean growth and reproduction (Sections 7.2.2.5 and 12.2). In the meantime, they are widely used to achieve immediate commercial objectives.

2.5 Disease

There are at least nine virus diseases of cultured crustaceans, six having been identified in penaeids and three in crabs. The viruses cause considerable mortalities within hatchery and nursery facilities and are readily transported from the wild to hatcheries, from nurseries to farms and from country to country with shipments of stock. No viral diseases have yet been reported from cultured *Macrobrachium*, lobsters or crayfish.

The bacterial disease Gaffkaemia causes serious losses among stored lobsters but vaccination is now possible. In juvenile shrimp and crayfish, however, most bacterial diseases are of secondary etiology (Alderman & Polglase 1988; Sindermann & Lightner 1988). Bacterial erosion of the exoskeleton can also cause significant financial loss in live storage operations and during the production of soft shelled crustaceans, due largely to the unsightly appearance of infected animals (Section 8.9).

The most serious of the fungal diseases is that caused in the wild by the crayfish plague fungus (*Aphanomyces astaci*) which is lethal to all native European and Australian crayfish. North American crayfish are generally resistant but can carry the disease and under stress will also succumb. Other fungal infections are mostly serious during hatchery and nursery phases of all species, while infestations with ectocommensal bacteria, blue-green algae or protozoa are frequently associated with high levels of dissolved organic matter accumulating in the water. Beyond the nursery phase, disease prevention rather than control is often the only course of action and medication seldom appears as an item in published costs (Section 10.9). Considerable financial losses during the on-growing phase do occur however as a result of diseases associated with poor diet, water quality or pond bottom chemistry (Sections 8.3, 8.9 and 12.2).

2.6 Genetics

Creditable progress in the study of heritable characteristics and cross breeding of crustaceans has been made in the last decade (Malecha 1983; Wickins 1984). Estimates of growth rate heritability for lobster (*H. americanus*), freshwater prawn (*M. rosenbergii*) and red swamp crayfish (*P. clarkii*) seem comparable (Lutz & Wolters 1989) and there may be some scope for improvement of favourable traits by selective breeding. However, greater improvements can often be gained by upgrading the culture environment or husbandry practices, and true domestication seems unlikely to be achieved in the near future (Section 12.10).

2.7 References

Alderman D.J. & Polglase J.L. (1988) Pathogens, parasites and commensals. In *Freshwater crayfish, biology, management and exploitation*, (Ed. by D.M. Holdich & R.S. Lowery), pp. 167–212. Croom Helm, London.

Bliss D.E. (Ed.) (1980–85) *The Biology of Crustacea*. Vol. 1–10, Academic Press, London.

Cobb J.S. & Phillips B.F. (Eds.) (1980) *The biology and management of lobsters*, Vol. 2, Ecology and management, pp. 333–84. Academic Press, London.

Dall W., Hill B.J., Rothlisberg P.C. & Sharples D.J. (1990) The biology of the Penaeidae. In *Advances in marine biology*, Vol. 27. (Ed. by J.H.S. Blaxter & A.J. Southward). Academic Press, London.

Holdich D.M. & Lowery R.S. (Eds.) (1988) *Freshwater crayfish, biology, management and exploitation*. Croom Helm, London.

Koshio S., Haley L.E. & Castell J.D. (1989) The effect of two temperatures and salinities on growth and survival of bilaterally eyestalk ablated and intact juvenile American lobsters, *Homarus americanus*, fed brine shrimp nauplii. *Aquaculture*, **76** (3/4) 373–82.

Lutz C.G. & Wolters W.R. (1989) Estimation of heritabilities for growth, body size, and processing traits in red swamp crawfish, *Procambarus clarkii* (Girard). *Aquaculture*, **78** (1) 21–33.

Malecha S.R. (1983) Crustacean genetics and inbreeding: an overview. *Aquaculture*, **33** (1–4) 395–413.

Radhakrishnan E.V. & Vijayakumaran M. (1984) Effect of eyestalk ablation in spiny lobster *Panulirus homarus* (Linnaeus): 1. On moulting and growth. *Indian J. Fisheries*, **31** (1) 130–47.

Sindermann C.J. & Lightner D.V. (Eds.) (1988) Disease diagnosis and control in North American marine aquaculture. *Developments in Aquaculture and Fisheries Science*, **17**, 1–431.

Wickins J.F. (1976) Prawn biology and culture. In *Oceanogr. Mar. Biol. Ann. Rev.* (Ed. by H. Barnes), **14**, 435–507.

Wickins J.F. (1984) Crustacea. In *Evolution of domesticated animals* (Ed. by I.L. Mason), pp. 424–28. Longmans, London.

Chapter 3
Markets

3.1 Overview

Interest in crustacean farming is usually stimulated by the desire to exploit markets for high value food products. Indeed, it is normally only the luxury prices obtained for crustaceans which makes their production through aquaculture a viable proposition in the first place. As demand grows for luxury foods, fuelled by expansion in western economies, and while yields from wild crustacean fisheries remain stable or increase only gradually, the potential market for new sources of crustaceans becomes increasingly apparent. All the same, it is advisable to view this potential in the context of the usual patterns that link supply and demand, remembering that high prices are usually only a reflection of limited supplies, and that they may well be depressed as a result of rapid increases in aquaculture production.

The ability of crustacean farming to influence market conditions is clearly illustrated by shrimp farming which, despite its short history, has already resulted in the oversupply of some global markets, with consequent price reductions. On the other hand, although this has reduced the profit margins for many farmers, lower prices have helped boost world consumption and shrimp are starting to compete more effectively in the mass market for everyday foods. As an example of this, seasoned and steamed shrimp prepared on the spot in 90 seconds are being sold as fast-food in locations such as New York and Los Angeles (Anon. 1990a).

Frozen shrimp accounts for 17–20% of world trade in fishery products and has established itself as a global commodity (Section 3.3.1). The situation, however, is markedly different with the other crustaceans that can be farmed. Most are dependent on local or national markets or on individually targeted export markets where demand for a particular species has been identified.

The difficulty of marketing farmed *Macrobrachium* at cost effective prices (Section 3.3.2) has severely restricted the development of large scale freshwater prawn farming despite the fact that this species is widely cultivated. Small producers can often obtain premium prices by supplying nearby restaurants and hotels. Success on a larger scale has been limited to countries where freshwater prawns are well established as a desirable food item, for example Thailand. However, even in this country, initial success has been tempered by falling prices resulting from excessive farm output. Freshwater prawns have shown some potential for bulk export in the form of frozen tails, but prepared in this way they must compete effectively in the well established market for marine shrimp.

The vast majority of crayfish farmed in the USA are consumed within the same southern states that produce them, with some exports directed at the high price Swedish market. Some of the crayfish production in Spain is sold to consumers in France and Sweden as well as home markets, and a number of Australian crayfish farmers are investigating export markets in Europe, the USA and Japan (Section 3.3.3).

Demand for clawed and spiny lobsters on world markets continues to be very strong. However, any future success with the marketing of farmed lobsters may rely on the successful development of new markets for small animals which, for clawed lobsters at least, is probably the only economically viable size for production by land-based culture systems (Section 3.3.4).

Farmed crabs are sold live on traditional local markets primarily in Asia. International trade does occur in processed crab meat, though this relies heavily on crabs from capture fisheries.

Although the great bulk of farmed crustaceans are consumed in Japan, the USA and Western Europe, the economies of several developing and newly industrialized countries are strengthening, particularly in Asia, and consumption of luxury crustacean foods in these countries can be expected to steadily increase. This has been the case in Taiwan, where high value aquaculture products are increasingly destined for home consumption as well as export. Home demand in Thailand, a country noted for its rapidly expanding economy, already supports the largest *Macrobrachium* farming industry in the world. Currently, shrimp consumption in major shrimp farming nations such as Indonesia and the Philippines is based on small low-value wild species which are often converted into products such as shrimp paste and shrimp crackers (Ferdouse 1990a). As farm output rises and shrimp prices decline in real terms, the farmed product will sell increasingly well within producer countries. Increasing tourism in many developing countries is also serving to boost demand from the hotel and restaurant trade.

Preferences for different types of crustaceans and different product forms within countries or regions remain very strong. In fact, observers note that there has been an overall trend towards increasing segmentation within crustacean markets and that one of the best strategies for processors is to make customized products that target individual market segments. Through product diversification and development an increasing variety of value-added crustacean foods are already being produced, with many finding ready acceptance with consumers. Some products are able to offer greater convenience and ease of preparation to the supermarket shopper, while others such as soft-shell crustaceans begin to open up novel markets (Sections 3.3.3.2 and 3.3.5). In all, they hold out considerable hope for generating increased overall sales volumes.

Aquaculture production is generally less variable than wild supply, and shrimp farming at least has demonstrated the potential to limit price fluctuations within often volatile markets. A related effect can be to reduce the need for large cold-storage holdings. In the USA over the period 1979–86, a more consistent supply of shrimp due to farming enabled average cold-storage holdings to be reduced from two to one month's supply (Siegel 1989). Nowadays when cold-storage holdings of shrimp build up to high levels, this is a sign of oversupply rather than insecurity about the availability of the product. In the late 1980s inventories climbed to record levels in Japan as buyers stockpiled shrimp awaiting improvements in prices.

Despite an overall improved price stability for shrimp, up-to-date marketing information can be very useful to producers and buyers alike. Regular publications with market analyses and forecasts include:

- Infofish International, Kuala Lumpur.
- LMR Shrimp Market Report, LMR Fisheries Research Inc., San Diego.

Plate 3.1 Thai women processing farmed shrimp. Note that in this picture the face masks are not always covering the nose properly.

- Seafood International, AGB Heighway Ltd., London.
- Seafood News, Heighway Ltd., London.
- Shrimp World Inc., 417, Eliza St., New Orleans, LA 70114.

3.2 Marketing crustaceans

3.2.1 Importance of correct handling and quality control

In general, crustacean flesh is rich in lipids, protein and free amino acids and has a tendency to perish very quickly. Thus, in order to ensure that products reach the consumer in good condition, attention to quality control and careful handling is essential right through all stages of harvesting, processing and marketing. Quality control measures can begin at the farm even before harvesting commences – for example by taking a sample of animals and checking for soft shells and any abnormalities. Soft shelled shrimp and prawns are liable to break up during harvesting and processing. For all farmed crustaceans the aim is to harvest when the majority are at mid inter-moult stage because at this time the water content and quality of the flesh are best.

The use of ice is vital when dealing with fresh product. Ice serves to prevent desiccation, retard bacterial growth and slow the rate at which flesh will spoil. In spite of this, Macintosh (1987), in his review of the handling of aquaculture products in south-east Asia, noted that many traders lacked knowledge in the proper use of ice. Common mistakes were the use of chunks rather than crushed or flaked ice (which is softer and has a much better contact area), the addition of ice only to the top of baskets of fish and shellfish, and the use of dirty ice which had been in contact with the market floor.

Careful handling of live animals is also critical and can greatly affect market value. Shipping and storage facilities provided for exports of live Canadian lobster must result in survival rates above 95% if premium prices are to be obtained (Shortall 1990). High survival rates for shipments of live *P. japonicus* in Japan are also used to indicate that the whole consignment is likely to be of premium quality.

Increasing consumer interest in lighter meals, balanced 'natural' diets and a general dislike for foods containing additives has led to widespread acceptance of crustacean flesh in many product forms as a health food. The low to moderate level of cholesterol found in crustaceans appears not to raise blood cholesterol levels (Tucker 1989), and indeed crustacean flesh is a rich source of n-3 fatty acids which can minimise or prevent certain heart and other diseases (Lands 1989; Piggot 1989). Farmed crustaceans are usually of equal or higher quality than equivalent products from the wild (most can usually be processed within an hour of harvest). However, if distinctions are made, consumer preference traditionally tends to favour wild sources as it already does in the case of salmon.

Great care is needed to guard the high quality, health food image of crustacean foods. Unfortunately shellfish (mainly molluscs) are responsible for serious outbreaks of food poisoning and their reputation is also vulnerable to fears about water pollution. Fourteen deaths due to *Shigella* in the Netherlands in 1984 were attributed to infected shrimp imported from Asia. In most countries strict microbiological specifications for imported foods are laid down by health authorities and shipments may be detained for inspection and destroyed if they are substandard.

Recommended international codes of practice for the handling of fish and fishery products, including crustaceans, are published jointly by the FAO and the WHO (FAO/WHO 1981; FAO/WHO 1984). Articles on quality standards, for example on frozen raw headless shrimp in the USA (Ramamurthy 1990), also appear in market publications such as Infofish International (Section 3.1).

Internationally traded crustacean products fetch different prices depending on their country of origin. The differences are based on established reputations for quality rather than the condition of each batch being handled. In this way the average price paid for white shrimp from China is lower than that for the equivalent Ecuadorian product.

Reputations for quality are difficult to build up but very easy to damage. The seizure of a single batch of home produced shrimp by Australian fisheries inspectors in April 1990, and subsequent declaration that it was unfit for human consumption, was considered potentially damaging to the whole of Australia's shrimp farming business (Ruello 1990a). Problems with maintaining a quality image for Malaysian shrimp exports have been attributed to a lack of confidence of investors in purchasing state-of-the-art processing machinery. Production has suffered from under-counting, with packages containing fewer shrimp than stated, and over-glazing (too much ice) of IQF (see Glossary) cooked and peeled product (Low 1988). The standing of individual farms can also affect the prices obtained for their products. Shrimp sold on Sydney

Fish Market from one farm in New South Wales fetched 25% more than shrimp sold the same day from a Queensland farm, the distinction being based on the perceived difference in quality from the two sources (Ruello 1990a).

Even with the concerted effort of a majority of farmers to raise standards, it is difficult to improve a country's reputation for quality, and action at a national level becomes essential. The Ocean Garden Products Corporation in the USA co-ordinates the sale of Mexican seafood on the US market and by enforcing quality control standards it has been able to keep the prices of Mexican shrimp above those of many other exporting countries.

3.2.2 Importance of reliable supplies

In many situations, farmed crustaceans can provide a more reliable and consistent supply than wild sources. This can be a distinct advantage when it comes to satisfying the needs of a processing industry as well as the final market. Indeed, despite some initial resistance, farmed black tiger prawns (*P. monodon*) have made a successful impact on world markets, because they now have a reputation for consistent supply as well as high quality. Aquaculture output has also had beneficial implications for the US crayfish market (Section 3.3.3.1). In contrast, pioneer freshwater prawn farms in the USA were unable to provide the regular supplies needed to develop a new market for their product and eventually closed down (Section 3.3.2).

If availability is predictable and prices are steady, it is much easier for restaurants to include crustaceans on their menus. Reliable supplies of Canadian lobsters on the UK market have made this product available for virtually 365 days of the year and allowed it to be included on the menu for sporting events and banquets (Shortall 1990). The former British Crayfish Marketing Association, realising the importance of predictable supply, established a UK crayfish season lasting from January to April and sets prices for signal crayfish at the beginning of each season to allow fixed menu pricing (Clarke 1989).

Some of the best prices for aquaculture products are obtained when seasonal demand cannot be met by wild catches. For example, demand for live kuruma shrimp (*P. japonicus*) in Japan peaks around new year and during the flower viewing season in April, and shrimp farmers are able to time their harvests to coincide with these periods when wild catches are typically low.

3.2.2.1 Harvesting strategies

In many situations the overall approach to harvesting will need to be determined by market considerations. Shrimp and prawn ponds and some types of crayfish pond may be harvested either in a single complete operation or as a series of partial steps. Single complete harvests of large shrimp ponds are only feasible when facilities for handling and processing bulk quantities are available, for example when supplying frozen shrimp for an export market. Multiple partial harvesting, however, is suitable for the supply of smaller sometimes local markets, e.g. hotels, where smaller batches are more acceptable than bulk quantities and can fetch higher prices. Live-sales in particular rely on this latter approach.

For a large farm it can be beneficial to design and manage the operation so as to obtain a steady flow of product. This helps to maintain a constant market presence and can be particularly useful for farms which are part of an integrated concern which operates its own processing plant. In one shrimp farm study (IFC 1987) it was estimated that 78 harvests per year would be required to achieve the desired level of production continuity and that these harvests could be provided by 30 ponds, assuming 2.6 cycles/pond could be obtained per year. While this is attractive on paper, the reality of continuous production is rarely achieved in outdoor ponds, not least because of the vagaries of weather conditions and fluctuations of water and feed quality.

3.2.3 Market development

The successful development of crustacean markets will involve both the creation of new markets for new products and product forms, and the expansion of existing markets to reduce or prevent problems of oversupply and falling prices. In this respect Filose (1988) recommends that the emphasis for aquaculture must switch away from a 'pure production mentality', in which huge tonnage increases are 'thrown' at buyers in a haphazard manner, and move towards the formulation of considered objectives and strategies to create new sales.

One obvious method for increasing the sales of a product is to extend the geographical range of its distribution, either within the producer country or through exportation. For live and frozen products, however, this does require that a network of suitable handling facilities are in place. Indeed, the rise of shrimp farming in regions like south-east Asia has only been possible through the evolution of infrastructure for cold-storage and transportation, to facilitate access to valuable city and export markets. In an analysis of market channels for aquacultured shrimp in Hawaii, Macaulay *et al.* (1983) identified the need for a greater number of more widely dispersed selling points, and for increased speed of delivery to shorten the time between orders being placed and deliveries being received.

Another approach to generating a greater volume of sales is to present products in more convenient and varied forms which gain ready acceptance through retail outlets. In many situations, people are only accustomed to eating shellfish in hotels and restaurants, so to encourage purchases for home consumption products need to be presented in prepared or easy-to-cook forms. This approach will, however, often require investment in new processing plant and the creation of a company market identity. Examples of alternative products to live crayfish include: frozen boiled whole crayfish; frozen peeled and un-peeled tails; and frozen or canned prepared products such as crayfish soup (Clarke 1989). Alternatives to live lobsters include cooked and frozen whole animals in brine, and blanched lobster sealed in a vacuum pouch. These product forms all have the benefit of an extended shelf-life and cheaper freight rates than live animals. The impact of value-added shrimp products in Europe has been reviewed by Lambert (1990).

Rather than produce a wide range of different products, one approach which may be especially suited to smaller organisations is to concentrate marketing efforts on just one particular segment of a market. Different market segments can be identified by geographic area and/or demographic factors which relate to a specific product usage (e.g. income level, age, sex, family size, social class, and occupation). Ideally a segment should be of a sufficient size, possess potential for growth, and not be over-occupied by competitors. Once a suitable segment has been identified promotional efforts can be targeted accordingly (Chaston 1983).

Consumers base their choice of seafoods on a range of factors including colour, cleanliness, size, apparent freshness, packaging and product brand, and some insight into this complex mix is usually required before new products can be developed and launched. Market research, either through direct test marketing or a market survey, can help establish likely levels of acceptability. Surveys often employ questionnaires and can be used to gauge the reactions and preferences of consumers, retailers or other people involved in the marketing process. Results can be interpreted and applied in the formulation of new marketing strategies before commitment is made to investment in costly large-scale production or processing facilities.

In addition to understanding likely reactions to new seafood products, it is often necessary to educate potential customers about the desirable features of crustacean products before a significant level of demand can be generated. This can be achieved through promotional activity and can be especially important if a luxury food image is to be projected. To reach a large number of consumers and promote a company or brand identity an advertising campaign can be launched; to communicate with a more limited audience of retailers, processors, or caterers a team of sales people may be more appropriate. Providing recipes, serving suggestions and cooking instructions on packages and during promotional campaigns can improve consumer uptake of new seafood products. The value of cooking instructions is illustrated by the case of freshwater prawns which can be easily spoiled through overcooking if prepared in the same way as the more familiar marine shrimp.

Pricing is another important aspect of marketing. It needs to reflect the general marketing strategy (perhaps involving price reductions to achieve greater market penetration), the production cost and quality of a product, and the level of promotional activity that may be required to maintain a high quality image.

Trade associations can play an important role in the co-ordination and execution of promotional efforts which may be beyond the means of individual producers or processors. They can also serve to lay down marketing standards. An example of one such organisation was the British Crayfish Marketing Association (BCMA) which had the objective of placing crayfish on the menu of leading UK hotels and restaurants. Before ceasing operations in 1991, it co-ordinated the marketing of 10 cm CL, hand-graded crayfish and established two grades of product: premium grade with undamaged claws at 15–19 kg^{-1} (approx. 50–70 g each) and standard quality with slight limb damage or smaller size (33 kg^{-1}). The marketing effort also involved encouraging regular buyers to install holding tanks (Clarke 1989; Richards 1988).

The BCMA was established as a co-operative venture with the help of a government grant, and clearly governments can play a useful role in supporting marketing organisations. In the developing world, institutional support can be especially valuable to small scale operators who do not possess adequate resources to efficiently market their products as individuals (Pillay 1977).

3.2.3.1 Strategies for farmed shrimp

The oversupply of small and medium sized shrimp as a result of aquaculture is becoming a constant feature within world markets (Infofish 1988). Various strategies have been proposed to overcome this problem which poses a serious threat to the economic viability of shrimp farming in some countries. Increasing competition is already favouring low cost producers like China and Indonesia over countries such as Taiwan and Thailand. The falling demand and export prices in Thailand became so acute in 1989 that trade associations and government departments began a co-ordinated promotional campaign to boost home shrimp consumption (Ferdouse 1990a).

At the level of the farm operator, a limited number of alternative strategies are available. One approach is to take advantage of the higher value of large shrimp and reduce competition with other aquaculture operations simply by growing larger animals. This is generally only possible through the use of lower stocking densities rather than prolonged on-growing (Figure 8.4). However, under pond conditions lower stocking densities usually result in reduced yields per hectare (Figure 10.9), and this may offset any increase in the value of the crop due to the larger size of individual shrimp. The reality of the economics of this 'trade-off' between yield and individual value is one of the main factors underlying the decision of most farmers to produce medium sized shrimp in the first place.

Diversifying production to include alternative shrimp species is another option, assuming that the requirements of the new species can be met and its market characteristics are born in mind. In the assessment of the culture potential of different shrimp species in Taiwan, aspects of market acceptability have been given considerable priority (Liao 1988; Liao & Chien 1990).

At the level of the processor, the output of more value-added products is helping to expand demand, particularly for the smaller sizes of shrimp. Pre-cooked and easy-to-cook products are increasingly favoured by consumers in place of fresh whole seafoods, both for their convenience and the absence of 'fishy' odours. Existing and proposed value-added products include:

- Cooked peeled and deveined shrimp in 200 g trays;
- Breaded or battered shrimp;
- Shrimp on seafood skewers (brochettes) with cuttlefish and paprika;
- Oriental stir fry dishes and ready-to-eat noodles;
- Shrimp in ready-to-eat consumer packs of sashimi;
- Small shrimp in tomato or cocktail sauce;
- IQF raw headless shrimp with fan tail on;
- Raw peeled and deveined headless shrimp with fan tail on and splayed flesh (butterfly form);
- Shrimp paste, soup and crackers.

Value-added products are being produced increasingly in developing countries, and joint venture deals are often established with foreign importers. Accompanying this trend is a movement away from exporting traditional 5 lb and 2 kg blocks towards the production of smaller consumer packs for retail sale.

To reduce reliance on any one particular market, diversification is advisable. For shrimp exporters, this may require selling product to Europe, the USA and Japan, but it reduces the impact of unforeseen problems specific to any one of these markets. Such problems in the past have included uncertainties in the US market due to holding orders being placed on shrimp imports from a particular country.

3.3 Review of world crustacean markets

3.3.1 Shrimp

The world market for shrimp (around 2.8 million mt) is dominated by Japan, the USA and Western Europe, which between them account for around 85% of the world trade. The remaining 15% is sold in lesser markets such as Australia, Singapore and Hong Kong or consumed within producer countries. A large section of the market (approximately 20–25%) is for cold-water species, of which *Pandalus borealis* and *Acetes japonicus* are the most significant. Their importance, however, varies greatly between the three main markets: around 50% of European shrimp consumption depends on cold-water species, but the equivalent figures for the USA and Japan are close to 10% and 5% respectively. Warm-water species, of which *P. chinensis* and *P. indicus* are the most important, account for the rest of shrimp consumption in these three markets.

Shrimp farms rely on warm-water species and in 1990 produced an estimated 663 000 mt, accounting for over 20% of world shrimp supplies. Three main species, *P. chinensis*, *P. monodon*, and *P. vannamei*, between them account for more than 80% of this farm output. In general, the great bulk of cultured shrimp flows from the developing tropical producer countries to the nearest of the three main markets in the developed world. Aquaculture now supplies more than 30% of US and Japanese shrimp imports (Anon. 1988). Japanese imports of Chinese shrimp have risen sharply in the late 1980s from around 10 000 mt in 1985 to over 37 000 mt in 1989 (Ferdouse 1990b). These figures, together with those for US imports of Chinese shrimp (up fivefold to 47 000 mt yr^{-1} from 1986–9), seem to testify to the remarkable expansion of shrimp culture in China, the scale of which has been doubted by some authorities (Forbes 1990).

Despite the increasing importance of aquaculture, shrimp prices are still determined to a large degree by the ex-vessel prices of wild-caught shrimp. This means that fuel prices are very significant since diesel fuel constitutes around 70% of the cost of operating a shrimp trawler. By contrast, in a semi-intensive shrimp farm, expenditure on fuel usually represents less than 10% of operating costs and in consequence these operations are thus much less sensitive to fluctuations in fuel prices (Section 10.9.1). Pricing is complicated by the fact that the size ranges of shrimp from fisheries and aquaculture do not coincide. While wild fisheries yield all sizes, aquaculture production centres on small and especially medium sized shrimp. The result is that, while supplies of large shrimp are limited by the wild catch and prices have remained high, supplies of small and medium sized shrimp have increased and there has been an overall downward trend in their prices through the 1980s. Figure 3.1 provides an example of the different prices that can be obtained for various size categories which are usually based on 'counts' – the number of whole shrimp or tails per kilogram or per pound respectively (Appendix II).

Prices also vary depending on the country of origin and the shrimp type. Important categories on the US market include Ecuador white, Mexico white, China white and Gulf brown (Gulf of Mexico). On the Japanese market major categories include Taiwan black tiger, Philippines black tiger, Thailand white, Indonesia white and China white. In addition, shrimp are categorized according to product form, such as peeled and deveined (P&D), peeled un-deveined (PUD), headless, whole, and individually

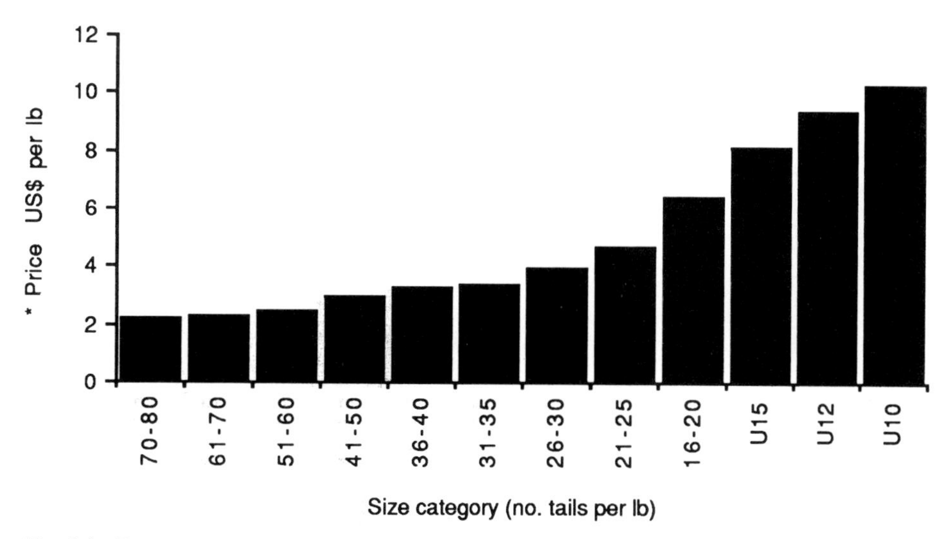

Fig. 3.1 Example of prices obtained for shrimp in different size categories (Appendix II). U = under. * Based on Ecuadorian export reference prices in El Universo, Guayaquil, July 1989.

quick frozen (IQF). Prices for categories with a relatively consistent supply, for example medium sized Ecuadorian white headless, are followed in trade journals (Section 3.1) for use as benchmarks and as indicators of overall price trends. Shrimp, therefore, is an extremely flexible product which has now become available in a wide variety of market forms (Figure 3.2).

3.3.1.1 Japan

The Japanese are probably the largest consumers of seafood in the world. They eat around 80 kg of seafood per person per year, of which 2 kg is shrimp. Most of the expensive seafoods in Japan are, however, suffering from dwindling wild supplies, and demand for imports continues to climb. Imports of shrimp totalled 261 000 mt in 1988, of which more than 98% were frozen, the remainder being live, fresh, chilled, dried, salted or in brine. Frozen product forms were raw headless (60–70%), shelled (10–15%), and whole (around 10%) (van Eys 1987; Yamaha 1989).

Demand for shrimp continues to increase, fuelled largely by continuous growth in personal disposable income and the associated increasing popularity of commercial eating and drinking establishments such as tempura shops, noodle shops and sushi bars. In addition, family-style restaurants are becoming more numerous and shrimp are now typically included in business lunches. Young people are increasingly consuming ready-to-eat food, and in response to people staying out late, food shops are staying open for extended periods.

Shrimp sales are also increasing on retail markets as a result of increased consumption in the home. More product is being made available in supermarkets in special consumer packs, prepared by defrosting large volume imports, repacking into smaller quantities, and refreezing. In fact, home consumption looks set to exceed consumption in restaurants and hotels. In the 1970s it accounted for 30% of sales but now estimates place the proportion at close to 50% (Yamaha 1989). This increase has

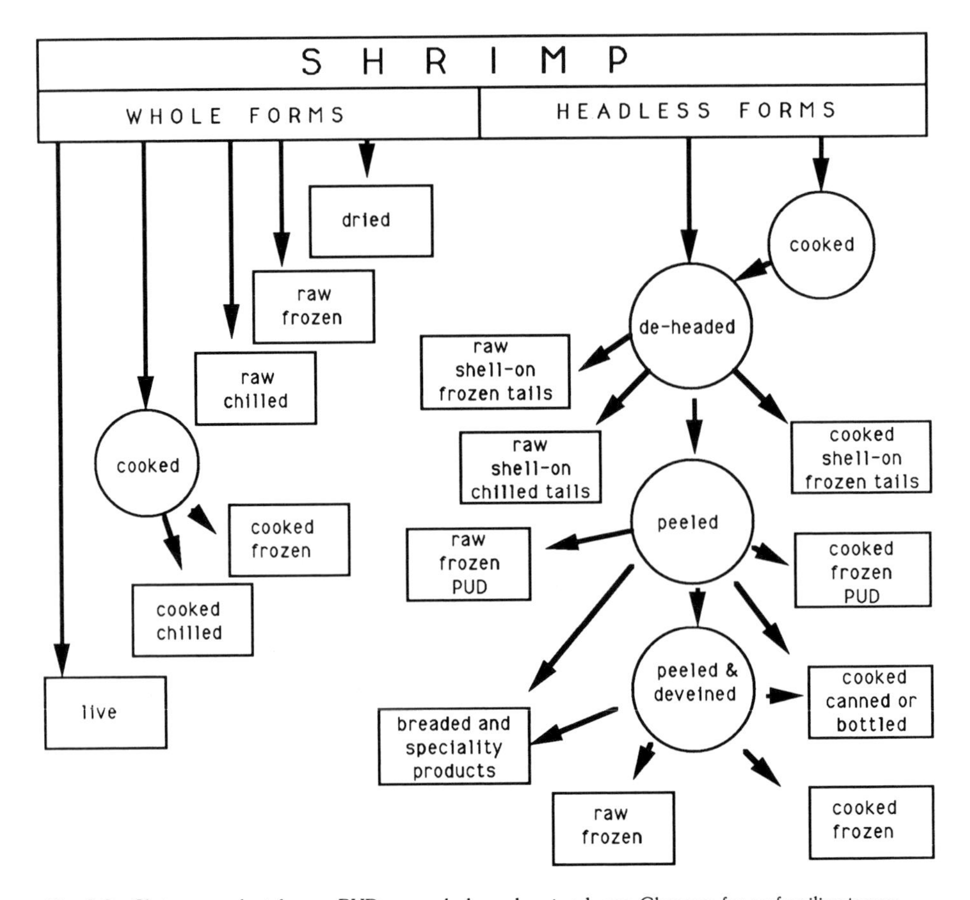

Fig. 3.2 Shrimp product forms. PUD = peeled un-deveined; see Glossary for unfamiliar terms.

been related to promotional campaigns successfully highlighting the health food image of shrimp, and to lower retail prices resulting from the fact that the small and medium sizes generally consumed at home are plentiful as a result of aquaculture. Large shrimp (16–20 tails per lb or fewer) on the other hand, which are consumed in commercial eating houses, come primarily from wild sources and are in more limited supply.

Patterns of consumption of different types of shrimp vary between regions and are based partly on consumer preference for shrimp of particular colours. While Osaka and Kyoto have become important areas for black tiger shrimp (mostly farmed *P. monodon*), southern Honshu shows a preference for economically priced Indian white shrimp, and Tokyo accepts all species.

Overall, the variety of imported species and product forms in Japan continues to increase. Sales of cold-water shrimp, including species from the North Atlantic, have a small but increasing share of the expanding market. The successful development of the Japanese market for black tiger shrimp has had much to do with the success of *P. monodon* farming. At first unfamiliar to the consumer because of its different colour and texture, its acceptance was slow, but reliability of supply from Taiwan, competitive pricing and high quality overcame the initial reluctance of consumers and importers alike. Although *P. monodon* production has suffered setbacks in Taiwan, this species dominates the output of shrimp aquaculture in south-east Asia, and Japan remains its principal world market. However, at the same time, the high quality of farmed *P monodon* is also helping to boost sales in US and European markets.

In general, since 1980 shrimp prices have become much more stable as a result of the development of the mass market. Yet the sensitivity of the market was recently highlighted by the death of Emperor Hirohito which depressed Japanese shrimp sales as celebrations were suspended and shrimp consumption slumped. As the value of the yen has increased against other hard currencies, such as the US dollar, imports of shrimp have been drawn towards the higher equivalent prices paid in Japan. Correspondingly any decline in the value of the yen is likely to see more shrimp diverted to either US or European markets.

Many Japanese importers and major retail chains have become involved in establishing processing plants, often as a joint venture, in tropical nations such as Malaysia (Singh & Jee 1988). As a result import channels are becoming more direct as these operations increasingly supply straight to retail outlets. Processors outside Japan are increasing their sales of value-added shrimp and seafood products which they are able to produce at much lower cost than Japanese-based operations. Home-based Japanese processors mostly buy peeled un-deveined shrimp (PUD) which are used for products such as shrimp croquettes or, in the case of small shrimp, as an ingredient in instant noodles. Strict adherence to quality guidelines is ensured by placing Japanese employees of importing companies on the production line of foreign processing plants. This approach to quality control has also enabled products to be custom-made to suit the needs of the Japanese market and has reduced the need to detain large amounts of shrimp in Japan for inspection by health officials (Sumner 1988).

A small but highly priced section of the Japanese shrimp market is for live *P. japonicus*. In fact, at US\$25–60 kg^{-1}, these live shrimp are 3.5–4 times more expensive than frozen shrimp and are the most costly shrimp for human food anywhere in the world. Colourful and with a delicious taste, they are served raw as sashimi in high class restaurants, sold to gourmet customers in tempura shops, or sold live in gift packages in department stores. If they are dead on arrival at the market their value is halved, so great attention is given to careful handling. After harvesting or capture they are dropped into an aerated cooling tank and the temperature is reduced in stages to 12°C. After 20–30 minutes, dead and soft-shelled shrimp are removed and the remainder sorted by size and packed into cardboard boxes between layers of dry chilled sawdust. Packed in this way they are able to survive for 10–30 hours (Yamaha 1989; Liao & Chien 1990).

Prices, however, fluctuate through the year and the highest demand is centred around the new year and the flower viewing season in April. Much of the supply is provided by wild fisheries but this premium value market also supports aquaculture operations in the southern islands of Japan and in Taiwan. The warmer, more southerly, locations of Taiwan, Kagoshima, Amami and Okinawa are able to maintain culture operations through the winter and take advantage of demand from December to May when there is a lack of supply from the fishery. Unfortunately, however, the colour of cultured *P. japonicus* is generally less intense than for wild specimens and there is a price difference, with the farmed product fetching 12–42% less (Liao & Chien 1990).

3.3.1.2 USA

Generally in the USA, as elsewhere outside Japan, shrimp is not regarded as an everyday food and is often reserved for special occasions and visits to restaurants. As a result of this luxury food status, growth in demand relies more heavily on growth in the level of personal disposable income and is thus strongly influenced by the state of the national economy. The image of shrimp as a health food is particularly strong in the USA and this is expected to boost demand in the short to medium term. The consumer attitude towards shrimp is expected to remain very positive as long as sufficient emphasis is placed on quality at all levels of production, processing and marketing.

US shrimp prices, particularly of the large sizes, are greatly influenced by the level of landings in the Gulf of Mexico. These vary seasonally and are usually at their greatest in the period from April to November. As in Japan, the bulk of product reaches the USA in the frozen raw headless form. Patterns of shrimp purchasing in the USA can be divided between different shrimp sizes (Appendix II) and regional consumer preferences (Filose 1988):

(1) *Large* (21–25 tails per lb or fewer, equivalent to whole shrimp of 27 g or more); consumed mostly in higher quality restaurants. Most supplies are obtained from the fishing fleets in Mexico, Panama and the USA. Mexico is the preferred source because it has an established reputation for high quality and so is able to set prices according to its landings.
(2) *Medium* (26–50 tails per lb, equivalent to whole shrimp of 14–26 g). Shrimp in this size range fetch considerably lower prices than large shrimp and represent an economical choice for households and restaurant owners. These sizes are produced by aquaculture, and the oversupply of white shrimp from Ecuador (mostly *P. vannamei*) and China (mostly *P. chinensis*) has resulted in reduced prices. This has partly resulted from Ecuadorian and Chinese producers by-passing traditional import channels. Filose (1988) identified a need to promote consumption of medium-sized shrimp in supermarkets and moderately priced restaurants.
(3) *Small* (more than 50 tails per lb, equivalent to whole shrimp of 13 g or less). As well as being consumed in restaurants and sold in supermarkets, these sizes are used for further processing. The smallest sizes (80 tails or more per lb) are peeled and sold as raw or cooked meats in a range of convenient products for home consumption which are distributed via supermarkets and other retail seafood stores.

Consumer preferences vary between regions and generally every major metropolitan region has its particular preference (Filose 1988). Premium quality white shrimp (mostly *P. setiferus*) are favoured on the west coast and in the north-east; brown shrimp (mostly *P. aztecus*) from the Gulf of Mexico is preferred in the middle section of the country and around Baltimore, Philadelphia and Washington DC; pink varieties (mostly *P. duorarum* and *P. notialis*) are preferred in the south-east.

As a general trend, Asian supplies to the US market are increasing. Chinese white shrimp in particular are making rapid progress. However, to achieve this the Chinese have had to offer their shrimp at prices 5–10% lower than other producers such as Ecuador which have an established record for consistent high quality. The image of

Chinese shrimp has been adversely affected by problems with soft shells, slow delivery schedules, and the sale of kilogram packs to a country where all dealings, computerized records, accounts, and indeed customers, are geared to buying shrimp by the pound (Anon. 1988).

US imports of Asian farmed *P. monodon* have been limited somewhat by the fact that Japan remains the main destination of this product and regular supplies, which are essential to make a sustained impact on US markets, have been hard to maintain. Other problems with farmed black tigers have included confusion over prices and quality because of product arriving from many different sources, and suppliers who often by-pass traditional import channels. In addition, consumer unfamiliarity with the term 'black tiger' has meant that it has been necessary to present the product in a cooked form or as a cocktail shrimp (Filose 1988).

A market niche in the USA has been reported for 'blue' tiger shrimp. These are farmed *P. monodon* which take on a bluish appearance rather than the usual black colour. The unusual colour is thought to relate to the diet and although at first affected product was discarded as substandard, it now appears to be saleable (Anon. 1990b).

US market channels have four basic levels: major importers, combined distributors/importers, pure distributors/wholesalers, and various types of end user. Shrimp farmers within the USA tend to adopt two different marketing strategies depending on the size of their operation. Small farms sell their product to coastal processors for packing and marketing, while larger farms contract a processor to pack under their own brand labels and then manage their own marketing (Chamberlain 1989).

3.3.1.3 Europe

European shrimp consumption is increasing rapidly; over the years 1982–7 alone it rose by 50%. There is thought to be considerable room for further expansion because annual per capita consumption at 0.5 kg still lags far behind the estimated 1.3 kg and 2.0 kg consumed in the USA and Japan respectively. As elsewhere in the world, shrimp in Europe benefit from their good, health food image.

The European market relies heavily on supplies of cold-water shrimp (and indeed their availability has a major influence on prices), but landings have either been stable or in steady decline. As a result, opportunities for the sale of warm-water shrimp have been steadily improving and countries such as India and Thailand have become major suppliers. In addition, an increasing part of the catch of cold-water shrimp is being diverted to the Japanese market and this too is stimulating European demand for shrimp from warm-water sources. No single country dominates European supplies and in fact more than 50 countries export shrimp to the EC, with developing nations accounting for two thirds of the supply. While Japan and US markets are dominated by frozen headless product, Europe prefers whole or cooked and peeled product.

The European market is characterized as being very price conscious, with importers readily altering their buying patterns to satisfy a price sensitive retail market. Increasing receptiveness to new seafood products is enabling competitively-priced, high quality shrimp from aquaculture to gain acceptance. However, warm-water shrimp in general will have to consistently out-perform established products in terms of price and quality if they are to greatly increase their market share. This may in part be due to the conservatism of many wholesale buyers, particularly in the UK, who tend not to think in terms of promoting or extending new markets but rather deal only in products they are able to sell on quickly. Even so, traditional European market channels are being increasingly by-passed and the number of links in the distribution chains are being reduced. In the UK, for example, at least two major retail organisations are buying direct from suppliers and eliminating the wholesale stage.

High quality black tiger shrimp, particularly from Indonesia, are starting to make an impact, and imports of white shrimp from Ecuador and China are also on the increase (Infofish 1989). The dark appearance of *P. monodon*, however, is limiting the acceptance of this product in countries like the Netherlands where light-coloured shrimp are preferred. Latin-American suppliers of white shrimp, predominantly Ecuador, are expected to turn increasingly towards European markets in response to strengthening European currencies and increasing competition from Chinese product on their traditional market in the USA. Also, by supplying Europe's market for whole shrimp rather than the US market for tails, they are able to obtain a 100% processing yield instead of only 57–68%.

Although the markets of countries in Western Europe are usually considered collectively as the European shrimp market, sharp distinctions in buying preferences exist between northern and Mediterranean regions and from one country to the next. In fact, an overall trend towards reinforcement of these distinctions and increased market segmentation has been observed. In countries bordering the Mediterranean the preferred market form is whole raw product which is cooked for dishes and meals requiring whole shrimp. Warm-water species are more important here than in northern Europe, and Cuba is a big supplier. Cold-water shrimp, notably from Argentina, are also consumed. In general, northern Europe prefers cooked and peeled shrimp mostly from cold-water sources.

Superimposed on this overall pattern are differences between countries. In Italy a big market exists for large pink or red shrimp and remains buoyant despite the increasingly high prices of the large sizes. Spanish consumers on the other hand prefer white shrimp (Nierentz and Josupeit 1988). In both Spain and Italy, farmed *P. japonicus* is readily accepted because its banded appearance is very similar to the locally available *P. kerathurus*. Production of *P. japonicus* within Europe, however, is no more than a few tonnes per year. France prefers cooked whole shrimp, while Denmark consumes more bottled or canned product (Evans 1990). West Germany and Belgium consume large amounts of tropical prawns which have been repacked in the Netherlands. However, Belgian consumers generally prefer the mild taste of cold-water *Crangon* species which is unlike that of tropical marine species. Belgium, France and Spain have the largest per capita shrimp consumption in Europe.

The UK market has a distinct preference for cooked and peeled cold-water shrimp which are usually eaten cold in salad dishes. Nonetheless, increasing UK demand for shrimp in Indian and Chinese restaurants is favouring imports of warm-water species which have a tougher texture more suitable for Oriental dishes. More than half UK sales are made to the catering trade and warm-water species now make up more than 50% of all shrimp consumption (Jamieson 1988). Around 5% of shrimp in the UK is destined for processing into value-added products.

3.3.1.4 Other markets

Significant shrimp importers include Singapore, which acquires product both for re-export and domestic consumption, and Hong Kong which imports around 70 000 mt annually, mainly from China for trans-shipment (Ferdouse 1990a). As already mentioned, demand for shrimp within tropical producer countries is also increasing, boosted in some areas by strengthening economies and expanding tourism.

Australia tends to export high quality shrimp from its own wild fisheries to the lucrative Japanese market, while importing cheaper and sometimes lower quality product from Asia (Low 1988; Ruello 1990b). Principally it buys cooked and peeled shrimp along with some raw meat. Some *P. stylirostris* farmed in New Caledonia has been flown to market in Sydney in cooked and uncooked forms and marketed under the name Paradise Prawns (Ruello 1990a). Australian supplies also come from a small but developing home-based shrimp farming industry (Hardman *et al.* 1990).

3.3.2 Freshwater prawns

Although the farming of *Macrobrachium rosenbergii* has become widespread, the scale of developments has generally not lived up to expectations and production has declined, particularly in Hawaii and Israel. One of the main reasons for this has been the difficulty of consistently selling prawns at profitable prices. The most reliable markets for *Macrobrachium rosenbergii* exist where freshwater prawns are traditionally eaten and have established an image as a desirable fishery product, for example in Thailand and Taiwan. Yet even in Thailand the prawn culture industry has suffered significant setbacks and output has recently declined. Elsewhere in the world, where well developed national markets do not exist, many small operations have been able to centre their marketing efforts on the local hotel and restaurant trade and nearby retail outlets. It is now recognised that this is probably the best short-term commercial strategy and may offer significant opportunities for development programmes to stimulate *Macrobrachium* production at the level of artisan farmers, especially where tourism is being encouraged (e.g. Cuba).

Some farming ventures which opted to produce freshwater prawns because they were seen as an easy species to culture, and which later encountered problems with marketing their product, have been accused of following a production orientated approach to species selection instead of matching output to meet market needs (Chaston 1983). Some farmers mistakenly considered *Macrobrachium* to be an ideal substitute for marine shrimp, when in fact the species is substantially different: it has a tougher shell and is more difficult to peel; its flesh has a different taste and a more delicate consistency; when cooked in the same way as marine shrimp the flesh becomes unappetising; it undergoes rapid deterioration if frozen prior to packing, and specialised freezing technology is required. In addition, the product cannot strictly be labelled 'shrimp' nor 'seafood' (although in fact, unfairly, crayfish sometimes are) and even if equivalent prices can be obtained for headless product the processing yield is 40% (or even less with large-clawed males) compared to 57–68% with marine shrimp.

As a result of these different characteristics, successful market development for prawns has relied on selling whole animals to a gourmet market, highlighting their differences to shrimp rather than their potential as a substitute product. Unfortunately, the US consumer is not familiar with head-on shrimp and prawns and the general preference for headless shrimp has resulted in consumer resistance to whole freshwater prawns. Test marketing of prawns in South Carolina showed good potential for sales but also recorded a preference for headless product (Liao & Smith 1981).

New (1990) noted that the economics of *Macrobrachium* farming would improve if freshwater prawns could be sold on a market which did not differentiate between shrimps and prawns. One such market may exist in Belgium, where freshwater prawns from Bangladesh are accepted because of their similarity in taste to the preferred cold-water species *Crangon* (Nierentz & Josupeit 1988).

Despite estimates of a potential US market for some 4500 mt of whole prawns per year, attempts to develop and supply this market have failed through an inability to provide a regular supply of prawns, and the poor acceptance of the product at cost effective prices (New 1990). US companies such as Amfac and General Mills have retreated from prawn projects in Hawaii and Honduras respectively, and one Texas operation abandoned farming essentially because 'prawn proved impractical to market'. Some Puerto Rican farms do, however, send shipments of prawns to the US mainland as well as to some other Caribbean countries.

In Thailand, which is the largest producer, exporter and market for *Macrobrachium rosenbergii*, prawn farming has contracted as a result of increased prices for rice paddy and strong competition from shrimp farming. While prawn prices have fallen as a result of over-production, shrimp fetch 40–70% more and are more easily marketed (New 1990). Despite these factors, around 6000–7000 mt of prawn tails were exported from Thailand to Europe in 1987 and 1988, and although a sound global commodity market for freshwater prawns has not yet been developed, New (1990) believes that these exports may herald some changes. If these materialize, low cost fishery production from Vietnam and Bangladesh is also likely to benefit.

An expanding market in Taiwan absorbs around 5000 mt per year of domestic farm production. Malaysia currently exports small quantities of prawns to Hong Kong and Singapore. A survey of Malaysian restaurants revealed a potential domestic market for 1000 mt of prawns per year, but also found that established retail fish traders generally regarded dealing in live *Macrobrachium* as too 'messy' to be of interest. This type of problem may hinder market development in tropical producer countries, except in places such as Indonesia and north-east Thailand where consumers are used to buying live fish directly from farms and live fish traders (Macintosh 1987).

Prawns produced in Hawaii are mostly sold on home markets live or freshly iced to ethnic Philippinos. The use of the label 'Hawaiian Prawns' has assisted the development of a premium speciality market but even so prawn production through the 1980s in Hawaii has been in gradual decline, primarily due to low pond yields and a limited domestic market (Brock 1988; New 1990).

Some prawns farmed in Israel have been sold in Europe, and a heads-on market has been located in Italy and Spain. Once again irregularity of supply has been a problem in developing the potential of these markets, and the generally unfavourable economics of prawn farming have resulted in Israeli freshwater aquaculture being redirected towards tilapia production (Karplus I., 1990, pers. comm.). Total prawn output from Israel has shrunk from 40 mt yr^{-1} to a mere 0.5 mt yr^{-1} in 1989, all of which is sold to local tourist hotels.

Although Thailand has begun to export frozen tails, prawns are usually sold live or fresh and whole on ice. Thai consumers prefer male prawns with short orange claws rather than blue claw males which give more claw waste. For live transport to nearby restaurants, prawns may need to be held in containers of aerated water as they have only a limited tolerance to air exposure (Liao 1988). Transport fresh on ice is feasible for short periods only because prawn flesh rapidly turns mushy under these conditions. To extend these periods the 'kill chill' process can be employed in which prawns are dipped in iced water and then blanched (pre-cooked) at 65°C for 15–20 seconds. Alternatively, flushing with carbon dioxide and storage at 4°C can also extend the life of fresh product. In processing prawns for tail meat, washing and de-heading both serve to reduce bacterial contamination and retard the spoiling of flesh.

3.3.3 Crayfish

Although international trade in crayfish is increasing, crayfish markets are most conveniently discussed in respect of the three main regions where they are fished and cultured: the USA, Europe and Australia.

3.3.3.1 USA

Crayfish find a ready market in the southern states of the USA where they are obtained from both fisheries and extensive farm systems. Louisiana accounts for 90% of US farmed crayfish and the bulk of its production is consumed in that state, where the appeal of crayfish rests largely on the popularity of cajun cooking, part of Louisiana's French heritage. Exports of US crayfish to Europe are expected to account for 3–5% of Louisiana production in the foreseeable future (Roberts & Dellenbarger 1989).

Production from the wild fisheries has always been highly seasonal and in Louisiana centres around March, April and May. Aquaculture has extended output to the period from November to May and this has assisted the trend towards expansion of sales in 'out-of-state' restaurant and retail markets. In order to compete more effectively with marine products, crayfish are often referred to incorrectly as seafood (Holdich 1990).

Traditionally most product has been sold live to local markets. However, general over-reliance on such markets has placed stress on the marketing system and has only been alleviated by an increased emphasis on alternative product forms. While approximately half the crop is still sold alive, most of the remainder is now processed for tail meat. This enables gluts of production, which would otherwise have oversaturated the market, to be utilised profitably. Currently though, in the absence of a commercially successful machine for meat removal, hand peeling is necessary and results in high labour costs of US\$2.75 kg^{-1} (Roberts & Dellenbarger 1989). Also, the processing yield is a mere 15% and although additional flesh is available in the claws, it is not considered economically viable to extract it.

In South Carolina most farmers minimize their marketing costs by selling their product directly to consumers (farm gate sales) who do not require delivery, continuous supply, or great attention to size grading (Liao 1984). This method of direct selling probably represents the most profitable strategy for small-scale producers but is only likely to be viable with the support of regular customers and a good passing trade. Larger crayfish farming operations in South Carolina make deliveries to restaurants and seafood retail stores.

The different product forms in which crayfish can be marketed are illustrated in Figure 3.3. For the restaurant trade market preference is for live product, while for domestic consumption both live product and processed meat are very important. Although overall consumer preference for live crayfish has been recorded in South

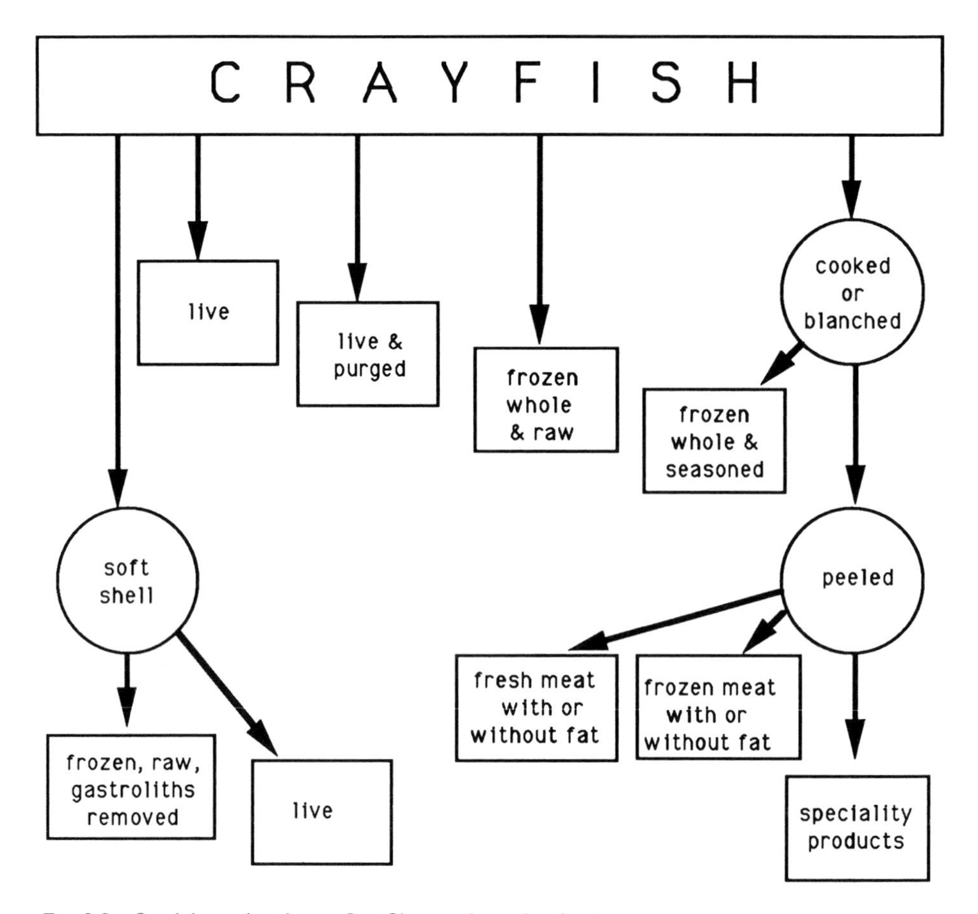

Fig. 3.3 Crayfish product forms. See Glossary for unfamiliar terms

Carolina, 40% of customers preferred to buy only cooked tail meat (Liao 1984). Negative impressions of crayfish were characterized by comments about the high price, and, for whole animals, the small meat yield and the difficulty of peeling. European markets, particularly in Scandinavia, are now being supplied with custom-made products such as graded whole crayfish packed in trays (Section 3.3.3.3). Often the larger animals (30–40 g) are destined for export (Holdich D., 1990, pers. comm.).

A number of significant constraints remain regarding the processing and marketing of crayfish meat in the USA. These result from the seasonality and price instability of the supply of crayfish; rancidity in tails packed with the fat (hepatopancreas included); a lack of standard size categories; and consumer ignorance about the product. However, significant improvements have been made in the areas of plant design; proper determination of cooking times; product stabilization; freezing techniques; quality control programmes; and packaging (Roberts & Dellenbarger 1989).

3.3.3.2 Soft shelled crayfish

A specialised section of the US crayfish market is for soft-shell product. This is a high value food with a price 15–20 times higher than hard shelled crayfish (Clarke 1989).

Most (90%) soft-shell crayfish fall in the size range 12.5–20 g (50–80 crayfish kg^{-1}) because larger animals are not available in sufficient quantities and smaller individuals are not efficient to produce (Culley & Duobinis-Gray 1989). Processing yield is excellent and there is no need for deveining or purging because crayfish do not eat in the period just before moulting. In fact, processing yield would be 100% but for the need to remove the two gastroliths – a pair of calcareous secretions located behind the rostrum – which reduces the yield to 92%.

Soft-shell crayfish are typically frozen in water (which serves to protect the delicate limbs) in 1 ℓ bags containing 454–680 g of product. New packaging methods include the use of vacuum shrink-wrap pouches, and some product is displayed on white plastic trays with six or 12 crayfish per pack. Only small amounts of soft-shell crayfish are sold live because of the likelihood of limb loss. Quality is a prime consideration. Buyers will not accept soft crayfish which have lost both claws, and will only take a limited number of one-claw animals. 'Paper-shelled' animals that have started to harden and turn leathery are also hard to sell, as are small specimens (>73 kg^{-1}) (Homziak 1989). Ice is commonly used to stop the process of shell hardening.

3.3.3.3 Europe

Among the most important crayfish importing countries in Europe have been France, West Germany, Belgium, Sweden and Switzerland, with the Turkish fishery for *Astacus leptodactylus* as the major supplier. However, production from this fishery has declined markedly as a result of industrialization and crayfish plague and it appears that much of the demand in European markets, estimated to total 10 000 mt per year (Huner *et al.* 1987), goes unsatisfied. Precise production figures are unavailable but at its peak Turkey may have produced an estimated 8000 mt per year. Now output is down to 15–20% of this figure (Clarke 1989) and supply shortfalls are probably greatest in France and Sweden.

Spain, however, has become an important exporter following the introduction of the plague resistant North American crayfish *Procambarus clarkii.* A wild fishery for this species became established and rapidly expanding output saturated domestic demand and provided material for export. In addition, almost the entire harvest of Kenyan crayfish, estimated at 250 mt per year (Goddard 1988), is exported to Europe because Kenyan domestic consumption is negligible (Huner 1988).

In Scandinavia it is traditional to eat crayfish during the summer, a custom upheld most strongly in Sweden and Finland. Crayfish can fetch as much as US$100 kg^{-1} during the first two weeks of the Swedish crayfish season in August. The market preference is for noble crayfish (*Astacus astacus*) but Louisiana has become a significant exporter of *Procambarus* to Sweden following the decline in the Turkish *Astacus* fishery. Small amounts of rusty crayfish, *Orconectes rusticus*, produced in Wisconsin, USA, are also exported to Sweden and because this species is similar in appearance to the noble crayfish it obtains better prices than *Procambarus* spp. (Clarke 1989).

The best prices for crayfish are obtained for live animals and Huner *et al.* (1987) note that in Europe there is little experience with products such as crayfish meat, soft shelled crayfish and unpeeled tails. This situation is gradually changing as more frozen product arrives from the USA. Some of it is processed in dill sauce and packed in specially shaped containers (Holdich 1990). Turkey exports both live and processed crayfish. Live shipments leave by airfreight in lightweight wooden boxes containing

5 kg of product, and processors sell cooked whole crayfish (18–25 kg^{-1}) packed in brine (Koksal 1988).

Quality requirements for edible crayfish on European markets whether alive or frozen include:

- Total length greater than 10 cm;
- Intact symmetrical appendages (especially chelae);
- Clean flexible shell;
- Well filled-out body (both meat and hepatopancreas);
- Uniform colour (orange/red preferred in cooked product) (Huner *et al.* 1987).

3.3.3.4 Australia

Crayfish, with their excellent flavour, are a much prized food in Australia where they are often prepared on barbeques or served as entrées in restaurants.

Although wild crayfish catches are generally in decline, erratic output from some fisheries, for example for the yabbie, still has the potential to swamp markets and depress prices (Sokol 1988). In Western Australia amateur fishing for marron is heavily regulated with strict size limits enforced to protect wild stocks. These regulations have, until recently, interfered with marron farming operations which need to be able to harvest and sell relatively small animals of 40 g. Although in captivity the marron is capable of growing much larger, animals of 40 g can be produced after one year of on-growing and are the most economical option for farmers. They are also considered to be suitable for restaurants (O'Sullivan 1988) and indeed smaller crayfish (<100 g) appear to have an enhanced flavour (Jones 1990).

In addition to the catering trade, market outlets for crayfish farmers within Australia include direct sales, retailers, wholesalers and auctions. Direct sales may be the best option for small operators who are located close to population centres or tourist routes because the requirements for consistent prices, regular sizes and supply are lower than for the catering trade. Wholesalers, retailers and auctioneers, on the other hand, are more able to deal with bulk quantities (O'Sullivan 1989). Alternative specialized markets include the aquarium trade, bait crayfish, and re-stocking (conservation) programmes.

Australian crayfish are mostly sold live. Within Australia, live yabbies of minimum size 40 g are supplied to top restaurants and served as entrées under the name baby lobsters. Asian markets, including Japan, also prefer live product and although crayfish in this form also get the best prices in Europe, European markets will accept cooked and frozen crayfish which incur lower shipping and maintenance costs and reduced risk of losses. Some live marron from Australia have already been air-freighted for sale in Europe, retailing in some fish shops at promotional prices four times lower than clawed lobster.

Some of the problems facing the marketing of Australian crayfish have been highlighted by O'Sullivan (1989). They include the fact that freshwater products are generally considered to have an inferior taste to their saltwater equivalents, even though in taste trials crayfish compare very favourably with marine lobster (Jones 1990). Also, if export markets are to be developed, particularly in Europe and the USA, Australian product needs to be made available in sufficiently large quantities to sustain the interest of buyers and must be able to compete with crayfish produced at relatively lower costs in extensive US systems (Section 10.9.2.3). To confront these problems O'Sullivan (1989) proposes an aggressive and co-ordinated marketing effort to provide regular supplies of high-quality, clean, intact and purged crayfish while emphasising the Australian, 'pollution-free' origin of the product.

Market development will benefit from the production of alternative product forms e.g. frozen boiled whole crayfish, frozen peeled and unpeeled tails, and frozen or canned prepared products such as crayfish soup (Clarke 1989). Studies carried out by Jones (1990) using a 30 member tasting panel are significant in this respect and address some basic questions relevant to processing and the development of recipes.

The tests centred on red claw (*C. quadricarinatus*) and established that for frozen product there was no deterioration in quality over six months of storage. However, frozen cooked whole crayfish had a marginally less acceptable flavour than frozen raw whole or headless product that was cooked after defrosting. Average tail meat yields for cooked product were established at 22.8%, 26.2% and 12.0% for red claw, marron and yabbie respectively. In the red claw an extra 4.5% yield was obtained by extracting claw meat from the larger (>100 g) specimens and although this yield was low, claw flesh was recognised as having a sweeter, more delicious flavour than tail meat. An optimal cooking method of seven minutes boiling in freshwater was established for whole red claw to retain the characteristic flavour and provide flesh with a tender, slightly resilient texture. Alternative methods of cooking i.e. boiling in saltwater, steaming or microwave cooking, had no noticeable effects on the flavour although the latter two techniques resulted in inferior meat yields.

Of particular significance to farmers are results that indicate that the flavour of red claw can be enhanced by growing them in saline water (up to 2.4% saline) rather than freshwater, and that the same effect on flavour could be induced in animals grown in freshwater by conditioning them for 48 hours in 3.0% saline water, immediately prior to harvest.

3.3.4 Clawed and spiny lobsters

The potential for profitable lobster aquaculture has always been associated with a well-established, high value market for the final product. Certainly demand for this luxury food continues to be very strong; indeed the word insatiable is often used to describe the world's appetite for lobster.

While the world supply of spiny lobster appears to be stable, clawed lobster catches have recently been increasing both through greater fishing effort and, in the case of Canada, through rigorous management of wild resources. Major suppliers of lobster apart from Canada are the USA, Australia and Cuba. Canadian landings of clawed lobster which account for around 25% of the total world lobster catch were 38 500 mt in 1989. The USA produces 20 000 mt from its own fisheries (also predominantly *Homarus americanus*), and to satisfy home demand, imports another 40 000 mt, much of it clawed lobster from Canada (Shortall 1990). US imports of spiny lobster account for one third of world output (Williams 1988) which is mostly traded internationally in the form of frozen tails. Demand for lobster in Europe and Japan is strong and growing more quickly than in the USA. France has the largest consumption of all European countries.

One of the main marketing problems facing the putative lobster aquaculturist is the fact that lobster markets are geared to receiving animals of the sizes supplied from wild fisheries, and the culture of animals to these sizes may not be economically viable. Size regulations are laid down by fishery authorities to protect wild stocks and enable lobsters to breed before they are captured. In the case of clawed lobsters, markets and fisheries usually deal with live animals of a minimum size of 350–500 g, yet lobster farms for *Homarus americanus* or *H. gammarus*, if they are to have a reasonable chance of making profits, may have to grow animals to only half this size (Sections 7.8.9 and 10.9.3.6). So it would probably be necessary to create a new market for small size lobsters while retaining the high market value through promotion of the product as a gourmet food item. Considerable market research would be needed to establish the acceptability of this or any other new crustacean product and some consumer resistance might develop against farmed product as it has done in the case of salmon. In some countries changes in legislation might also be required in order to exclude cultured lobsters from existing regulations. In Britain, however, it would be a defence to show that the undersized lobsters in question were reasonably believed to have been produced by farming (Howarth 1990). In the case of clawed lobster, a smaller product would probably find ready market acceptance and would complement rather than compete with the market for larger wild specimens. In a parallel example, some success has been achieved in marketing small Australian crayfish at 40 g in an established market for sizes in excess of 100 g (Section 3.3.3.4).

3.3.4.1 Clawed lobsters

Although the best prices are obtained for live lobster, and in this form they can be air-freighted for sale in markets throughout the world, increased emphasis on alternative product forms may be needed in the future. This may be especially important if culture operations ever develop on a significant scale and large volumes of product need to be marketed. Output from the Canadian wild fishery is already sold in a range of different forms. Half is cooked whole or processed for meat and half is marketed live. Large lobsters are sold live and divided into two size categories: 'markets' of 1–3 lb (454–1362 g) and 'jumbos' of over 3 lb (>1362 g). Smaller specimens of $\frac{1}{2}$–1 lb (227–454 g) are either:

(1) cooked alive, sealed in a vacuum pouch ('popsicle pack') with brine then frozen for sale in six size grades;
(2) cooked, shelled, packed into cans and frozen for sale as either high-grade tail and claw meat or lower grade broken meat.

Live lobster is mostly destined for the catering trade, and frozen lobster products are mainly supplied to the retail market. Whole frozen lobsters are a popular product in Europe, and the USA takes more than 90% of the Canadian exports of frozen canned meat.

The occurrence of genetically determined red and blue colourmorphs of the American clawed lobster *Homarus americanus* may provide opportunities to develop the culture of animals with a unique appearance. This could enhance the marketability of the live product: red, for example, is considered by many consumers to be the 'natural' colour of lobsters (Aiken & Waddy 1989).

The farming of lobsters would provide an ideal opportunity to capitalise on seasonal price fluctuations caused by uneven supply and demand. While most landings are made during the spring, summer and autumn, particularly during periods of calm weather, year-round demand causes prices to rise in the winter. Demand in Europe is particularly high around Christmas. The use of holding pounds, however, as a way of retaining fished lobsters for more profitable sale out of season, has become popular and has tended to reduce price fluctuations somewhat. Overall, the prices obtained by fishermen in the UK have not kept pace with inflation in the past decade as a result of increased supply from home waters and imports from Canada. Canadian lobsters have generally made a good worldwide impact because of consistent landings but there is a preference and price differential in favour of European lobster (*H. gammarus*) on European markets, based on a perceived difference in flavour (Esseen M., 1990, pers. comm.)

3.3.4.2 Spiny lobsters

Small quantities of spiny lobster are produced in on-growing operations which rely on supplies of wild-caught juveniles or sub-adults. In Taiwan such operations produced 400 000 lobsters in 1987, for domestic consumption. However, mass imports of Chinese spiny lobsters or crawfish threaten to undermine prices and the profitability of on-growing units. In Taiwan most if not all lobsters are sold live. They are transported for up to 24 hrs in cardboard boxes containing 16 kg of lobster in four layers with chilled sawdust. A valuable market exists for juvenile lobsters used to stock the on-growing operations, and animals above 25 g are sold for US$42 kg^{-1}. There are no size restrictions on the capture of wild lobsters in Taiwan (Chen 1990).

Some trade takes place in live spiny lobsters flown into Europe from Australia. Careful handling and low temperatures (<12°C) are essential to ensure product quality. There is a potential gourmet market for soft-shell spiny lobster of 40–60 g, which could be produced from wild caught pueruli (see Glossary).

3.3.5 Crabs

The crabs produced by aquaculture are sold live to markets based on traditional local consumption of the same species caught in the wild. International trade in crab is mostly reliant on processed meat from wild-caught crabs. When the need arises, however, there is little to impede transport of live crabs over long distances. Mud crab (*Scylla serrata*), for example, can survive out of water for one week if sprinkled occasionally with water to keep them moist (Section 7.10.6).

Cultured crabs are a highly valued product in certain Asian markets. In Taiwan, female mud crabs which are filled with orange coloured roe are especially prized and fetch far more than males or immature females. After spawning, their value drops by 90%. Harvested mud crabs are bound with thick rope interwoven with rice straw and kept moist during transport to market. While roe-bearing females are sold by number, other mud crabs are sold by weight and the weight of the rope is included in the price (Chen 1990). Female portunid crabs with full ripe ovaries are a luxury food item in Japan. Although they are not farmed, restocking programmes have been aimed at enhancing the Japanese fishery for *Portunus trituberculatus*. Again price is very sensitive to condition: if the female is berried rather than roe-bearing her value falls by 60–70% (Cowan 1983).

The viability of soft-shell crab production in shedding systems in the USA (Section

7.10.9) relies on the high processing yield and the high market value of this gourmet seafood. More than 1400 mt of soft shelled blue crab *Callinectes sapidus* are produced each year from Maryland to Georgia, for sale as a local delicacy or distribution to top-line restaurants throughout the eastern USA. Shipments of live crabs and dressed IQF product are also made to Japan. Other markets may exist in Hong Kong, Singapore and Taiwan.

3.3.6 Analogue products

The sales of crustacean analogue products are increasing worldwide and have important implications for some traditional crustacean products. Surimi, the most common product, can be produced in various forms such as crab-sticks and shrimp tails to take advantage of the high prices obtained for crustaceans. It is made from mechanically de-boned, washed and stabilised white fish flesh, produced mainly in the USA and Japan usually using Alaska pollock as the raw material. The washing removes odorous substances and the whole process produces a bland white product which can be combined with flavourings and other additives. Various shapes such as curled shrimp tails can then be produced by extrusion. The advantages of surimi are its relatively low price, consistent quality and availability, and the fact that it can use the desirable label 'seafood'.

Surimi crab analogues in the form of sticks, flakes and chunks have made a big impact on the US crab market where they are able to fill lower priced market niches in which the natural product is unable to compete (Vondruska 1986). Shrimp imitations have been less successful (Holmes 1988), although surimi does compete effectively with breaded and battered shrimp. Shrimp analogues do not sell well in restaurants because the small savings that customers can make do not outweigh the preference for the genuine article rather than an imitation product. Another use for imitation shrimp products is in toppings for pizzas, but surimi cannot compete with the gourmet image of shrimp and its impact will continue to be restricted because of this.

3.3.7 By-products

The potential of various crustacean by-products has been investigated as a way of generating additional revenue from processing operations. By-products from crustacean processing include heads and shells of shrimp and crayfish. When US crayfish are processed for peeled tail meat a massive 85% becomes residue. There may be some potential for the extraction of the carotenoid pigment astaxanthin (Meyers 1987a) and recovery of meat from the chelipeds (Huner 1988). Crustacean waste also represents a potential source of high protein meal rich in calcium, chitin and pigments, for use in animal feeds. Shrimp heads from some processors are sold to be dried and converted into meal for inclusion in compound shrimp diets, and chitosan, a derivative of chitin which has haemostatic and wound healing properties, can be extracted from shrimp shells (Brzeski 1987; Anon. 1989a). Meyers (1987b) successfully extracted the meat from shrimp heads and used it to make a novel product – shrimp sausages.

3.4 Markets for aquaculture technology, products and services

As the crustacean farming industry expands, opportunities for the provision of goods and services also increase. Already demand for a large variety of supplies and various technical and support services has persuaded some companies to diversify into crustacean aquaculture and others to start up and dedicate themselves exclusively to supplying the needs of the sector. In 1987 the UK Department of Trade and Industry reviewed the market opportunities for intensive marine aquaculture technologies and services (DTI 1988). The market was valued at approximately £850m annually, and a number of opportunities for UK companies were identified. Markets relevant to crustacean farming were:

- feedstuffs*
- speciality food components, e.g. pigments*
- micro-encapsulated and spray-dried algal diets for larvae*
- feeding equipment
- harvesting pumps and graders
- security systems for farms
- predator scaring devices*
- pump technology for farms
- recirculation and biofiltration systems
- aeration equipment
- heat pumps
- water filtration and UV sterilisation equipment
- equipment for live transport
- water quality monitoring equipment
- consultancy for new farms, in turnkey projects, environmental monitoring
- insurance and site assessment*
- biotechnological approaches, e.g. transgenic manipulation and cryo-preservation*
- lobster culture technology
- diagnostic aids based on poly- and mono-clonal antibody systems*
- new vaccines*
- education, training and associated instructional aids and software*
- insurance services*
- venture capital for foreign based-projects
- specialised processing plant for shellfish

* indicates technology and services already well developed in the UK, and therefore offering the best opportunities for export.

3.4.1 Supplies

Feed companies have been quick to respond to the growth in crustacean farming and have developed a range of specialised diets. Of these, shrimp diets are the most significant and are now among the most important of all aquaculture feeds, which together make up the fastest growing sector of the global market for animal feeds (Ratafia & Purinton 1989). Shrimp diets are produced in, and exported from Japan, the USA and Europe, but their output is becoming increasingly centred in the major shrimp farming nations themselves. In some of these countries the governments limit imports of feed to protect home-based production. Partly as a result of this and partly to circumvent trade barriers, foreign feed companies are often encouraged to set-up overseas joint ventures.

Accompanying the production of crustacean feed is the demand for its various ingredients, notably fishmeal but also soybean meal, wheat, and various additives including binding agents, pigments and vitamin and mineral mixtures (Section 8.8.2). Consumption of fishmeal for shrimp diets in 1988 has been estimated at 183 000 mt (Pike 1989) and large increases in demand are anticipated. New and Wijkstrom (1990) predicted a requirement for around 350 000 mt for Asian shrimp culture in the year 2000 and Pike (1989) estimates that for all aquaculture, demand in the same year will total 1.5 million mt.

On a global scale, the fishmeal industry expects to be able to cope with the increasing requirements of aquaculture even though fishmeal production is unlikely to rise. Aquaculture currently accounts for around 10% of consumption, with the remainder destined for poultry and mammalian feeds. During the formulation of animal as opposed to shrimp feeds, greater reliance can be placed on terrestrial sources of protein (such as soybeans and wheat) and this can release an increasing portion of the available fishmeal to aquaculture (Pike I., 1990, pers. comm.). Even so, increasing demand overall will push up the price of fishmeal and adversely effect the economic viability of some intensive and semi-intensive crustacean farms (McCoy 1990). A point of concern raised by New and Wijkstrom (1990) is that shrimp farming within Asian nations has increased the demand for home-produced fishmeal and local supplies of fish for its manufacture. This has reduced the availability of low cost fish for human consumption, and also inflated the price of important local foods such as poultry which partly rely on fishmeal for their diets (Section 11.2).

Other aquaculture supplies include specialized feeds for use in hatcheries. Live feeds such as *Artemia* are consumed in nearly all operations that rear crustacean larvae and early post-larvae. Live algae for penaeid larvae are grown on the spot in hatcheries but there is a small market for sterile starter cultures. Specialised artificial diets have been developed to reduce dependence on live feeds and satisfy demand for more convenient forms of feed (Section 8.8.1). They are typically highly priced (sometimes in excess of US$100 kg^{-1}), and because of the relatively small volumes required they are easily shipped around the world. Several companies based either in Europe, Japan or the USA, currently compete for a leading position in this expanding market. Most of the diets are size-graded compounded microparticles, although innovations include micro-encapsulated diets, heterotrophically grown and spray-dried algae, nutrient-enriched dried yeast, and special nutrient mixtures for feeding to *Artemia* and rotifers to boost their nutritional value. The global market for larval feeds (primarily shrimp larvae) has the potential to rise to around 2000 mt in the present decade.

A significant component of the cost of trapping crayfish in the USA is bait. Trash fish is traditionally used, but, recognizing the need for an effective and more stable product with good storage properties, some 13 US companies now market artificial baits.

Other supplies used by crustacean farmers include chemical products for ponds, such as inorganic fertilizers; the fish eradicators rotenone and saponin; and various treatment and conditioning products such as bentonite, zeolite and lime. In hatcheries smaller quantities of chemicals are used, including nutrients for algae culture, disinfectants and various therapeutants. In the future, markets for diagnostics and alternative therapeutants such as vaccines may expand and reduce dependence on antibiotics. Diagnostic reagents can enable the rapid recognition of diseases and already there are at least 17 companies worldwide who are involved in aquaculture diagnostics. Medications can either be administered to the culture water or added to feeds, but manufacturers of new therapeutants for aquaculture need to conform with the drug and chemical registration processes (Ratafia & Purinton 1989). Although there is a great need for disease diagnosis and advice, usually only the services of vets or fish disease specialists are available. The number of crustacean disease specialists in the whole world is probably fewer than 10.

3.4.2 Equipment

A sizeable market exists for the equipment used to establish farms and hatcheries (Tables 10.11 and 10.8). There are several aquaculture supply companies which specialize in providing a wide range of equipment and some can offer complete start-up packages. Other companies provide specialized or customized equipment in addition to the regular items needed during production.

A large part of the investment in a crustacean processing plant is in its equipment. Most units require cold storage facilities, refrigeration plant and ice machines and often additional specialized equipment depending on the type of processing to be carried out (Section 10.9.5). Equipment needs are greatest for the production of value-added foods and these are becoming increasingly important in world crustacean markets (Section 3.1). The various processes involved in feed production (Figure 8.7) also require large amounts of hardware and again offer opportunities to the suppliers of specialized equipment to sell and install their systems, and provide spare parts and back-up services.

3.4.3 Broodstock and nauplii

Significant opportunities exist to satisfy the hatchery demand for broodstock, and in the case of shrimp, for nauplii. Demand for penaeid broodstock in south-east Asia has stimulated substantial international trade, some of which is carried out in contravention of legal restrictions. Demand for shrimp nauplii can be measured in billions. One study estimated that hatcheries in five leading shrimp farming nations (China, Ecuador, Taiwan, Indonesia and Thailand) use a total of 197 billion nauplii per year. Estimated average prices for the nauplii varied greatly between US$0.035 and US$1.25 per thousand, and a total market value of more than US$30m was identified for these five countries. Some nauplii are traded internationally (Section 7.2.3) and the potential of world markets has encouraged experimentation with nauplii cryo-preservation techniques to facilitate storage and long-distance transport.

Markets also exist for juvenile and broodstock crayfish. In Europe, for example, more than 1 million crayfish juveniles are sold each year, either as second or third stage hatchlings or as one summer old juveniles (Huner *et al.* 1987). Once the problems of rearing the larvae of spiny lobsters have been solved, there is a potential market for juveniles to stock on-growing operations, and if clawed lobsters can be successfully ranched there could also be a sizeable market for their juveniles.

3.4.4 Services

Basic services required by shrimp farms can include pond cleaning and harvesting, for which extra manpower and special equipment may be called for. Professional services required by the crustacean farming business as a whole include market analysis, financial planning and appraisal, accounting, legal services and insurance (Sections 9.5.2 and 11.6).

In addition many different forms of technical service are in demand. Complete turnkey systems usually comprise comprehensive engineering, design and management packages (Section 9.5.4). The services of consultants or consulting companies may be hired for a range of different subjects, including the preparation of prefeasibility and feasibility studies, project supervision, training and technical assistance (Section 9.5.2).

The specialized aquaculture expertise of highly trained technical advisors and consultants is easier to sell when advanced and intensive technology is required. At the operational level most extensive culture systems need only basic management skills (MacPherson & Mackay 1987). This is one reason why Westerners, and technology sellers in general, tend to encourage the development and installation of intensive systems since these are most likely to require their input. This should not however be allowed to distort the choice of culture systems in situations where extensive and semi-intensive methods are clearly the most appropriate (Section 11.2).

Demand for up-to-date information on developments in research, culture technology, markets and general industry news, supports the publication of a whole range of books, newsletters and periodicals. Some of these are dedicated to crustaceans, while most are concerned with aquaculture in general or with aquaculture in a particular country. Sources of aquaculture information are listed in McVey *et al.* (1989) and Turnbull (1989), and the latter includes a section on computerised information searches.

Many managers now make use of computers and specialized commercial software to improve the production efficiency of hatcheries and farms. Once software systems become established, future revenue for software companies will rely on the development of updated and improved packages. Using computers to monitor and control pond or hatchery water conditions, however, can easily lead to problems when technicians accept the data returned as accurate without question. The system that can distinguish between a fouled probe and deteriorating water quality has yet to be invented.

Finally, there are many opportunities to provide or sell educational programmes or technical courses for students and other interested parties such as businessmen and entrepreneurs. Popular short courses on shrimp aquaculture are run in the USA especially for people who are new to crustacean aquaculture. They combine hands-on experience with a basic introduction and training in techniques, and provide background information on subjects such as economics and marketing. Lists of practical and academic courses relating to aquaculture have been compiled for the UK and Ireland (Anon. 1990c), and in Anon. (1989b) (principally covering the USA).

3.5 References

Aiken D.E. & Waddy S.L. (1989) Culture of the American lobster, *Homarus americanus*. In *Cold-water aquaculture in Atlantic Canada* (Ed. by A.D. Boghen), pp. 79–122. The Canadian Institute for Research on Regional Development, Moncton.

Anon. (1988) The world shrimp industry. In *Shrimp '88, Conference proceedings*, pp. 1–6, Bangkok, Thailand, 26–28 Jan. 1988. Infofish, Kuala Lumpur.
Anon. (1989a) Chitosan plant set up. *Infofish International*, (5) 8.
Anon. (1989b) Universities and institutions offering aquacultural courses. 18th annual buyers guide, *Aquaculture Magazine*, pp. 40–49.
Anon. (1990a) 'Fast shrimp' in the USA. *Seafood International*, June 1990, p. 14.
Anon. (1990b) Black and blue tigers. *Seafood International*, June 1990, p. 11.
Anon. (1990c) At a glance course guide. *Fish Farmer*, **13** (2) 52–6.
Brock J.A. (1988) An overview of factors contributing to low yields and diseases in the prawn (*Macrobrachium rosenbergii*) culture in Hawaii. In *Proc. 1st Australian Shellfish Aquacult. Conf.*, Perth, 1988 (Ed. by L.H. Evans & D. O'Sullivan), pp. 117–46. Curtin University of Technology.
Brzeski M.J. (1987) Chitin and chitosan – putting waste to good use. *Infofish International*, (5) 31–3.
Chamberlain G.W. (1989) *Status of shrimp farming in Texas*. Presented at 20th meeting World Aquacult. Soc., Los Angeles, Feb. 12–16 1989.
Chaston I. (1983) *Marketing in fisheries and aquaculture*. Fishing News Books, Blackwell Scientific Publications, Oxford.
Chen L.-C. (1990) *Aquaculture in Taiwan*. Fishing News Books, Blackwell Scientific Publications, Oxford.
Clarke S. (1989) Freshwater crustacean farming: the world scene. *SAFISH* **13** (4) 10–12. South Australian Dept. Fisheries.
Cowan L. (1983) Crab farming in Japan, Taiwan and the Phillipines. *Information Series Q 184009*, Queensland Department of Primary Industries, Australia.
Culley D.D. & Duobinis-Gray L. (1989) Soft-shell crawfish production technology. *J. Shellfish Res.*, **8** (1) 287–91.
DTI (1988) Feasibility study on the technology of mariculture, Vol. 2. Review of technologies and services. Dept. of Trade and Industry, *Resources from the Sea Programme, report 87/27*. Aberdeen University Marine Studies Ltd, Aberdeen.
Evans S. (1990) Seafood trends – The changing role of the importer. *Proc. 21st Ann. Shellfish Conf.*, 15 and 16 May 1990; pp. 36–41. Shellfish Assoc. of Great Britain.
FAO/WHO (1981) Codex standards for fish and fishery products. *Codex alimentarius*, Vol. 5, 1st edn. FAO/WHO, Rome.
FAO/WHO (1984) Recommended international code of practice for shrimps and prawns. *Codex alimentarius*, VB, 2nd edn. FAO/WHO, Rome.
Ferdouse F. (1990a) Asian shrimp situation. *Infofish International* (1) 32–8.
Ferdouse F. (1990b) Shrimp from China. *Infofish International* (3) 24–6.
Filose J. (1988) The North American market perspective. In *Shrimp '88, Conference proceedings*, pp. 20–25, Bangkok, Thailand, 26–28 Jan. 1988. Infofish, Kuala Lumpur.
Forbes A. (1990) The shrimp industry in the People's Republic of China. *Aquaculture Magazine*, **16** (1) 44–8.
Goddard J.S. (1988) Food and feeding. In *Freshwater crayfish, biology and exploitation* (Ed. by D.M. Holdich & R.S. Lowery), pp. 145–56. Croom Helm, London.
Hardman P.J.R., Treadwell R. & Maguire G. (1990) *Economics of prawn farming in Australia*. Presented at International Crustacean Conference, Brisbane, 2–6 July, 1990.

Holdich D. (1990) 'Crawfish state' Louisiana plays host to astacologists. *Fish Farmer*, **13** (4) 32–3.
Holmes K. (1988) Impact of analogue products. In *Shrimp '88, Conference proceedings*, pp. 44–9, Bangkok, Thailand, 26–28 Jan. 1988. Infofish, Kuala Lumpur.
Homziak J. (1989) Producing soft crawfish: Is it for you? *Aquaculture Magazine*, **15** (1) 26–32.
Howarth W. (1990) *The law of aquaculture*. Fishing News Books, Blackwell Scientific Publications, Oxford.
Huner J.V. (1988) *Procambarus* in North America and elsewhere. In *Freshwater crayfish, biology, management and exploitation* (Ed. by D.M. Holdich & R.S. Lowery), pp. 239–61. Croom Helm, London.
Huner J., Gydemo R., Haug J., Jarvenpåa T. & Taugbøl T. (1987) Trade, marketing and economics. In *Crayfish culture in Europe*. Report from the workshop on crayfish culture, 16–19 Nov. 1987, Trondheim, Norway.
IFC (1987) *Marine shrimp farming: A guide to feasibility study preparation*. Aquafood Business Associates, International Finance Corp., P.O. Box 16190, Charleston, SC 29412, USA.
Infofish (1988) Shrimp: imports growing but demand is slow. Globefish market summary, *Infofish International* (6) 33.
Infofish (1989) Shrimp. Globefish market summary, *Infofish International* (5) 25.
Jamieson D. (1988) The European shrimp market. In *Shrimp '88, Conference proceedings*, pp. 37–44, Bangkok, Thailand, 26–28 Jan. 1988. Infofish, Kuala Lumpur.
Jones C.M. (1990) The biology and aquaculture potential of the tropical freshwater crayfish *Cherax quadricarinatus*. *Information series QI 90028*, Queensland Department of Primary Industry, Australia.
Koksal G. (1988) *Astacus leptodactylus* in Europe. In *Freshwater crayfish, biology, management and exploitation* (Ed. by D.M. Holdich & R.S. Lowery), pp. 365–400. Croom Helm, London.
Lambert R. (1990) Value-added shrimp products in Europe. *Infofish International* (4) 11–14.
Lands W.E.M. (1989) Fish and human health: a story unfolding. *World Aquaculture*, **20** (1) 59–62.
Liao D.S. (1984) Market analysis for crawfish aquaculture in South Carolina. *J. World Maricult. Soc.*, **15**, 106–7.
Liao D.S. & Smith T.I.J. (1981) Test marketing of freshwater shrimp *Macrobrachium rosenbergii* in South Carolina. *Aquaculture*, **23** (1–4) 373–9.
Liao I.-C. (1988) History, present status and prospects of prawn culture in Taiwan. In *Shrimp '88, Conference proceedings*, pp. 195–213, Bangkok, Thailand, 26–28 Jan. 1988. Infofish, Kuala Lumpur.
Liao I.-C. & Chien Y.-H. (1990) Evaluation and comparison of culture practices for *Penaeus japonicus*, *P. penicillatus*, and *P. chinensis* in Taiwan. In *The culture of cold-tolerant shrimp* (Ed. by K.L. Main & W. Fulks), pp. 49–63. The Oceanic Inst., Honolulu.
Low J. (1988) Market outlook for marine shrimps. *Proc. Sem. Marine Prawn Farming in Malaysia*, pp. 85–91. Malaysian Fisheries Soc., Serdang, Malaysia.
Macaulay P.J., Samples K.C. & Shang Y.C. (1983) Assessing the performance of market distribution channels for aquacultured products: The case of Hawaii-

produced marine shrimp. In *Proc. 1st Int. Conf. on Warm Water Aquaculture – Crustacea* (Ed. by G.L. Rogers, R. Day & A. Lim), pp. 34–42. Brigham Young University, Hawaii, 9–11 Feb. 1983.

Macintosh D.J. (1987) *Aquaculture production and products handling in ASEAN.* ASEAN Food Handling Bureau, Kuala Lumpur.

MacPherson N. & MacKay T. (1987) *Worldwide market opportunities for mariculture technology and services.* Phase 1 report for Aberdeen Univ. Mar. Studies, Mackay Consultants, Inverness, Scotland.

McCoy D. (1990) Fishmeal – the critical ingredient in aquaculture feeds. *Aquaculture Magazine,* **16** (2) 43–50.

McVey E.M., Hanfman D.T., Smith M.F. & Townsend Young A. (Eds.) (1989) *Practical aquaculture literature II.* A bibliography, Bibliographies and Literature of Agriculture, No. 75, US Dept. Agriculture, National Agricultural Library, Aquaculture Information Centre.

Meyers S.P. (1987a) Crawfish – total product utilization. *Infofish International* (3) 31–2.

Meyers S.P. (1987b) Sausages from shrimp wastes. *Infofish International* (5) 36.

New M.B. (1990) Freshwater prawn culture: A review. *Aquaculture,* **88** (2) 99–143.

New M.B. & Wijkstrom U.N. (1990) Feed for thought. *World Aquaculture,* **21** (1) 17–23.

Nierentz J.H. & Josupeit H. (1988) The European shrimp market. *Infofish International,* (5) 14–18.

O'Sullivan D. (1988) The culture of the marron (*Cherax tenuimanus*) in Australia: A review. *Program and Abstracts, East meets West,* p. 60. 19th Annual Conference and Exposition, World Aquacult. Soc., Hawaii.

O'Sullivan D. (1989) *Marketing freshwater crayfish.* Presented at Symp. on Trop. Aquacult., N. T. University, Australia, 5–6 Aug. 1989 (mimeo).

Piggot G.M. (1989) The need to improve omega-3 content of cultured fish. *World Aquaculture,* **20** (1) 63–8.

Pike I. (1989) Fish meal industry should meet farm demands. *Fish Farming International,* **16** (7) 86–8.

Pillay T.V.R. (1977) *Planning of aquaculture development – an introductory guide.* FAO, Fishing News Books, Blackwell Scientific Publications, Oxford.

Ramamurthy V.D. (1990) US quality standards for frozen raw headless shrimp. *Infofish International,* (1) 50–52.

Ratafia M. & Purinton T. (1989) Emerging aquaculture markets. *Aquaculture Magazine,* **15** (4) 32–46.

Richards M. (1988) Co-operative marketing of signal crayfish in the United Kingdom. In *Proc. 1st. Australian Shellfish Aquacult. Conf.* (Ed. by L.H. Evans & D. O'Sullivan), pp. 326–30. Perth, 1988, Curtin University of Technology.

Roberts K.J. & Dellenbarger L. (1989) Louisiana crawfish product markets and marketing. *J. Shellfish Res.,* **8** (1) 303–7.

Ruello N. (1990a) Prawn market update. *Austasia Aquaculture Magazine,* **4** (10) 20.

Ruello N. (1990b) Indonesian farmed prawns in Sydney. *Austasia Aquaculture Magazine,* **4** (1) 23.

Shortall D. (1990) Canadian lobster – resource management and market development. *Proc. 21st. Ann. Shellfish Conf.,* pp. 21–31, 15 and 16 May 1990. Shellfish Assoc. of Great Britain.

Siegel R.A. (1989) The growing influence of aquaculture in US seafood markets: salmon and shrimp. In *Aquaculture a Review of Recent Experience*, pp. 296–313. OECD, Paris.

Singh T. & Jee A.K. (1988) Marine prawn farming in Malaysia. Present problems, future prospects. *Malaysian Fisheries Soc., Occasional Publication No. 1*, Universiti Pertanian Malaysia, Selangor, Malaysia.

Sokol A. (1988) The Australian yabby. In *Freshwater crayfish, biology, management and exploitation* (Ed. by D.M. Holdich & R.S. Lowery), pp. 401–25. Croom Helm, London.

Sumner J. (1988) Import regulations and trade barriers. In *Shrimp '88, Conference proceedings*, pp. 50–58, Bangkok, Thailand, 26–28 Jan. 1988. Infofish, Kuala Lumpur.

Tucker B.W. (1989) Sterols in seafood: a review. *World Aquaculture*, **20** (1) 69–72.

Turnbull D.A. (1989) *Keyguide to information sources in aquaculture*. Mansell Publishing, London.

van Eys S. (1987) Japanese shrimp market – price and quality above all. *Infofish International*, (4) 19–21.

Vondruska J. (1986) *Blue crab markets and analog products*. National Marine Fisheries Service, Prepared for National Blue Crab Industry Assoc. Annual Meeting, New Orleans, Louisiana, Feb. 26–28, 1986.

Williams A.B. (1988) *Lobsters of the world – An illustrated guide*. Osprey Books, Huntingdon, NY.

Yamaha (1989) Prawn culture. *Yamaha fishery journal*, No. 30, Yamaha Motor Co., Shizuoka-ken, Japan.

Chapter 4
Candidates for cultivation

4.1 Introduction

It is now technically possible to culture a large number of commercially valuable crustaceans through all or at least a major part of their life cycle, in many cases at a considerable distance from their home range. However, it is not yet feasible to culture many of these animals profitably because of low market value, slow growth rate or because of the need to create an expensive culture environment. Such considerations have ruled out several superficially attractive species for aquaculture at least for the foreseeable future. Among these are: *Nephrops norvegicus* (scampi) unsuitable, along with several other nephropid lobsters, because of a pronounced burrowing habit or a preference for offshore environments; most caridean prawns (other than two or three *Macrobrachium*, and perhaps one *Pandalus* species) on the basis of small size, slow growth and low fecundity; and the majority of crabs, on account of low value and sometimes lack of data on their requirements during culture.

Recent Japanese advances in the laboratory culture of spiny lobster larvae are exciting but at the time of going to press only a few individuals have survived to reach the juvenile stages and commercially viable mass rearing techniques have yet to be developed. Despite these erstwhile drawbacks, some of the larger crustaceans, notably *Homarus* and some crabs and spiny lobsters, may be suitable for sea ranching provided problems concerning ownership can be resolved (Sections 7.8.11, 7.10.8 and 11.6.3).

The primary goal of most (but not all) crustacean aquaculture enterprises is to make money as quickly as possible. In the simplest terms an 'ideal' species must therefore command a high price in an established market, it must grow rapidly to a marketable size on inexpensive, readily available diets, be resistant to disease while not having stringent environmental or biological requirements, and be readily available for culture from wild-caught or hatchery-reared stock. The only species to possess the majority of these attributes are to be found among the penaeid shrimp, which is of course why they form the bulk of farmed crustaceans (Table 1.1). There are no known freshwater species possessing so many favourable characteristics although there is currently (1992) some speculation that the Australian red claw crayfish (*Cherax quadricarinatus*) might one day supersede *M. rosenbergii* in popularity among aquaculturists.

For some species or locations several quite different approaches to culture methodology are worth considering; in other situations a species may be grown profitably using methods modified only to take account of increasing intensification. This chapter examines the factors governing the selection of species to be cultured in relation to environment, biology, product marketability and other aquatic organisms (polyculture). At the same time possible candidates for 'catch crops' or off-season cultures that might be employed to improve cash flows or increase marketing opportunities (Liao 1988) are considered, together with some other potentially cultivable species.

4.2 Location

Compatibility with the local environment and the availability of brood- or seedstock (post-larvae or juveniles) are perhaps the two most important reasons for choosing to culture a species within its home range. Table 4.1a shows the broad environmental requirements of important cultivable crustaceans while Table 4.1b indicates their natural geographic distribution and gives examples of areas into which they have been transplanted for the purposes of cultivation.

When a species is cultured outside or near the limits of its normal distribution but within a comparable climatic zone, the farmer must establish reliable supplies of juveniles or broodstock and/or maintain broodstock production facilities locally. Reliable importation of juveniles is seldom achievable in the long term and becomes particularly difficult and expensive as scale increases. If present trials with cryopreserved penaeid eggs and nauplii prove successful, the technique may provide a useful, though probably expensive (Section 3.4.3), means of buffering fluctuations in availability or quality of seed (Section 12.2). A species that reproduces readily in captivity in the new environment is therefore more suited to transplantation than one which does not. The early development of *Macrobrachium rosenbergii* farms in Hawaii, the Caribbean Islands and Israel, and of *Procambarus* and *Pacifastacus* culture in Europe, provide good examples of this.

Transportation of crustaceans to new locations carries with it the risks of transferring diseases, and ecological problems may arise when the introduced species escapes and becomes established outside the farm (Section 11.4). Increasingly legislation is being invoked or adapted to control such movements in an attempt to protect the industry and local fisheries (Sections 11.4.2 and 11.6.3.2).

Under some circumstances (for example where proximity to western markets is a

Table 4.1a The environmental habitats of the major species and groups of cultivable crustaceans.

	Marine/brackish	Freshwater
Tropical	*Penaeus indicus* *P. merguiensis* *P. monodon* *P. stylirostris* *P. vannamei* *Metapenaeus* spp. *Panulirus* spp. *Scylla serrata*	*Macrobrachium* spp. *Cherax quadricarinatus*
Sub-tropical/ warm temperate	*Penaeus chinensis* *P. japonicus* *P. penicillatus* *Jasus* spp. *Panulirus* spp. *Portunus* spp.	*C. destructor* *C. tenuimanus* *Procambarus clarkii*
Temperate	*Pandalus platyceros* *Homarus gammarus* *H. americanus*	*Astacus leptodactylus* *Pacifastacus leniusculus*

Table 4.1b The geographic distribution of the major species and groups of cultivable crustaceans with examples of known transplantations.

Crustacean	Natural range	Transplantations
Penaeid shrimp		
Penaeus chinensis	Korea, China, Gulf of Bohai, Northwest Pacific	New Zealand, France, USA
P. japonicus	Japan, Korea, Western Central Pacific	Mediterranean, Brazil France
P. monodon	Indo-Pacific, Australia	Middle East, West Africa, Middle America, Italy
P. stylirostris	Eastern Central Pacific	USA, Caribbean Islands, Tahiti, Hawaii, Taiwan
P. vannamei	Eastern Central Pacific	USA, Caribbean Islands, Tahiti, Hawaii, Taiwan
Caridean prawns		
Macrobrachium rosenbergii	Indo-Pacific, Australia	Caribbean Islands, USA, Hawaii, Israel, USSR, Southern Africa
Pandalus platyceros	Eastern Central Pacific	—
Crayfish		
Astacus leptodactylus	Turkey	UK, Italy, France, Switzerland, Austria
Cherax destructor	South-eastern Australia	—
C. tenuimanus	Western Australia	USA, Caribbean Islands Panama, South Africa, China, Malaysia, Japan, France, New Zealand
C. quadricarinatus	Northern Australia, (Queensland)	USA, Caribbean Islands
Pacifastacus leniusculus	North-western USA	Scandinavia, UK, USSR, Italy, Spain, Holland, Denmark, East Europe
Procambarus clarkii	Southern USA	East Africa, Spain, Caribbean Islands, Central and South America, Japan, Taiwan, France, South Africa, China, Thailand
Lobsters		
Homarus americanus	Canada, Atlantic USA	Italy, Japan, Pacific USA
H. gammarus	Europe, Mediterranean	—
Spiny lobsters		
Jasus spp.	Australia, New Zealand, South Africa	France, Japan
Panulirus spp.	Japan, Australia, India, Africa, Caribbean, Eastern Central Pacific	Japan
Crabs		
Portunus spp.	Japan, Korea	—
Scylla serrata	Indonesia, Thailand, Australia	—

prime advantage) it may be thought desirable to culture a species in a climatic region to which it is not well adapted. Attempts may also be made to increase growth rate of temperate water species such as the shrimps *Hymenopenaeus mulleri*, *Artemesia longinaris* and *Sclerocrangon boreas* by raising the temperature of the water, for example through direct or indirect use of heated effluent or geothermal waters (Wickins 1982). In addition to the problems of animal supply, these situations require the creation and maintenance of some kind of heat exchange or controlled-environment system. Nevertheless, such approaches have been attempted commercially (Rosenberry 1982; McCoy 1986; Anon. 1987; Campbell 1989) and the tolerance of several species to controlled-environment systems evaluated (Forster & Beard 1974; Hanson & Goodwin 1977; Wickins & Beard 1978; Morrissy 1984; Beard *et al.* 1985; Kowarsky *et al.* 1985; Cogan 1987; Waddy 1988; O'Sullivan 1990). However, no controlled-environment on-growing systems are currently (1992) known to be commercially viable.

4.3 Broodstock

4.3.1 Seasonal availability

Security and predictability of broodstock supply are prerequisites for any successful crustacean hatchery enterprise. Where there is an established fishery for the species to be cultured, or where natural breeding populations are abundant, ripe impregnated or egg bearing females may be caught for at least part of the year. Special dispensation

Table 4.2 Broodstock categories.

Species that mature, mate and spawn readily in captivity
Shrimps *Penaeus indicus, P. japonicus, P. merguiensis*
Prawns *Macrobrachium rosenbergii, Macrobrachium* spp.
Crayfish *Astacus leptodactylus, Cherax destructor, C. quadricarinatus, C. tenuimanus, Pacifastacus leniusculus, Procambarus clarkii*
Species requiring environmental or hormonal manipulation to breed reliably
Shrimps *Penaeus chinensis, P. monodon, P. stylirostris, P. vannamei*
Prawns *Pandalus platyceros*
Clawed lobsters *Homarus americanus, H. gammarus*
Spiny lobsters *Jasus* spp., *Panulirus* spp. ?
Crabs *Scylla serrata* ?

to fish for breeding animals may be required in some areas at certain times of the year, while in other areas a complete ban may be enforced. When assessing a new or expanding project it is wise to consider potential competition for broodstock supplies, particularly in areas where aquaculture developments are being encouraged. Alternative sources of supply should also be investigated.

4.3.2 Ease of establishing and maintaining a broodstock

A number of species mature, mate and spawn readily in production ponds, while others must experience specific seasonal photoperiod and temperature regimes or be induced to breed through surgical or dietary manipulations which act on the hormonal system. In Table 4.2 important cultivable species are categorised according to the ease with which breeding populations may be maintained. Details of the environmental and nutritional requirements for inducing maturation, courtship, mating, spawning and incubation in captivity are described in Chapter 7.

A very different problem arises with *Pandalus platyceros* whose aquaculture potential in temperate regions has been investigated (Wickins 1972; Oesterling & Provenzano 1985). This prawn is a protanderous hermaphrodite. It matures and functions first as a male, but 1.5 to 2.5 years later matures and functions as a female. This habit, combined with a low fecundity (Table 4.6f) poses formidable difficulties in the management of captive broodstock.

4.4 Larvae

4.4.1 Duration and complexity of larval life

The duration and complexity of the larval phase as well as the fecundity of cultivable crustaceans have a major impact on hatchery design, running costs and the technical skills required to maintain a predictable output of post-larvae. However, a lengthy

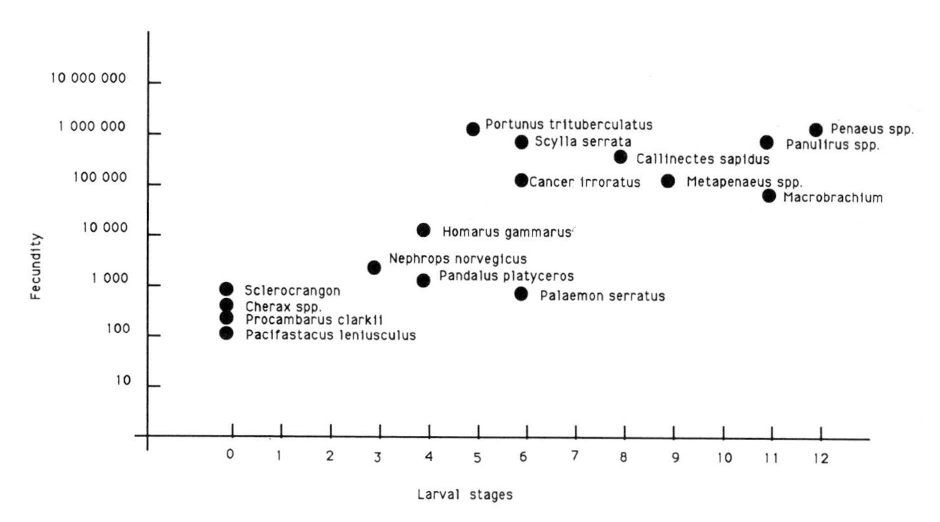

Fig. 4.1 The relationship between fecundity and the number of free-swimming larval stages in selected crustaceans.

larval phase is not always associated with greater complexity, although abbreviated, less complex larval development is generally associated with reduced fecundity (Figure 4.1). For example, penaeid larvae pass through three quite distinct morphological stages, each with different modes of nutrition, in half the time *M. rosenbergii* larvae take to reach their first significant metamorphosis. Fecundity is inevitably low (generally less than 1000 eggs) in freshwater crayfish and the large cold water shrimp *Sclerocrangon boreas*. This is because sufficient yolk must be deposited in each egg to sustain complete development so that juveniles hatch directly from the eggs. In contrast, spiny lobsters incubate large numbers of small eggs with very little yolk but have an enormously protracted larval life.

4.4.2 Resistance to disease

All larvae seem susceptible to disease and infestations but most mortalities are believed to be the result of, or exacerbated by, stresses arising from poor water, water management or food; the longer the larval life the greater the difficulty of maintaining good culture conditions. Penaeid hatcheries seem particularly vulnerable to introduced diseases due to the worldwide trade in broodstock and nauplii (Section 8.9).

4.5 Post-larvae and juveniles

4.5.1 Availability from the wild

Many thriving extensive culture operations (for example penaeid farms in south-east Asia and Ecuador) are based on supplies of naturally occurring post-larvae or juveniles. With intensification and expansion, such industries become increasingly vulnerable to natural fluctuations in seed supply and the price changes that follow (Section 10.9.1). While hatcheries certainly contribute to overcoming deficits in supply, they seldom completely fulfil the demand (Table 4.3). Nurseries stocked with either wild or hatchery produce can go some way to maximising the use of the available seed, and a high tolerance of a species to crowding, compounded diets and, in some cases, recirculated water is a distinct advantage.

In the absence of a proven hatchery technology, attempts to initiate spiny lobster fattening operations based on the collection of wild pueruli or juveniles have often been frustrated by the difficulty of locating reliable supplies or by the costs involved in collection (Ting 1973). Recent advances in the understanding of juvenile ecology and in collector design, however, have rekindled interest in the prospects for fattening and ranching operations (Sections 7.9 and 12.8).

4.5.2 Nursery

Following metamorphosis or capture, crustacean post-larvae benefit from special attention (weaning) as they pass through a period of transition from epipelagic to benthic life. Fundamental physiological and behavioural changes occur at this time as they are weaned away from the diets and environmental conditions they enjoyed as larvae, to compounded diets, sometimes (for example during *Macrobrachium* culture) with a concurrent change of salinity. The rate at which weaning can be successfully completed and the juveniles grown to a size suitable for stocking in on-growing ponds

Table 4.3 Hatchery capacity and the demand for post-larvae.

Country	Number of hatcheries	Hatchery output	Anticipated demand (wild and hatchery)	Year
		(Millions of post-larvae per year)		
China				
Penaeus	10 govt. + backyard	72 443	72 443	1987
	340	42 000	—	1989
Ecuador				
Penaeus	70	3000	6000	1986
	85 (+ 30 not in production)	6000	11 000	1989
	40 producing	4320	15 000	1990
Indonesia				
Penaeus	38 + 85 planned	34–110	1750	1987
	93	1460	7300	1988
Macrobrachium	5	3.6	—	1987
Japan				
Portunus	17	16	—	1982
Korea				
Penaeus	3 govt. + backyard	78.3	—	1989
Malaysia				
Penaeus	32	539–600	—	1987
Philippines				
Penaeus	56	100	—	1987
Taiwan				
Penaeus	1200	4500	4500+	1987
Macrobrachium	20	25	—	1987
Thailand				
Penaeus	5 major + >1000 backyard	41 18.8	152 180 (*P. monodon*)	1984 1985
	5 major + >1000 backyard	6000	6000+	1989
Macrobrachium	44	80	—	1987
USA				
Penaeus	6	138 (not yet achieved)	800–1000	1989

Sources Cowan 1983a; Chan 1988; Juario & Benitez 1988; Wickins 1988; Anon. 1990; Buddle 1990; Main & Fulks 1990.

Table 4.4 Growth rate and survival of crustaceans during the nursery phase of culture.

Species	No. m^{-2}	Period (days)	Survival (%)	Growth		Reference
				from	to	
Penaeus spp.						
P. chinensis	25	14	100	0.15 g	0.97 g	Forster & Beard 1974
	166	28	92	0.13 g	2.21 g	Forster & Beard 1974
	150–300	45	86–100	7 mm TL	50 mm TL	Main & Fulks 1990
	135–234	18–23	33–46	6–7 mm TL	0.06–0.21 g	Main & Fulks 1990
P. indicus	25	14	93	0.26 g	0.61 g	Forster & Beard 1974
	166	28	68	0.23 g	1.08 g	Forster & Beard 1974
	50–200	32–80	>60	PL5	1 g	AQUACOP 1984
P. japonicus	25	14	100	0.2 g	1.13 g	Forster & Beard 1974
	166	28	99	0.2 g	1.62 g	Forster & Beard 1974
P. monodon	25	14	100	0.15 g	1.27 g	Forster & Beard 1974
	166	28	100	0.15 g	1.87 g	Forster & Beard 1974
	50–200	35–85	<50	PL5	1 g	AQUACOP 1984
	50–150	45–50	47–85	PL10–15	0.5–1.8 g	Tacon A. 1986 pers. comm.
	50–100	30–40	60	PL5	1–2 g	Juario & Benitez 1988
	125–1000	35	83–98	0.013 g	0.25–0.87 g	Briggs M. 1990 pers. comm.
P. stylirostris	50–200	30–45	<50	PL5	1 g	AQUACOP 1984
	150–250	45–50	55–85	PL4–6	0.8 g	Tacon A. 1986 pers. comm.
P. vannamei	50–200	45–50	>60	PL5	1 g	AQUACOP 1984
	200–300	45–50	69–95	PL4–6	0.8 g	Tacon A. 1986 pers. comm.
Prawns						
M. rosenbergii	90–193	60–120	17–76	PL5	1 g	AQUACOP 1983
	25	14	100	0.13 g	0.33 g	Forster & Beard 1974
	166	28	98	0.15 g	1.04 g	Forster & Beard 1974
	40	35	65–83	0.04 g	0.38 g	Mulla & Rouse 1985
Pandalus platyceros	N/A	60	98	0.63 g	2.8 g	Oesterling & Provenzano 1985
Crayfish						
Astacus astacus	800	112	49	0.07 g	0.63 g	Keller 1988
A. leptodactylus	20–30	120	85–90	0.03 g	3.34 g	Tcherkashina 1977
A. leptodactylus	130	90	51	0.04 g	0.84 g	Koksal 1988
Cherax destructor	263	42	70	0.10 g	0.75 g	Geddes *et al.* 1988
Cherax destructor	154	50	40	0.02 g	0.65 g	Jones 1990

Table 4.4 (contd.)

Species	No. m^{-2}	Period (days)	Survival (%)	Growth		Reference
				from	to	
Cherax quadricarinatus	1278	36	52	0.02 g	0.24 g	Jones 1990
Pacifastacus leniusculus	200	100	42	0.03 g	1.18 g	D'Abramo *et al.* 1985
Procambarus clarkii	200–300	30	80–90	0.02 g	0.50 g	Huner & Barr 1984
Lobsters						
Homarus americanus	N/A	60	95	0.44 g	1.48 g	Norman-Boudreau & Conklin 1983
H. gammarus	400	30	80	Sta. 4	0.16 g	Beard *et al.* 1985

varies between species from a few days to a month or two. A short period is generally advantageous although some on-growing strategies are dependent on indoor nursing throughout a winter period (*Macrobrachium* culture in the southern USA and Israel) in order to be able to stock ponds with juveniles as soon as it is warm enough for growth outdoors. Examples of the survival and growth rate of selected species during the nursery phase of culture are given in Table 4.4.

Some species (*Homarus*) are so cannibalistic that severe losses and mutilation occur if they are not placed in individual rearing containers soon after metamorphosis. Various attempts to rear juvenile homarids communally for six to 12 months have been made (Henocque 1983; Waddy 1988) and some workers have suggested it might be worthwhile to periodically remove or immobilize the claws to reduce aggression and fighting during this stage (Aiken & Young-Lai 1981; Kendall *et al.* 1982; Karplus *et al.* 1989). The treatment however, increases risks of disease, incurs post-operative losses and would not be considered humane in some countries.

4.6 On-growing

4.6.1 Growth rate and size distribution

Rapid growth rate is one of the most important attributes of a candidate for aquaculture in that it maximises cash flow, minimises the period in which crop loss would be financially most damaging, and minimises time taken to recover from crop failure. The farmer, however, is interested not only in average growth rate but also in the proportion of the crop that fetches the highest price. This is usually that which contains the largest animals (Table 4.5), but in some crab and crayfish operations it may be ovigerous or soft shelled animals (Sections 7.5.7, 7.10.4 and 7.10.9). Both sex and stocking density can affect the final distribution of sizes in the crop, and here again *Macrobrachium* presents particular problems in that dominant males grow substantially larger than females and hierarchies develop which encourage the further spread of sizes. Specific stocking and harvesting strategies are employed to minimize these

Table 4.5 Examples of the proportion (%) of harvested crustaceans reaching marketable sizes.

Penaeus monodon, Thailand, examples of seasonal variations.

Size (g)	<25	26	29	32	36	>38
Jan.	1.3	0	27	5.5	32.6	25.4
Feb.	5.7	0.8	31.4	10.8	23.6	22.6
Mar.	7.4	38.1	14.7	6.1	20	10.8
Apr.	9.5	15.6	19.5	0.2	16	29.9
May	45	2.7	22.5	5.9	9	6.9
June	50.5	1.5	22.7	3.1	6.4	5.3
July	32.6	0.7	22.5	9.4	7.8	13.8

P. monodon, Philippines.

Size (g)	<22.1	24.2	32.1	52.6	>52.7
%	5.5	5.6	54.1	34.3	0.2

P. monodon, India.

Size (g)	9	12	15	19	24	29	34	41
%	20.3	25.3	7.0	7.9	8.3	10.4	12.4	8.3

P. stylirostris, Belize.

Size (g)	<11	11	13	16	19	22	26
%	5.8	2.9	11.6	39.1	26.1	11.6	2.9

Macrobrachium rosenbergii, Israel (Karplus *et al.* 1986).

Size (g)	<20	20–25	25–30	30–35	35–40	40–45	45–90
%	9.9	6.6	8.5	14.5	18.4	35.5	6.6

Macrobrachium rosenbergii, S. Carolina, USA (Smith *et al.* 1982).

Tails kg^{-1}	>154	111–154	78–110	56–77	<56
Pond					
Alcolu 1	16	26	40	14	4
Alcolu 3	5	0	25	60	10
Cameron	6	4	27	32	31
Moncks C.	14	22	35	24	5
Murrells	5	9	30	33	23
Ridgeland	23	37	27	10	3
Sumter	8	27	44	18	3
Walterboro	21	43	21	11	4

Macrobrachium rosenbergii, Jamaica.

Count lb^{-1}	8–12	13–15	16–20	21–25
%	20	15	35	30

Cherax destructor, Australia. / *C. tenuimanus*, Australia.

Size (g) (after 1 yr)	<40	>40	>40 (*C. tenuimanus*)
%	56–64	36–44	70

Homarus gammarus, UK (Change in percentage reaching 85 mm CL with time).

Years	<2	2.25	2.5	2.75	3	3.25	3.5
%	3	18	38	22	12	6	1

effects (Section 7.3.7) and research on the production and usefulness of mono-sex populations is in progress (Section 12.3). Even when lobsters are held individually, the range of sizes that develops may mean that only a proportion of the crop reaches a legally saleable size at any one time (Table 4.5). In the absence of behavioural interaction the variation may be due to both genetic and husbandry components (Section 12.10).

4.6.2 Tolerance to water quality changes

Species that survive and grow well under culture conditions are generally those that are adapted to a shallow water, naturally changeable environment (Figure 4.2). Nowhere is this more apparent than in the cases of inshore and estuarine penaeids, *Macrobrachium*, the red swamp crayfish (*P. clarkii*) and perhaps the yabbie (*C. destructor*). Species from more stable or deep oceanic environments (e.g. spiny lobster larvae and clawed lobsters) do not usually perform so satisfactorily and often require more stringently controlled culture conditions.

4.6.3 Resistance to disease

A high degree of resistance to disease is a feature of many successfully cultured species but three features are worthy of note in the context of assessing a culture project. Firstly no species is immune to disease and serious losses can occur if the animals or their environment are overstressed. The near collapse of the Taiwanese shrimp culture industry in 1988 following the unprecedented demands made by the rapidly expanding industry provides a timely warning of this (Lin 1989). Secondly, the use of seed from outside the farm carries with it the risk of importing disease, particularly if the exporting hatchery itself imports broodstock from elsewhere. Thirdly, European and Australian crayfish are all susceptible to the crayfish fungus plague likely to be present wherever North American species are found (Sections 8.9 and 11.4). Substantial investment in cultivating these groups outdoors in such areas may be unwise.

4.6.4 Other factors

Among other factors that may influence the choice of a culture species or an assessment of a project are the ability of an animal to escape from ponds and the ease and efficiency with which it may be harvested. *Pacifastacus leniusculus* and some species of *Cherax* and *Macrobrachium* can travel easily over damp terrain, and anti-escape fences may be necessary. Burrow-dwelling species of crayfish, for example *P. clarkii*, have to be harvested by trapping rather than pond draining, with the result that an unknown number may remain unharvested. The distance of the farm from the market and the ability of an animal to withstand the rigours of live transportation and storage are of importance if valuable speciality live sales are sought (Section 3.2.1). Lobsters, crayfish, crabs and spiny lobsters are traditionally sold live, but most shrimp and prawns are sold frozen or on ice.

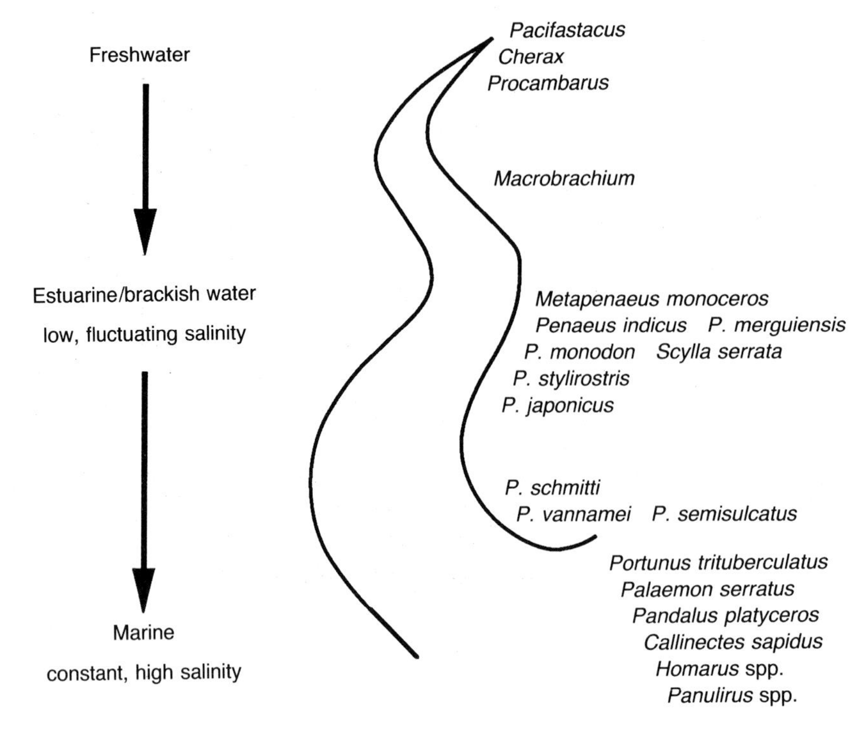

Fig. 4.2 Habitat preferences of selected crustaceans.

Two morphological features of decapod crustacea are of interest when choosing a species to culture. These are the presence of large chelae, typically indicative of a territorial or predatory species, and the proportion of the body that is eaten, usually the tail and claw muscle. *Homarus*, *Nephrops*, most *Macrobrachium* species and probably most crayfish are naturally territorial or solitary, and when crowded together are likely to become unduly aggressive. The density at which fighting and cannibalism becomes a problem varies between species, with crayfish being far more tolerant of crowding than lobsters. Limb loss, shell blemishes and increased size variation are the chief manifestations of overcrowding, and all of them seriously affect the market value of the crop.

Although both head and tail are eaten in some parts of south-east Asia, and during the consumption of soft shelled crabs and crayfish, most markets are based on the edible muscle of tail and claws. The amount of edible muscle per unit weight of animal is greater in penaeids (57–68%) than in *M. rosenbergii* (35–40%) and crayfish (10–40%) (Tables 4.6e–i).

The appearance and coloration of the farmed product becomes increasingly important as competition increases between farms and between farmed and fished crustaceans. The ability to influence coloration and flesh composition during culture – through, for example, tank background colour, light regimes and diet – is available for a range of species and could be especially useful for those that would be grown under a reasonable degree of environmental control (Section 3.2).

4.7 Comparison of species

The most significant biological and husbandry aspects that affect the main operational phases of crustacean farming are presented for broodstock (Table 4.6a), larvae culture (Table 4.6b), nursery (Table 4.6c) and on-growing (Table 4.6d). These tables show the broad differences and similarities between the major groups or species of interest and give an indication of the range in survival rates or yields that might be anticipated. Wherever possible, data judged to represent performances obtainable under commercial conditions were selected. In other cases, ranges of values from pilot or laboratory trials were used. The average values can be used to compare general performance attributes of the groups, but it is emphasised that the following examples should not be used in financial calculations; they are shown here simply to give a sense of perspective when considering which species to culture.

As an example, rough estimates of the numbers of harvestable penaeids that can be reared from one female can be obtained by multiplying the mean fecundity by broodstock productivity (600 000 × 0.55, Table 4.6a) which gives 330 000 nauplii which, when multiplied by the estimated overall survival (0.36, Table 4.6d) gives 118 800 harvestable shrimp. Estimates for the remaining groups are:

- *M. rosenbergii* 7020
- temperate crayfish 37
- tropical crayfish 237
- clawed lobsters 3187
- spiny lobsters* 210 animals

* assuming broodstock productivity is 30%

No figure is calculated for crabs because of the widely different estimates of broodstock performance between species.

Similar, very approximate, comparative estimates may be calculated for the area of on-growing pond or tank bottom required to produce 1 tonne of market sized crustaceans. For example, survival multiplied by average initial stocking density (0.7 × 15, Table 4.6d) yields 10.5 shrimp of 30 g each or 0.315 kg m^{-2}; thus 1000/0.315 = 3174 m^2 of pond bottom are needed to produce 1 tonne of shrimp. This figure equates roughly to a yield of 3.15 mt ha^{-1} per harvest, typical of well-managed semi-intensive farms (Section 10.9.1.5). Comparative values for the other species/groups, in m^2, again assuming a good water supply and competent husbandry, are:

- *M. rosenbergii* (single batch harvest) 8888
- temperate crayfish 9524
- tropical crayfish 2778
- lobsters (single layer battery) and spiny lobsters 351
- crabs 2538

The low figures for clawed and spiny lobsters are due in part to their large size at harvest (about 350 g); the high value for crabs harvested at a similar size is a reflection of poorer survivals during nursery and on-growing phases.

Table 4.6a Generalised data for captive broodstock.

Species/group	Maturation frequency (yr^{-1})	Fecundity range	Source	Facility	Productivity (%)*
Penaeids	3–6	0.2–1.0 million	Wild, ponds or tanks	Indoor spawning tanks 0.1–40 mt.	30–80
Macrobrachium	3–6	20 000–80 000	Ponds	Covered incubation/ hatching tanks 1–10 mt with hides	50–80
Crayfish					
Temperate	1–2	60–260	Ponds	Covered incubation/ hatching tanks or cages, hides, mesh floor.	50–80
Tropical	3–5	140–1000	Wild or ponds	as above	80
Clawed lobsters	0.5–1.0	5000–80 000	Wild	Covered incubation tanks, claws banded	30
Spiny lobsters	1	80 000–200 000	Wild	Experimental, covered incubation tanks	30E
Crabs	3–4	50 000–3 million	Wild or ponds	Covered tanks, sand bottom	4–80

* After allowing for female mortality, egg and hatching losses. E = Estimated.

Table 4.6b Generalised data for larvae culture.

Species/group	Duration of larval life (days)	Density (no. ℓ^{-1})	Feed	Rearing facility	Expected survival (%)
Penaeids	12–14	30–200	Algae, rotifers, *Artemia*, flakes, micro-particulates, micro-capsules	Static water tanks 0.1–200 mt	50–80
Macrobrachium	20–40	30–200*	*Artemia*, compounded diets, sieved fish/ invertebrates	Static or recycled brackish water, 1–20 mt tanks.	30–60
Crayfish					
Temperate	0	—	—	—	—
Tropical	0	—	—	—	—
Clawed lobsters	9–14	25–50	Fresh/frozen shrimp, mollusc flesh, adult *Artemia*, mysids.	40–80 ℓ Kreisels (Hughes *et al.* 1974)	20–60
Spiny lobsters	90–230	5–15	*Artemia*	Experimental	<1
Crabs	23–30	6–50	Algae, rotifers, *Artemia*, sieved bivalve flesh	Static 10–200 mt tanks	4–70

* two phase culture (Section 7.3.3)

Table 4.6c Generalised data for nursery culture.

Species/group.	Duration of nursery phase (days)	Stocking density (no. m^{-2})	Rearing facility	Expected survival (%)
Penaeids	10–60	50–200	Small (0.04–1 ha) ponds, open and covered tanks/ raceways	70–90
Macrobrachium	7–90	100–5000	Static or recycled covered tanks, small ponds	70–90
Crayfish				
Temperate	70–120	2–10	Shallow troughs, numerous hides	40–80
Tropical	30–90	2–15	As above, shaded	60–80
Clawed lobsters	20–30	350–400	Individual cells	70–80
Spiny lobsters	90E	200E	Experimental, shaded tanks with hides	>60
Crabs	14–28	2000–3000 max	Shaded tanks, hanging net shelters	20–70

E = Estimated

Table 4.6d Generalised data for on-growing.

Species/group	Duration of on-growing phase (months)	Stocking density (no. m^{-2})	Basic facility	Size at harvest (g)	Expected survival (%)	Estimated mean overall survival (hatch to harvest %)
Penaeids	3–4	5–25	Earth ponds, 0.1–100 ha, 1 m deep	15–45	60–80	36
Macrobrachium	3–5 or continuous	5–10	Earth ponds, 0.1–1.0 ha, 1 m deep	15–45	40–60	18
Crayfish						
Temperate	24–36	3–4	Earth channels, 10–60 m long, 5–12 m wide, 1.5 m deep	40–60	30–90	36
Tropical	12–18	3–5	As above, and ponds 0.1–0.5 ha	40–200	60–90	52
Clawed lobsters	24–30	90–100*	Individual cells, transfer to larger cells periodically	345–400	80–90	25
Spiny lobsters	18–30	—	Experimental, shaded tanks with hides, ponds	350+	80–90E	0.5
Crabs	3–8	0.5–4	Shaded tanks, ponds	200–500	30–70	8

* $9\,m^{-2}$ at harvest. E = Estimated

Table 4.6e A comparison of attributes to be considered when choosing a species to culture: marine and brackishwater shrimp.

Attribute	*P. chinensis*	*P. indicus*	*P. japonicus*	*P. merguiensis*	*P. monodon*	*P. penicillatus*	*P. semisulcatus*	*P. stylirostris*	*P. vannamei*
Age at first maturity (months)	9(v)	7(v)	9(a)	6–7(y)	10–12(v)	—	—	10(v)	6–12(g,v)
Size at first maturity(g)	M 20–30(a,v) F 30–60(a,v)	M 6(v) F 8(v)	M 15–20(a,v) F 20–25(a,v)	M 6–15 F 6–15(z)	M 30–40(m,v) F 60–70(m,v)	M 22–40(a) F 35–60(a)	—	M 35–40(v) F 35–55(v)	M 30–40(v) F 35–45(v)
Control of breeding	With and without ablation(v)	Ablation not necessary(v)	With and without ablation(v)	With and without ablation(y)	With ablation (v)	With and without ablation(v)	With and without ablation(s)	With ablation (v)	With ablation (v)
Interval between spawning (days)	5–30(c,v)	1–9(y)	60(v)	—	4–20(l,v)	4–7(f)	12–15(s)	3–19(y)	3–40(d,g,v)
Fecundity									
captive	100 000–300 000(a,v)	23 000–75 000	200 000(v)	10 000–90 000(y)	150 000–300 000(v)	50 000–200 000(f)	58 000–284 000(s,ad)	70 000–400 000	80 000–250 000(g,v)
wild	400 000–500 000(e)	—	up to 1 000 000	—	400 000–1 000 000(m)	100 000–500 000(f)	—	—	100 000–500 000(g)
Incubation period (hours)	—	—	10–14(a,d)	18(y)	13(m)	—	—	12–18(g)	12–18(g)
No. of larval stages	12(m)	11–12	12(ac)	11–12	12(m,x)	11	12(ac)	11	11
Duration of larval phase (days)	17(a)	—	12–13(a)	10–12(z)	8–10	9–14	11–14(ae)	8–12(g)	8–12(g)
Age stocked for on-growing (days post metamorphosis)	10–30(b)	3–>40(q,t)	50(m,n)	42(z)	25–55(i,j,m)	5–10(m)	43–51(r)	45–60(ab)	33–75(g,t)

Size stocked for on-growing(g)	0.005–0.05 (b)	<0.01–1.48 (q,t)	0.5–0.8(m)	0.2(z)	0.5(l)	0.7–1.0cm (m)	0.17–2.4(r)	0.5–1.0(ab)	0.6–2.0 (h,t,u)
Density stocked (no. m^{-2})	4–20(b)	2.5–515(q,t)	1–400(n,v)	4–162(o,z)	2–66(i,l)	20–330 (a,m)	3–45(p,r)	6.5(t)	3–110 (h,t,u)
On-growing period (days)	126–190(b)	70–319(q,t)	122–183 (m,n)	76–112(z,aa)	80–225(i,l)	95–141(a)	49–162(r)	252(t)	67–164(t)
Size at harvest (g)	9.4–43.5(b)	4–11(t)	10–25(v)	7–12.5(z,aa)	21–33(i)	9–21(a)	7–21(r)	17–28(t)	7–23(t,u)
Survival (%)	25–55(b)	32–91(q)	40–70(m)	47–73(z)	30–80(i)	45–90(a)	44–87(r)	5–70(g,t)	40–90(g,h)
Yield (kg ha^{-1} $crop^{-1}$)	314–2308 (b)	231–36 000 (p,t)	300–30 000 (n,v)	200–5850 (o,z)	500–14 500 (i,l)	3400–12 300 (a)	135–2740 (p,r)	2500–50 000 (t,af)	490–20 000 (t,u)
Meat yield (%)	57(k)	—	—	65(k)	—	more than other penaeids (a,m)	—	—	63–68% (k,u)
Crops yr^{-1}	1	1–2+(q,t)	1–2(m)	—	1–3(i,l)	1–2	1(r)	1(t)	1–3(w)

(a) Liao & Chien 1990; (b) Table 10.2; (c) Wang & Ma 1990; (d) Rho 1990; (e) Chen 1990a; (f) Hu 1990; (g) Lee D.O'C. (unpublished data); (h) Wyban *et al.* 1989; (i) Chen *et al.* 1989; (j) IFC 1987; (k) Rosenberry 1989; (l) Chiang & Liao 1985; (m) Chen 1990b; (n) Spotts 1984; (o) Maguire 1979; (p) Nandakumar 1982; (q) Gopalan *et al.* 1982; (r) Issar *et al.* 1988; (s) Browdy & Samocha 1985; (t) AQUACOP 1984; (u) Hirono 1986; (v) AQUACOP undated report; (w) Chamberlain 1989; (x) Platon 1978; (y) Wickins, J.F. (unpublished data); (z) Wickins & Beard 1978; (aa) Briggs 1988; (ab) Pretto 1983; (ac) Samocha *et al.* 1989; (ad) Samocha 1980; (ae) Samocha & Lewinsohn 1977; (af) New 1988.

Table 4.6f A comparison of attributes to be considered when choosing a species to culture: marine and freshwater prawns.

Species/group	*Macrobrachium rosenbergii*(a)	*M. acanthurus*	*M. birmanicum*	*M. carcinus*	*M. malcolmsoni*	*Pandalus platyceros* (b,c)	*Palaemon serratus*(b)
Age at first maturity	4–7 mo	4–7 mo**	4–7 mo**	4–7 mo**	4–7 mo**	24–36 mo*	8 mo
Size at first maturity	20–30 g	—	—	—	102 mm TL	25–30 g	3–3.5 g
Control of breeding	yes, not widespread	not practised	not practised	not practised	not practised	no	not practised
Interval between spawning	3–4 mo	—	—	—	—	12 mo	4 mo
Fecundity	80 000	2000–13 400 (d)	2100–42 000 (h)	10 000–80 000 (d)	3500–63 000 (j)	2500–4500	1–2000
Incubation period	19–21 days	18 days	—	—	—	4–5 mo	30–40 days
No. of larval stages	11	10(e)	5(g)	12(i)	16(k)	4	6
Duration of larval phase (days)	20–40	32–40(e)	20–25(g)	32(e)	28(l)	15–24	18–35
Age stocked for on-growing (days post-metamorphosis)	30–60	30–60**	30–60**	30–60**	30–60**	1 mo	2 mo
Size stocked for on-growing (days post-metamorphosis)	0.5–1.0 g	15–20 mm TL(f,g)	—	13–30 mm TL (f)	20–50 mm TL (m,n)	0.5	0.5

Density stocked (no. m^{-2})	5–10	7–11(f,g)	—	1–6(f)	3–6(m,n)	50–100	50–100
On-growing period	3–5 mo, or continuous	125–188 or continuous (f,g)	—	166–180 or continuous(f)	120–150 or continuous (m,n)	18 mo	18–24 mo
Size at harvest	25–45 g	74–121 mm TL(f,g)	—	28–80 mm TL (f)	124–180 mm TL(m,n)	10–18 g	5–8 g
Survival (%)	40–60	8–15(f,g)	—	5(f)	30–70(m,n)	10–85	15–75
Yield (kg ha^{-1} $crop^{-1}$)	1–2500	—	—	—	534–925(m,n)	492 g m^{-2}*	315 g m^{-2}*
Meat yield (%)	37, peeled, raw	—	—	—	—	35	35
Crops yr^{-1}	1–3 or continuous	—	—	—	—	0.7	0.5–0.7

(a) New 1990; (b) Wickins 1972; (c) Oesterling & Provenzano 1985; (d) Coelho *et al.* 1982; (e) Choudhury 1970; (f) Dobkin *et al.* 1974; (g) Khan *et al.* 1984; (h) Prakash & Agarwal 1985; (i) Choudhury 1971; (j) Ibrahim 1962; (k) Kewalramani *et al.* 1971; (l) Sankolli & Shenoy 1980; (m) Rao *et al.* 1986; (n) Rajyalakshmi *et al.* 1980. * laboratory tanks (Wickins & Beard 1978). ** estimated probable value.

Table 4.6g A comparison of attributes to be considered when choosing a species to culture: freshwater crayfish.

Species/group	*Astacus leptodactylus*	*Pacifastacus leniusculus*	*Procambarus clarkii*	*Cherax destructor*	*Cherax tenuimanus*	*Cherax quadricarinatus*
Age at first maturity	2–3 yr(g)	1.5–3 yr(b)	3–9 mo	<12 mo	1–2 yr(b)	6 mo(e)
Size at first maturity	82 mm TL(g)	100 mm TL(b) (15 g)	45–125 mm TL	35–45 mm CL	30–50 mm CL(b)	40 mm CL(h)
Control of breeding	Yes, not widespread	Yes, not widespread	Yes, not widespread	Possible	Possible	Yes
Interval between spawning (months)	12	12(b)	2–5(h)	<6	12(b)	1–2(e)
Fecundity	200–400(g)	70–260(b)	100–700	124–960(h)	200–800(b)	150–250(e) 60–600(i)
Incubation period	5–6 months(g)	4–5 months(b)	1–3 weeks	20–30 days(h)	4–10 weeks(b) 12–16 weeks(f)	56–71 days(i)
Age stocked for on-growing (from hatching)	3–4 months	3–4 months	mature adult	28 days	30–60 days	50–60 days(i)

Size stocked for on-growing (grams)	0.5(g)	0.5	mature adult	0.1–0.5	0.2–1	0.5–1(i)
Density stocked (no. m^{-2})	5–10	2–10	25–100 kg ha^{-1}	5–10	5–15(c)	2–10(i)
On-growing period	1–2 year	1–2 year	1 year	4–12 months	12–24 months	6–24 months(i)
Size at harvest (grams)	60–100	60–100	30–80	50–100	40–120(c)	40–100
Survival (%)	60(g)	30–40	47–88(j)	18–36(j)	60	49–94(k)
Yield (kg ha^{-1} $crop^{-1}$)	500–1000(g)	500–1000(b)	50–3000(l)	1500–2000	1000–2500(c)	1000–2000(i)
Meat yield (%)	15–23(g)	15–25E	10–26(a)	25(h) 7.7–17.4(i) (cooked)	22–30(i) (cooked)	14–40 mean 22(i) (cooked)
Crops yr^{-1}	0.5	0.5	1	1–2.5	0.5–1	0.5–2

(a) Moody 1989; (b) Wickins 1982; (c) Morrissy 1988; (d) Jones 1988; (e) Sammy 1988; (f) Holker 1988; (g) Koksal 1988; (h) Sokol 1988; (i) Jones 1990; (j) Mills & McCloud 1983; (k) O'Sullivan 1988; (l) de la Bretonne & Romaire 1989. E = Estimated.

Table 4.6h A comparison of attributes to be considered when choosing a species to culture: lobsters and crabs.

Species/group	*Homarus americanus*	*Homarus gammarus*	*Panulirus, Jasus* spp.	*Scylla serrata*	*Portunus trituberculatus*
Age at first maturity	3–4 years(a)	3–6 years(a)	3–5 years(a)	4–6 months	12 months
Size at first maturity	200–700 g	400 g	N/A	10 cm CW 200 g(h)	10 cm CW
Control of breeding	Yes, experimental	Partial	Partial, experimental	Yes, not widespread	Possible, not widespread
Interval between spawning	10–24 months	12–24 months	2 months(f)	2–3 months(g)	3–6 weeks(b)
Fecundity	5000–50 000	5000–10 000	0.5–1 million(c)	0.8–1.5 million(e)	1–3 million(b)
Incubation period	4–7.5 months	4–9.5 months	4 weeks(c)	16–17 days(e)	20–25 days(b)
No. of larval stages	4–5	4	6–11	5–6(e)	5(e)
Duration larval phase (d)	9–14	10–14	90–250	20–24(e)	23
Age stocked for on-growing (post metamorphosis)	2 weeks	2 weeks	6–12 months	2–4 weeks	1–4 weeks

Size stocked for on-growing	5–6 mm CL	5–6 mm CL	2–5 cm TL	1.5–3 cm CW(h)	1–2 cm CW
Density stocked (no. m^{-2})	90–100*	90–100*	N/A	0.5–1(e)	Release at sea(b)
On-growing period	1.5–2.5 years	2–3 years	2–3 years(c)	3–6 months(e)	1–2 years
Size at harvest(g)	345–400	345–400	350	8–9 cm CW 200–250(h)	10–15 cm CW
Survival %	60–80	60–80	N/A	40(e) 50–70(h)	20–40(b)
Yield (kg ha^{-1} $crop^{-1}$)	24 000	24 000	N/A	8000 crabs ha^{-1}(e) 55–1800(h)	3–12% recaptures
Meat yield %	30–40 cooked	30–40 cooked	40–45(d) cooked	25–30(d) cooked	25–30(d)
Crops yr^{-1}	0.5	0.5	0.3–0.75	1–2(h)	0.5–1

(a) Wickins 1982; (b) Cowan 1983a; (c) Oesterling & Provenzano 1985; (d) Jones 1990; (e) Cowan 1983b; (f) Chittleborough 1974; (g) Ong 1966; (h) Macintosh 1982.
* Moved to larger compartments 9 m^{-2} at harvest.

Table 4.6i Culture attributes of selected crustacean groups. Best = *****, worst = *.

Species/group	Control of breeding	High fecundity	Short larval life	Rapid growth	Tolerant of crowding	High meat yield	Simple technology	Compounded diets available	Commercial viability
Penaeids	****	****	*****	*****	*****	*****	*****	*****	*****
Macrobrachium	*****	***	***	****	**	****	****	*****	****
Crayfish									
Temperate	***	*	*****	***	***	**	*****	**	****
Tropical	****	*	*****	****	***	***	****	***	***
Lobsters									
Clawed	****	**	*****	**	*	****	*	****	*
Spiny	**	****	*	**	***	****	*	**	*
Crabs	***	*****	***(a)	***(*)	*	***	***	**	**

(a) varies with species.

A further simple calculation (yield per unit area multiplied by number of crops per year) serves to illustrate the potential annual productivity of the land. It must, however, be emphasized that seasonally variable factors such as temperature, rainfall, seed or broodstock availability may severely curtail the useful growing season. In fact, few open air farms are able to realize year-round production. However, to highlight further the differences between the groups of cultivable crustaceans, the hypothetical annual yields calculated from Table 4.6d, in mt ha^{-1} are:

- Penaeids 9.45
- *M. rosenbergii* 2.69
- Crayfish: temperate 0.31
- Crayfish: tropical 2.52
- Clawed & spiny lobsters 11.40
- Crabs 5.91

The superficially attractive value for clawed and spiny lobsters belies the difficulties of high rearing facility costs (clawed lobsters) and of unsolved mass larvae rearing techniques (spiny lobsters).

Additional species specific information is presented for penaeid shrimps (Table 4.6e), caridean prawns (Table 4.6f), freshwater crayfish (Table 4.6g), and clawed lobsters, spiny lobsters and crabs (Table 4.6h). In Table 4.6i the most important features are compared in order to further assist in the choice of species for the various culture options described in chapter 5.

Penaeid shrimp clearly have the best biological attributes for culture but it is by no means certain that *Macrobrachium* will be able to retain their second place if trials presently being made with tropical crayfish, notably the Australian red claw (*C. quadricarinatus*), are successful. Red claw seems to have three major advantages over *Macrobrachium*: no independent larval existence, no requirement for saline water, and a possible greater tolerance to crowded conditions. Disadvantages are poorer fecundity and the potential risk from plague fungus disease. It is extremely difficult to compare objectively the disease susceptibility of different crustaceans, since all may succumb to infection or infestation under stressful culture conditions. Unfortunately, it seems likely that all Australian crayfish are susceptible to the crayfish fungus plague which has taken such a heavy toll of natural and cultured European crayfish populations.

Cultivable crabs and temperate water crayfish rank similarly, according to the criteria used in Table 4.6i (total number of stars), although market considerations and difficulties of rearing some crab larvae probably account for the differences in the extent to which the two groups are farmed (Section 4.4).

The two main biological features that militate against the farming of clawed and spiny lobsters are the prolonged larval life of spiny lobsters and the need to rear clawed lobsters in individual compartments to avoid cannibalism. As a direct result of these features it has not yet been possible to develop a mass culture method for larval spiny lobsters, or an economic battery rearing system for clawed lobsters (Section 10.9.3.4).

4.8 References

Aiken D.E. & Young-Lai W.W. (1981) Dactylotomy, chelotomy and dactylostasis: methods for enhancing survival and growth of small lobsters (*Homarus americanus*) in communal conditions. *Aquaculture*, **22** (1–2) 45–52.

Anon. (1987) Tobacco firm starts smolt production and launches in lobsters. *Fish Farming International*, **14** (9) 18–20.

Anon. (1990) Seed shortage faces US shrimp farms. *Fish Farming International*, **17** (4) 25.

AQUACOP (1983) First results of a 10 ha *Macrobrachium* farm in Tahiti. In *Proc. 1st Int. Conf. on Warm Water Aquaculture – Crustacea*, 9–11 Feb. 1983 (Ed. by G.L. Rogers, R. Day & A. Lim), pp. 179–87. Brigham Young University, Hawaii.

AQUACOP (1984) Review of ten years of experimental penaeid shrimp culture in Tahiti and New Caledonia (South Pacific). *J. World Maricult. Soc.*, **15**, 73–91.

AQUACOP (undated) *Commercial shrimp culture: Selection of species and rearing techniques*, p. 25. Centre Oceanologique du Pacifique, BP 7004, Taravao, Tahiti.

Beard T.W., Richards P.R. & Wickins J.F. (1985) The techniques and practicability of year-round production of lobsters, *Homarus gammarus* (L.) in laboratory recirculation systems. *Fish. Res. Tech. Rep. 79*, MAFF Direct. Fish. Res., Lowestoft.

Briggs M. (1988) Techniques and constraints of shrimp farming in Thailand. *Report on Shrimp Team Visit*. 2–21 May 1988, Institute of Aquaculture, Stirling, Scotland.

Browdy C.L. & Samocha T.M. (1985) Maturation and spawning of ablated and nonablated *Penaeus semisulcatus* de Haan (1844). *J. World Maricult. Soc.*, **16**, 236–49.

Buddle R. (1990) Ecuador's shrimp hatcheries aim for top quality. *Fish Farmer*, **13** (2) 36–7.

Campbell M. (1989) Prince Edward Island. Lobster culture technology on hold. *Atlantic fish farming*, 20 Aug. 1989 (page unknown).

Chamberlain G.W. (1989) *Status of shrimp farming in Texas*. Presented at 20th. Mtg. World Aquacult. Soc., Los Angeles, 12–16 Feb. 1989.

Chan H.H. (1988) The status of marine prawn seed production in Malaysia. *Proc. Sem. Marine Prawn Farming in Malaysia*, 5 March 1988, pp. 21–35. Malaysian Fisheries Society, Serdang, Malaysia.

Chen J. (1990a) Shrimp culture industry in the People's Republic of China. In *The culture of cold-tolerant shrimp* (Ed. by K.L. Main & W. Fulks), pp. 70–76. The Oceanic Institute, Honolulu, Hawaii.

Chen L.C. (1990b) *Aquaculture in Taiwan*. Fishing News Books, Blackwell Scientific Publications, Oxford.

Chen J.C., Liu P.C. & Lin Y.T. (1989) Culture of *Penaeus monodon* in an intensified system in Taiwan. *Aquaculture*, **77** (4) 319–28.

Chiang P. & Liao I.C. (1985) The practice of grass prawn (*Penaeus monodon*) culture in Taiwan from 1968 to 1984. *J. World Maricult. Soc.*, **16**, 297–315.

Chittleborough R.G. (1974) Western rock lobster reared to maturity. *Aust. J. mar. Freshwat. res.*, **25**, 221–5.

Choudhury P.C. (1970) Complete larval development of the palaemonid shrimp *Macrobrachium acanthurus* (Weigmann 1836) reared in the laboratory. *Crustaceana*, **18**, 113–32.

Choudhury P.C. (1971) Complete larval development of the palaemonid shrimp *Macrobrachium carcinus* (L.) reared in the laboratory (Decapoda, Palaemonidae). *Crustaceana*, **20**, 51–69.

Coelho A.P., Porto M.R. & Soares C.M.A. (1982) Biologia e cultivo de camarones del agua doce. *Serie Aquicultura No. 1*, pp. 1–55. Recife, Pernambuco.

Cogan P. (1987) Marron battery study proves encouraging. *FINS*, **20** (3) 5–6, Fishing Industry News Service, Perth, Western Australia.

Cowan L. (1983a) Hatchery production of crabs in Japan. In *Proc. 1st Int. Conf. on Warm Water Aquaculture – Crustacea*, 9–11 Feb. 1983, (Ed. by G.L. Rogers, R. Day & A. Lim), pp. 215–20. Brigham Young University, Hawaii.

Cowan L. (1983b) *Crab farming in Japan, Taiwan and the Philippines*. Information series Q184009, Queensland Dept. of Primary Industries.

D'Abramo L.R., Wright J.S., Wright J.H., Bordner C.E. & Conklin D.E. (1985) Sterol requirement of cultured juvenile crayfish, *Pacifastacus leniusculus. Aquaculture*, **49** (3/4) 245–55.

de la Bretonne L.W. Jr. & Romaire R.P. (1989) Commercial crawfish cultivation practices: a review. *J. Shellfish Res.*, **8** (1) 267–75.

Dobkin S., Azzinaro W.P. & Montfrans J. van, (1974) Culture of *Macrobrachium acanthurus* and *M. carcinus* with note on the selective breeding and hybridization of these shrimps. *Proc. 5th Ann. Mtg. World Maricult. Soc.*, pp. 51–62, Charleston, S. Carolina.

Forster J.R.M. & Beard T.W. (1974) Experiments to assess the suitability of nine species of prawns for intensive cultivation. *Aquaculture*, **3** (4) 355–86.

Geddes M.C., Mills B.J. & Walker K.F. (1988) Growth in the Australian freshwater crayfish, *Cherax destructor* Clark, under laboratory conditions. *Aust. J. mar. Freshwat. res.*, **39**, 555–68.

Gopalan U.K., Purushan K.S., Santhakumari V. & Meenakshikunjamma P.P. (1982) Experimental studies on high density, short term farming of shrimp *Penaeus indicus* in a 'pokkali' field in Vypeen Island, Kerala. *Proc. Symp. Coastal Aquacult.* 1982, (1) 151–9. Mar. Biol. Assn. of India, Cochin, India.

Hanson J. & Goodwin H. (Eds.) (1977) *Shrimp and prawn farming in the western hemisphere*. Dowden, Hutchinson and Ross Inc., Stroudsberg, Pennsylvania.

Henocque Y. (1983) Adaptability and propagation of lobster seedlings transplant experiments in Japan. *Symp. fr-japon Aquacult.*, Montpellier 16 Dec. 1983, **1**, 123–8.

Hirono Y. (1986) Shrimp pond management. In *Acuacultura del Ecuador*. Cámara de Productores de Camarón, Guayaquil, Ecuador (in Spanish and English).

Holker D.S. (1988) Marron – *Cherax tenuimanus*. Unpublished report.

Hu Q. (1990) On the culture of *Penaeus penicillatus* and *P. chinensis* in southern China. In *The culture of cold-tolerant shrimp* (Ed. by K.L. Main & W. Fulks), pp. 77–91. The Oceanic Institute, Honolulu, Hawaii.

Hughes J.T., Shleser R.A. & Tchobanoglous G. (1974) A rearing tank for lobster larvae and other aquatic species. *Progv. Fish Cult.*, **36** (3) 129–32.

Huner J.V. & Barr J.E. (1984) *Red swamp crawfish: biology and exploitation*. Sea Grant Publication No. LSU-T-80-001, Louisiana Sea Grant College Program, Louisiana State University, Louisiana, USA.

Ibrahim K.H. (1962) Observations on the fishery and biology of the freshwater prawn *Macrobrachium malcolmsonii* (Milne-Edwards) of the River Godavari. *Indian J. Fish.*, **9**, 433–67.

IFC (1987) *Marine shrimp farming: A guide to feasibility study preparation*. Aquafood Business Associates, International Finance Corporation, P.O. Box 16190, Charleston, SC 29412.

Issar G., Seidman E.R. & Samocha Z. (1988) Preliminary results of nursery and pond culture of *Penaeus semisulcatus* in Israel. *Fisheries and Fishbreeding in Israel*, **21** (4) 27–37.

Jones C.M. (1988) Aquaculture potential of *Cherax quadricarinatus*: research objectives and preliminary studies. In *Proc. 1st. Australian Shellfish Aquacult. Conf.*, Perth, 1988, (Ed. by L.H. Evans & D. O'Sullivan), pp. 73–8. Curtin University of Technology.

Jones C.M. (1990) *The biology and aquaculture potential of the tropical freshwater crayfish* Cherax quadricarinatus. Information Series QI 90028, Queensland Department of Primary Industries.

Juario J.V. & Benitez L.V. (Eds.) (1988) *Perspectives in aquaculture development in South-east Asia and Japan*. South-east Asian Fisheries Development Centre, Aquaculture Department, Iloilo, Philippines.

Karplus I., Hulata G., Wohlfarth G.W. & Halevy A. (1986) The effect of size-grading juvenile *Macrobrachium rosenbergii* prior to stocking on their population structure and production in polyculture. 1. Dividing the population into two fractions. *Aquaculture*, **56** (3/4) 257–70.

Karplus I., Samsonov E., Hulata G. & Milstein A. (1989) Social control of growth in *Macrobrachium rosenbergii*. 1. The effect of claw ablation on survival and growth of communally raised prawns. *Aquaculture*, **80** (3/4) 325–35.

Keller M. (1988) Finding a profitable population density in rearing summerlings of European crayfish *Astacus astacus* L. In *Freshwater Crayfish* 7 (Ed. by P. Goeldlin de Tiefenau), pp. 259–66. Musée Zoologique Cantonal, Lausanne, Switzerland.

Kendall R.A., Van Olst J.C. & Carlberg J.M. (1982) Effects of chelae immobilisation on growth and survivorship for individually and communally raised lobsters, *Homarus americanus*. *Aquaculture*, **29** (3–4) 359–72.

Kewalramami H.G., Sankolli K.N. & Shenoy S. (1971) On the larval history of *Macrobrachium malcolmsonii* (Milne Edwards) in captivity. *J. Indian Fish. Assoc.*, **1**, 1–25.

Khan S., Khanam S.F. & Shahadet A. (1984) Development of early larval stages of *Macrobrachium birmanicum* (Schenkel 1902) (Crustacea, Decapoda, Palaemonidae). *Bangladesh J. Zool.*, **12**, 79–89.

Koksal G. (1988) *Astacus leptodactylus* in Europe. In *Freshwater crayfish, biology, management and exploitation* (Ed. by D.M. Holdich & R.S. Lowery), pp. 365–400. Croom Helm, London.

Kowarsky J., Gazey P. & Rippingale R. (1985) Intensive culture potential of freshwater crayfish – a research update (March 1985). *Marron Growers Bulletin*, **7** (1) 8–15.

Liao I.C. (1988) Feasibility study for alternative culture species in Taiwan – *Penaeus penicillatus*. *J. World Aquacult. Soc.*, **19** (4) 227–36.

Liao I.C. & Chien Y.H. (1990) Evaluation and comparison of culture practices for *Penaeus japonicus*, *P. penicillatus*, and *P. chinensis* in Taiwan. In *The culture of cold-tolerant shrimp* (Ed. by K.L. Main & W. Fulks), pp. 49–63. The Oceanic Institute, Honolulu, Hawaii.

Lin C.K. (1989) Prawn culture in Taiwan. What went wrong? *World Aquaculture*, **20** (2) 19–20.

Macintosh D.J. (1982) Fisheries and aquaculture significance of mangrove swamps, with special reference to the Indo-West Pacific region. In *Recent advances in aquaculture* (Ed. by J.F. Muir & R.J. Roberts), pp. 3–85. Croom Helm, London.

Maguire G.B. (1979) *A report on the prawn farming industries of Japan, the Philippines and Thailand.* Brackish Water Fish Culture Research Station, C/-P.O., Salamander Bay, New South Wales, 2301.

Main K.L. & Fulks W. (Eds.) (1990) *The culture of cold-tolerant shrimp. Proceedings of an Asian–US workshop on shrimp culture.* The Oceanic Institute, Honolulu, Hawaii.

McCoy H.D. II. (1986) Intensive culture systems past, present and future, Pt 1. *Aquaculture Magazine*, **12** (6) 32–5.

Mills B.J. & McCloud P.I. (1983) Effects of stocking and feeding rates on experimental pond production of the crayfish *Cherax destructor* Clark (Decapoda: Parastacidae). *Aquaculture*, **34** (1/2) 51–72.

Moody M.W. (1989) Processing of freshwater crawfish: a review. *J. Shellfish Res.*, **8** (1) 293–301.

Morrissy N. (1984) Assessment of artificial feeds for battery culture of a freshwater crayfish, marron (*Cherax tenuimanus*) (Decapoda: Parastacidae). *Dept. Fish. Wildl. West. Aust. Rept. No. 63*, 1–43.

Morrissy N.M. (1988) Marron farming – current industry and research developments in Western Australia. In *Proc. 1st. Australian Shellfish Aquacult. Conf.*, Perth, 1988, (Ed. by L.H. Evans & D. O'Sullivan), pp. 59–72. Curtin University of Technology.

Mulla M.A. & Rouse D.B. (1985) Comparisons of four techniques for prawn (*Macrobrachium rosenbergii*) nursery rearing. *J. World Maricult. Soc.*, **16**, 227–35.

Nandakumar G. (1982) Experimental prawn culture in coastal ponds at Mandapam Camp. *Proc. Symp. Coastal Aquaculture 1982, (1)*, pp. 103–11. Marine Biological Association of India, Cochin.

New M.B. (1988) Shrimp farming developments in other areas. In *Shrimp '88, Conference proceedings*, 26–28 Jan. 1988, Bangkok, Thailand, pp. 102–22. Infofish, Kuala Lumpur, Malaysia.

New M.B. (1990) Freshwater prawn culture: a review. *Aquaculture*, **88** (2) 1–44.

Norman-Boudreau K. & Conklin D.E. (1983) Purified protein in artificial diets for juvenile lobster. In *Proc. 1st Int. Conf. on Warm Water Aquaculture – Crustacea*, 9–11 Feb. 1983, (Ed. by G.L. Rogers, R. Day & A. Lim), pp. 343–51. Brigham Young University, Hawaii.

Oesterling M.J. & Provenzano A.J. (1985) Other crustacean species. In *Crustacean and mollusk aquaculture in the United States* (Ed. by J.V. Huner, & E. Evan Brown), pp. 203–34. AVI Inc., Westport, Connecticut.

Ong K.S. (1966) The early developmental stages of *Scylla serrata* Forskal (Crustacea: Portunidae), reared in the laboratory. *Proc. Indo-Pacif. Fish. Coun.*, **11** (2) 135–46.

O'Sullivan D. (1988) Marron farmers anticipate a bright future. *FINS*, **21** (1) 21–5. Fishing Industry News Service, Perth, Western Australia.

O'Sullivan D. (1990) Intensive freshwater crayfish system tested. *Austasia Aquaculture Magazine*, **5** (4) 3–5.

Platon R.R. (1978) Design, operation and economics of a small-scale hatchery for the larval rearing of sugpo, *Penaeus monodon* Fab. *Aquaculture Extension Manual No. 1*, SEAFDEC, Tigbauan, Iloilo, Philippines.

Prakash S. & Agarwal G.P. (1985) Studies on the effect of non-pollutional interference on freshwater prawn (*Macrobrachium birmanicum*) choprai fishery of middle stretch of River Ganga. *Uttar Pradesh J. Zool.*, **5**, 42–8.

Pretto R. (1983) Penaeus shrimp pond grow-out in Panama. In *Handbook of Mariculture, Vol. 1 Crustacean aquaculture* (Ed. by J.P. McVey), pp. 169–78. CRC Press, Boca Raton, Florida.

Rajyalakshmi T., Reddy O.R., Rao A. & Ramakrishna R. (1980) Growth and production of riverine prawn *Macrobrachium malcolmsonii* (H. Milne Edwards) in pond culture, Andhra Pradesh, India. *Proc. 1st Nat. Symp. on shrimp farming*, Aug. 1978, p. 101, Bombay, India.

Rao K.G., Reddy O.R., Rao P.V.A.N.R. & Ramakrishna R. (1986) Monoculture of Indian freshwater prawn, *Macrobrachium malcolmsonii* (Milne Edwards). *Aquaculture*, **53** (1) 67–73.

Rho Y.G. (1990) Present status of Kuruma prawn (*Penaeus japonicus*) seed, in Korea. In *The culture of cold-tolerant shrimp* (Ed. by K.L. Main & W. Fulks), pp. 36–41. The Oceanic Institute, Honolulu, Hawaii.

Rosenberry R. (1982) Shrimp Systems International Ltd., *Aquaculture Digest*, **7** (12) 1–5.

Rosenberry R. (1989) *World Shrimp Farming 1989*. Aquaculture Digest, San Diego, USA.

Sammy N. (1988) Breeding biology of *Cherax quadricarinatus* in the Northern Territory. In *Proc. 1st. Australian Shellfish Aquacult. Conf.*, Perth, 1988, (Ed. by L.H. Evans & D. O'Sullivan), pp. 79–88. Curtin University of Technology.

Samocha T.M. (1980) *Developmental aspects of* Penaeus semisulcatus *de Haan, 1844 (Crustacea, Decapoda) under laboratory and pond conditions*. Extract from PhD. Thesis, Tel-Aviv University, Israel.

Samocha T. & Lewinsohn C. (1977) A preliminary report on rearing penaeid shrimps in Israel. *Aquaculture*, **10** (3) 291–2.

Samocha T.M., Uziel N. & Browdy C.L. (1989) The effect of feeding two prey organisms, nauplii of *Artemia* and rotifers, *Brachionus plicatilis* (Muller), upon the growth and survival of larval marine shrimp, *Penaeus semisulcatus* (de Haan). *Aquaculture*, **77** (1) 11–20.

Sankolli K.N. & Shenoy S. (1980) *Macrobrachium malcolmsonii* a prospective competitor for the jumbo prawn *M. rosenbergii*. *Proc. 1st Nat. Symp. on shrimp farming*, Aug. 1978, pp. 151–3. Bombay, India.

Smith T.I.J., Sandifer P.A., Jenkins W.E., Stokes A.D. & Murray G. (1982) Pond rearing trials with Malaysian prawns, *Macrobrachium rosenbergii*, by private growers in South Carolina, in 1981. *J. World Maricult. Soc.*, **13**, 41–55.

Sokol A. (1988) The Australian Yabby. In *Freshwater crayfish, biology, management and exploitation* (Ed. by D.M. Holdich & R.S. Lowery), pp. 401–25. Croom Helm, London.

Spotts D. (1984) The development of Shigueno-style shrimp culture in southern Japan. *Aquaculture Magazine*, **10** (4) 26–31.

Tcherkashina N.Y. (1977) Survival, growth and feeding dynamics of juvenile crayfish (*Astacus leptodactylus*) in ponds and the river Don. In *Freshwater Crayfish 3* (Ed. by O.V. Lindqvist), pp. 95–100. University of Kuopio, Kuopio, Finland.

Ting K.Y. (1973) Culture potential of spiny lobster. *Proc. 4th Ann. Wkshop World Maricult. Soc.*, 23–26 Jan., pp. 165–70, Monterrey, Mexico.

Waddy S.L. (1988) Farming the homarid lobsters: state of the art. *World Aquaculture*, **19** (4) 63–71.

Wang K. & Ma S. (1990) Advances in larval rearing techniques for *Penaeus chinensis* in China. In *The culture of cold-tolerant shrimp* (Ed. by K.L. Main & W. Fulks), pp. 42–8. The Oceanic Institute, Honolulu, Hawaii.

Wickins J.F. (1972) Experiments on the culture of the spot prawn *Pandalus platyceros* Brandt and the giant freshwater prawn *Macrobrachium rosenbergii* (de Man). *Fish. Invest., London, Ser. 2*, **27** (5).

Wickins J.F. (1982) Opportunities for farming crustaceans in western temperate regions. In *Recent advances in aquaculture* (Ed. by J.F. Muir & R.J. Roberts), pp. 87–177. Croom Helm, London.

Wickins J.F. (1988) Crustacean culture and technology. In *Feasibility study on the technology of mariculture. Vol. 2, Review of technologies and services*, pp. 43–106. Department of Trade and Industry Resources from the Sea Programme, Report 87/27. Aberdeen University Marine Studies Ltd., Aberdeen.

Wickins J.F. & Beard T.W. (1978) *Prawn culture research*. Lab. Leafl. (42), Ministry of Agriculture, Fisheries and Food. Direct. Fish. Res., Lowestoft.

Wyban J.A., Sweeney J.N., Kanna R.A., Kalagayan G., Godin D., Hernandez H. & Hagino G. (1989) Intensive shrimp culture management in round ponds. In *Proceedings of the 1989 South-east Asia shrimp farm management workshop*, Philippines, Indonesia, Thailand, 26 Jul.–11 Aug. (Ed. by D.M. Akiyama), pp. 42–7. American Soybean Association, Singapore.

Chapter 5
Culture options

Once a crustacean develops beyond the hatchery and nursery phases of culture, a range of options may exist for the longer and thus more expensive operation of on-growing. For convenience these options are grouped in relation to the climatic zone that provides the most suitable temperatures for growth; operations in some regions, however, may be seasonally constrained by variations in broodstock availability, seed supply or water quality.

The production of soft shelled crustaceans and the options for stock enhancement, sea ranching or other forms of hatchery-based fisheries are discussed independently of geographical location towards the end of this chapter.

5.1 Tropical climates

In the tropics, where temperatures are generally suitable for year-round growth, methods for the fattening of juveniles may be usefully categorized under four headings, based broadly on expected yields and husbandry practices (updated from Wickins 1976; 1986). It must be emphasized, however, that these categories cannot be defined precisely, and caution should be exercised if they are used in business negotiations.

5.1.1 Extensive culture

Typically, penaeid shrimp and crabs are farmed extensively and produce yields of up to 1000 kg ha^{-1} when harvested, with up to three harvests per year (Table 5.1). Much reliance is placed on natural food production within the pond, which may be enhanced by fertilization with animal manure or chemical fertilizers. The ponds are stocked with naturally occurring juveniles but with little prospect of controlling density. There is usually some tidal water exchange of up to 5–10% per day, although this may be restricted to spring tide periods. Pond sizes range from about 1 to 100 ha.

5.1.2 Semi-intensive and intensive culture

The crustaceans farmed at higher densities are predominantly penaeid shrimp, *M. rosenbergii*, crabs, and recently Australian crayfish (*Cherax* spp.) Cultures yield from 500 to 15 000 kg ha^{-1} each year, and frequently from 2 to 2.5 crops per year (Tables 5.2a and b). Mainly compounded foods are used, occasionally in conjunction with fertilization to enhance natural production. Performance depends on controlled stocking with wild and/or hatchery reared post-larvae. Good yields are only likely with carefully managed tidal or pumped water exchange of up to 30–40% per day, and the provision of supplementary aeration towards the end of the culture period. Pond sizes range from around 0.2 to 2 ha, and modern farms may have square or

Table 5.1 Examples of tropical extensive aquaculture: penaeids, crabs.

Species	Location	Annual yield range ($kg\,ha^{-1}$)	Crops yr^{-1}	Live weight g	Reference
Penaeus monodon	India	1000–1500	2–3	26–39	Sundararajan *et al.* 1979
Penaeus monodon	India	200–1000	2–3	21–26	Chakraborti *et al.* 1986
P. vannamei & *P. stylirostris*	Ecuador	240–1200	2–3	19–22	CPC 1989
Scylla serrata	Philippines	399	1–2	>200	Cowan 1983b

rectangular purpose-built earthen ponds with separate inlet and outlet channels to assist water management. In some localities the ponds are lined with clay, concrete or butyl linings (Sections 8.1.4 and 10.9.1.4). Productivity of many semi-intensive ponds is limited to a maximum of about 3 000 $kg\ ha^{-1}$ per harvest by low and fluctuating oxygen levels, poor phytoplankton management and irregular water exchange.

Culture of crustaceans to marketable size in cages is a relatively recent development involving shrimp, spiny lobsters and crabs (Table 5.2c). Cage culture may be employed when land is scarce or to avoid some of the difficulties caused by poor water exchange. However, when unrestricted development occurs, the number and proximity of the cages can seriously impede natural water circulation within and between the cages (Section 11.2.5).

5.1.3 Super-intensive culture

This category of cultivation is dominated by penaeid shrimp which can provide yields of over 10 000–30 000 $kg\ ha^{-1}$, with two to four crops per year. Battery farming of lobsters, while not yet commercially viable, has the potential to provide similar yields (Table 5.3). Super-intensive culture demands almost total reliance on compounded feeds, precisely controlled feeding rates, and the use of hatchery-reared, nursed juveniles (see Glossary) stocked at specific densities. Continuous exchange of new water, generally over 50–200% per day, as well as aeration, is necessary. Some farms also recycle a proportion of the culture water. Shrimp are reared in concrete, plastic or fibreglass tanks and raceways ranging in size from 0.03 to 0.2 ha. In the case of *P. japonicus* in Japan, circular concrete tanks with a sand substrate on a false floor are used (Shigueno tanks – Section 7.2.6.6).

5.2 Warm temperate and Mediterranean climates

Warm temperate or Mediterranean climates have well defined growth seasons. Most farmers must therefore obtain and stock juveniles as soon as the water temperature is high enough for growth. Alternatively they must provide additional cover, heating or a controlled indoor environment to sustain stock, and in some cases growth, throughout the cool season. Cultures are typically extensive or semi-intensive and utilise shrimp, prawns and crayfish (Table 5.4). Very often, juveniles only begin to occur naturally in sufficient numbers for culture after the temperature has risen beyond the minimum

Table 5.2a Examples of tropical semi-intensive culture: penaeids, *Macrobrachium*, crayfish, crabs.

Species	Location	Annual yield range (kg ha^{-1})	Crops yr^{-1}	Live weight g	Reference
Penaeus monodon	South-east Asia	2044	2–6	28	IFC 1987
P. stylirostris	New Caledonia	3000	1 rotated with *P. monodon*	18–25	AQUACOP 1984
P. vannamei	Ecuador	1150–1725	2–3A	15–25A	Dueñas *et al.* 1983
P. vannamei & *P. stylirostris*	Panama	1023–2273	2–3A	18–20	Pretto 1983
Macrobrachium rosenbergii	Hawaii	1396–2206	Continuous	>30	Malecha 1983
M. rosenbergii	Thailand	2000–2500	Continuous	36	Kwei Lin & Boonyaratpalin 1988
Cherax quadricarinatus	Australia	2000–4000	0.5–1	40–70 (1 yr) 100–200 (2 yr)	Jones 1990
Scylla serrata	Taiwan	up to 1800	2–3	200–300	Chen 1990

A = Assumed.

Table 5.2b Examples of tropical intensive pond culture: penaeids, spiny lobster.

Species	Location	Annual yield range ($kg\,ha^{-1}$)	Crops yr^{-1}	Live weight g	Reference
Penaeus monodon	Thailand	4000–12 000	2	33	Briggs 1989
P. monodon	Taiwan	8700–13 700	1, but two crops possible	23–33	Chen *et al.* 1989
P. penicillatus	Taiwan	4298–12 286	1, generally rotated with *P. monodon*	9.5–12.3	Liao & Chien 1990
Panulirus homarus	Taiwan	Up to 150 000 lobsters/1.3 ha	0.5–1	600	Chen 1990

Table 5.2c Examples of tropical marine cage culture: penaeids, spiny lobster, crabs.

Species	Location	Annual yield range ($kg\,ha^{-1}$)	Crops yr^{-1}	Live weight g	Reference
Penaeus merguiensis	Singapore	20–30 000	2	12	Lovatelli 1990
P. monodon	Thailand	32–48 000	2 experimental	25E	Lovatelli 1990
Panulirus polyphagus	Singapore	up to 45 $kg\,m^{-3}$	2–3	fattening from 100 to 300 g	Lovatelli 1990
P. homarus & P. versicolor	India (floating cages)	up to 1800	1–1.5	220–310	Kuthalingam 1990
Scylla serrata	Singapore	9–10 $kg\,m^{-3}$	10–20 day fattening	200–300	Lovatelli 1990

E = Estimated.

Table 5.3 Examples of super-intensive cultures: penaeids, tropical crayfish, lobsters.

Species	Location	Annual yield range (kg ha^{-1})	Crops yr^{-1}	Live weight g	Reference
Penaeus japonicus	Japan	20 000–30 000	1	20	Spotts 1984
P. vannamei	Tahiti	17 000–22 000	1–2	14–20	AQUACOP 1989
P. vannamei	Hawaii	up to 45 000	3	15.7	Wyban & Sweeney 1989 Wyban *et al.* 1989
P. vannamei	Chicago, USA (stacked trays)	1.07–1.51 kg m^{-2}	Continuous (2–3)	20	Rosenberry 1982 McCoy 1986
*Cherax quadricarinatus**	Australia (pilot battery)	32 kg per module of four trays	Potentially continuous	70 g	O'Sullivan 1990
Homarus americanus	PEI Canada (battery)	6–7 kg m^{-2}	Continuous (0.3–0.5)	350–450	Waddy 1988 Campbell 1989

* Pilot system operated with *Cherax tenuimanus* but also applicable to *C. quadricarinatus*. Tray dimensions not reported.

Table 5.4 Examples of crustacean culture in warm temperate and Mediterranean type climates: penaeids, *Macrobrachium*, crayfish.

Species	Location and culture type	Annual yield range ($kg\,ha^{-1}$)	Crops yr^{-1}	Live weight g	Reference
Penaeus chinensis	N. China	1123–2308	1	19.9–43.5	Zhang 1990
P. japonicus	Japan	4200	1	15–25	Spotts 1984; New 1988a
	France	210–590	1	15–25E	New 1988a
	Portugal	200–700	1	—	Arrobas 1989
	Spain	150–250	1	15–20	Cañavate & Sanchez 1989
	Korea	615	1	20	Kim 1990; Park 1990
	Italy	189–525	1	25–30	Lumare *et al.* 1989
P. monodon	Italy	465	1	16.9	Ponticelli *et al.* 1989
P. semisulcatus	Israel	739–2740	1	7–21	Issar *et al.* 1987
P. vannamei	Texas, USA.	1800–3500	1	17–22	Chamberlain 1989
Macrobrachium rosenbergii	Israel	up to 2800	1	20	New 1988b
	S. USA	up to 6700	1–2	15–25	New 1988a
Procambarus clarkii	Spain	350	1	40–80E	Lorena 1986
	S. USA	50–3000	1–2	40–80	de la Bretonne & Romaire 1989
Cherax destructor	Australia (simulation)	2000–3800	1	55	Staniford 1989
	Australia	1500	1	20–80	Villarreal 1988
Cherax tenuimanus	Australia	80–710	0.5–1	40–120	Villarreal 1988

E = Estimated.

needed for growth to begin. In these circumstances and where only hatchery reared juveniles are available, it may be advantageous to produce young from broodstocks maturing at the end of the summer, nurse the young through the winter in an indoor controlled-environment system, and stock them outside as soon as temperatures rise.

An added advantage occurs if the juveniles have increased in size during winter. It may be possible to offset nursery heating costs in the winter through the use of industrial thermal effluents or geothermal waters. Where the quality, temperature or quantity of the heated water is sub-optimal it is technically possible to transfer the heat to the culture water using heat exchangers or heat pumps (Herdman 1988) (Section 8.4.4). Interestingly, one company is reported to be experimenting with the production of individually confined lobsters (presumably *H. americanus*) in Hawaii. A culture temperature of 22°C is maintained by mixing the warm surface seawater with cold seawater pumped from the depths (Rosenberry 1990).

Hatcheries and over-wintering nurseries are seldom used in the farming of the red swamp crayfish *P. clarkii*, since the majority of populations are self-sustaining. Increasing attention is being paid, however, to the use of hatcheries as the culture of *Cherax* species intensifies and spreads beyond Australia.

The culture of crustacea in conjunction with other valuable species (for example molluscs like abalone) either in polyculture or during alternate seasons, represents another possible option for warm temperate environments (Wickins 1982). Since, however, the necessary level of investment is likely to be higher in these regions than it is for polyculture in the tropics, it is usually best to ensure that the culture of each species is economically viable in its own right, rather than relying on obtaining optimal performance from both to ensure success.

5.3 Temperate climates

There are far fewer culture options and species available for crustacean farmers in cool climates than there are for mollusc and fish farmers (Table 5.5). This is because cultivated molluscs are lower in the food chain than crustaceans (they filter or graze natural algae) and require neither additional feed nor daily attention. The majority of fish successfully farmed in cool waters are far more tolerant of crowding than are crustaceans, and many species can utilize the whole water column rather than just the floor area. They are thus able to make more effective use of the culture facility than crustaceans. The slow growth of temperate water crustaceans, coupled with their need for daily attention and feeding, and the comparatively low yields obtainable per unit floor area, effectively reduce the choices available to extensive or semi-intensive crayfish culture and, potentially, to indoor controlled environment or 'battery' farming (Wickins 1982).

Even so, it is worth mentioning that several attempts to rear clawed lobsters in coastal waters have been reported (Van Olst *et al.* 1977; Wickins 1982) although none achieved commercial viability. The systems included groups of floating individual containers, sea bed cages serviced by divers, and a fixed sea bed-to-surface cage system serviced at the surface by means of an access pier. In some cases hatchery reared lobsters were to be reared to market size, while in others partly grown lobsters were held until they grew large enough to command a higher price. The principal advantages of offshore systems were thought to be lower capital, water treatment and

Table 5.5 Examples of crustacean culture in temperate climates: penaeids, *Macrobrachium*, crayfish.

Species	Location and culture type	Yield range	Crops yr^{-1}	Live weight g	Reference
P. japonicus	UK (power stn. effluent ponds)	Experimental	2 estimated	10–15	Wickins 1982 Anon. 1988
	Italy (power stn. effluent ponds)	Experimental	Winter catch crop	11–15	Palmegiano & Saroglia 1981
Macrobrachium rosenbergii	USSR (power stn. effluent ponds)	up to 1000 $kg\,ha^{-1}\,yr^{-1}$	1	6.5–60	Khmeleva *et al.* 1989
	New Zealand (geothermal water)	Experimental	1–2	up to 50	Anon. 1989
Pacifastacus leniusculus	extensive and semi-intensive lakes, ponds	3 crayfish/metre of bank	0.5	70–100	Alderman & Wickins 1990

pumping costs. The disadvantages were lack of environmental control, difficulty of inspection, feeding and costly maintenance and predator control.

The use of heated industrial effluent waters for on-growing, either directly or following heat exchange, has been considered from both biological and financial standpoints on a number of occasions (Van Olst *et al.* 1980; Aston 1981; Johnston & Botsford 1981; Palmegiano & Saroglia 1981; Tiews 1981; Wickins 1982; D'Abramo & Conklin 1985). The conclusions have led to a number of experimental and pilot scale trials with crustaceans as well as fish, which have included attempts to grow juveniles both outdoors and indoors in raceways, ponds or tanks. To the best of our knowledge, however, most of these did not maintain the reliability in flow and quality of supply necessary to justify significant investment in crustacean culture facilities. Many of the problems seemed to arise from the disparity in size and the operational priorities of the heat generating industry and the aquaculture unit (see also Anon. 1988).

Battery farming or controlled-environment culture of crustacea has also been investigated (Wickins 1982; McCoy 1986; Cogan 1987; Rosenberry 1989a; O'Sullivan 1990). Optimum temperatures for growth are maintained throughout the year and animals are stocked and harvested from the unit at regular intervals regardless of season. Implicit in the concept are complete control over broodstock and juvenile supply and the ability to maintain water quality, typically by recirculation through suitable treatment plant.

Even though it is a high-risk operation, it has attracted a number of entrepreneurs because of its similarity to a manufacturing process. The controlled production allows a consistent product quality and the ability to alter the appearance of the product (through diet or illumination) to suit the customer. The ability to produce continuously permits cost-effective processing, distribution and marketing operations. There seems little doubt that it is technically feasible to grow lobsters, some shrimp and crayfish in such systems, but the high capital and labour costs involved seem likely to render the operation uneconomic or at least unduly sensitive to market forces.

5.4 Polyculture

In a polyculture system, penaeid shrimp, crabs or, less commonly, palaemonid prawns are grown as an addition to a crop of fish such as milkfish, tilapia or carp, which are grown for local sale (Table 5.6). Occasionally, because of juvenile scarcity or unseasonable temperatures, the species may have to be grown consecutively in a form of 'crop rotation'. The cultured crustacean provides a useful supplementary income and may even be the most valuable crop. Other polyculture prospects include crayfish with ducks, the latter being either for food or for sport, and marine prawns, such as the spot prawn *P. platyceros* grown with abalone. Herbivorous fish, such as grass carp, are sometimes added to ponds containing crustaceans to control plant growth (they may be called 'sanitary' fish) but this is not considered to be true polyculture even though the fish may have value when harvested.

5.5 Production of soft shelled crustaceans

Most if not all the large edible crustaceans (40 g or over) are potentially marketable as a gourmet food item immediately after moulting while the shell is still soft. At present, however, only crabs and crayfish are marketed in significant quantities in this form

Table 5.6 Examples of polyculture: penaeids, palaemonids, crabs.

Species	Location	Annual yield range (kg ha^{-1})	Crops yr^{-1}	Live weight g	Reference
Penaeus indicus, P. monodon, Metapenaeus monoceros, M. dobsoni	India	480–960	1–2	0.8–37	Pai *et al.* 1982
Metapenaeus dobsoni, M. monoceros, P. indicus, P. monodon with fish.	India	200 (shrimp and fish)	1 (rotation with rice)	0.8–12.5	Nagaraj & Neelakantan 1982
Penaeus spp. with clams	Hawaii	2000 expected	1–2	15–25E	York 1983
Macrobrachium rosenbergii with carp	Israel	96–312	1	>45	Cohen *et al.* 1983
M. rosenbergii with bait fish	USA	533	1	17–29	Scott *et al.* 1988
Pandalus platyceros with coho salmon in cages	USA	74 g m^{-2} of cage net surface	1	8.6	Rensel & Prentice 1979
Scylla serrata with shrimp or milkfish.	Philippines	340–500	1–2	200–230	Macintosh 1982 Cowan 1983b

E = Estimated

(Table 5.7) but prices are high and there would seem to be considerable potential for the development of markets for soft shelled spiny lobsters, clawed lobsters and Australasian crayfish.

Soft shelled crustaceans (Sections 3.3.3.2 and 3.3.5) must be eaten or processed within the few hours between casting the shell and what is known as the 'paper shell' stage, when water enters the tissues and spoils the texture. Crabs and crayfish about to moult are identified from catches and placed in shallow trays of clean, running water for up to about seven days. Inspections may be made as often as every 15 minutes, so harvesting and any initial processing will be continuous during the season (Sections 7.5.7 and 7.10.9).

Soft shelled crustaceans may be produced from wild or cultured stock, although in practice selection of pre-moult animals from ponds without repeatedly disturbing the stock would be difficult. It is not yet practicable to induce synchronous moulting within a pond-reared population, although it would be relatively easy to harvest newly moulted individuals from a 'battery' culture system.

5.6 Hatchery supported fisheries, ranching

5.6.1 Restocking and stock supplementation

A number of hatcheries in the Far East, North America and Europe are employed to produce juvenile crustaceans to restock inland fresh or coastal marine waters as part of

Table 5.7 Examples of soft shelled crustacean production operations: *Macrobrachium*, crayfish, crabs.

Species	Location	Holding system	Wild or cultured	Product features	Animal size	Reference
Macrobrachium rosenbergii	USA	Crayfish shedding trays	Cultured	Experimental study	25–35 g (assumed)	Lutz *et al.* 1990
Procambarus clarkii	South USA	Wooden trays	Wild and cultured	Vacuum packed	16 g	Culley & Duobinis-Gray 1989
Callinectes sapidus	USA	Wooden trays floating boxes	Wild	Dressed, IQF	80–140 mm CW	Oesterling & Provenzano 1985

national programmes aimed at restoring or enhancing depleted fisheries. Such operations are known by a variety of names, the most common being restocking, stock enhancement or supplementation and hatchery supported fishery. In at least one country (Taiwan) effort is directed towards increasing the availability of wild broodstock penaeids for hatchery use (see Table 5.8).

Comprehensive ecological and hydrodynamic surveys are prerequisites of any release programme and must indicate suitable habitat and season, both for release and subsequent growth, as well as provide data on predators and natural recruitment. Fluctuations in catches caused by varying recruitment, environmental conditions and weather patterns mask the effects of stock enhancement programmes in open waters. These factors make it difficult to judge the returns accurately and an element of faith is usually necessary when justifying large national restocking programmes.

Production of the juveniles for release may be contracted from private or publicly funded hatcheries but in most cases the released animals will be indistinguishable from their wild counterparts when captured. This means that fishermen are unlikely to contribute willingly towards the cost of their production, and also that no private restocking schemes will be implemented without the security of ownership rights. The management of large national programmes may involve a blend of levies on the fishermen and subsidies for the producers, or a system of transferable share quotas (Arnason 1989). Once a restocking scheme is started, public opinion may make it difficult to stop even though there may be little or no scientific evidence that enhancement of the stock or fishery has occurred (Hughes J., 1980, pers. comm.). It seems likely that in many fisheries greater production would result from environmental or habitat improvements than from releases of hatchery reared juveniles.

5.6.2 Ranching and habitat modification

Some types of extensive crayfish farm, for example the wooded and marsh pond farms in the southern USA, the 'stock and forget' farms in the UK and some farm dams in Australia (Sections 7.5–7.7), more closely resemble private fisheries or ranches than conventional aquaculture operations (Table 5.8). Broodstock or juveniles are stocked once, or over just three to five years, and become self-sustaining provided they are not over-fished. Very little management is involved but ownership is clearly defined and any investment can be protected.

Attempts to ranch marine or brackish water crustaceans are only likely to be of commercial interest in areas where legislation permits leasing or ownership and the stock can be maintained within a defined boundary or indisputably distinguished from wild stocks (Section 11.6.3). Examples include the use of net fences to confine hatchery reared juveniles (embayment) or the introduction of an easily identifiable exotic species such as *Homarus americanus* to Japanese waters (Kittaka *et al.* 1983).

A variation on the ranching theme involves investment in the creation or modification of seabed habitat in areas where adults occur naturally, so that fishing becomes more efficient (Fee 1986). The changes made to the seabed are most likely to concentrate crustaceans such as spiny and clawed lobsters and crabs in a convenient locality for fishing. To protect such investment, access must be regulated by suitable legislation. It is argued that concentrating the stock does not in fact increase natural productivity but does increase mortality of the stock due to fishing. However, unlike young clawed lobsters, spiny lobsters (*Panulirus* spp) do not appear to construct

Table 5.8 Examples of crustacean stock supplementation and ranching projects: penaeids, *Macrobrachium*, crayfish, crabs, clawed and spiny lobsters.

Species	Location	Culture system	Releases	Returns/results	Reference
Penaeus chinensis	China	Hatchery	4 billion released	4.6–8.2% est. 4800 mt, worth US$24 million	Rosenberry 1989b Shang 1989
	Korea	Hatchery/nursey	8–11 million annually	Unknown	Park 1990
P. japonicus	Japan	Hatchery/ nursery	300 million annually	5–8% recapture at market size	Oshima 1984 New 1988a
P. monodon	Taiwan	Ponds	Tagged juveniles	15–35% return of broodstocks	Chiang & Liao 1985
Macrobrachium rosenbergii	Thailand	Hatchery/nursery	3 million over 3 years	2% recaptured	NACA 1986
	Malaysia	Hatchery/nursery	'large numbers'	Aim to improve broodstocks	Juario & Benitez 1988
Astacus astacus	Europe	Hatchery, artificial incubation, adults	Various, $<100\,m^{-1}$ of bank	45–60% survival	Arrignon 1981; Holdich & Lowery 1988; Keller 1988

Pacifastacus leniusculus	Europe	Adults, hatchery	Various, 200/site for 5 yrs	Self-sustaining populations	Furst 1989; Westman *et al.* 1989
Procambarus clarkii	USA	Adults	50–60 kg ha^{-1}	Self sustaining populations	de la Bretonne & Romaire 1989
Homarus americanus	USA	Hatchery	500 000 yr^{-1}	Unknown, additions made to existing fishery	Syslo 1986
	Japan	Hatchery	<5000	Experimental studies	Kittaka *et al.* 1983
H. gammarus	UK	3 hatchery/ nursery projects	10 000 yr^{-1} for 5 yrs	Experimental, returns of tagged lobsters from existing fishery	Bannister *et al.* 1989; Bannister & Wickins 1989
	France	3 hatchery projects	250 000 yr^{-1} for 10 yrs	Unknown, additions made to existing fishery	Henocque 1983
Panulirus argus	Caribbean	Habitat modification	Some juveniles transplanted	Unknown, projects within existing fisheries	Miller 1983; Fee 1986; Eggleston *et al.* 1990
Portunus trituberculatus	Japan	Hatchery	10–27 million annually	3–12% estimated	Cowan 1983a

burrows or modify existing crevices. This means that in years when spiny lobsters become particularly abundant, the carrying capacity of a ground for these species may become limited by the number of available crevices.

In his review of habitat modification methods employed to enhance fisheries for spiny lobsters (*Panulirus* spp), Miller (1983) argued that a real increase in survival, and therefore natural production, might be induced in areas where food was abundant, by providing additional habitat (crevices) for the natural population. Taking the concept a step further, survival of juveniles might also be enhanced if species-specific habitats were designed and deployed in natural nursery areas (Fee 1986). This advance has now been made and investigators in Cuba and New Zealand are trying to determine the effectiveness of transplanting the juvenile spiny lobsters caught in the artificial nursery habitats to seabed areas that are either naturally more favourable for on-growing or are made so by the deployment of artificial habitat.

It may also be possible to ranch marine crustaceans with low migratory instincts (e.g. *Homarus gammarus*) by creating or modifying seabed habitat in such a way that it becomes suitable for occupation by hatchery reared juveniles (Sheehy 1982). If the area created does not receive natural settlement, and if currents carry away any larvae eventually produced by the occupants, the population will not become self-sustaining in the long term and will require regular replenishment from a hatchery. Ownership or lease arrangements would allow protection of commercial investment but such a scheme might also be of interest to organisations concerned with the disposal of non-toxic, solid industrial wastes. Suitably configured and deposited, such material could, with maturity, provide the ecosystem needed to support new, hatchery-based fisheries. Indeed, the maintenance of a hatchery might be imposed as a condition when granting permission for the disposal of such wastes.

Studies currently being conducted in Britain on artificial reefs of fly ash blocks (Collins *et al.* 1990) revealed rapid natural immigration and colonization by lobsters from the surrounding area. This observation lends support to the hypothesis that, if deployed in suitable areas, reefs of such waste materials may have the potential to revitalise impoverished seabed areas and sustain a worthwhile level of fishing even without repeated releases of hatchery reared juveniles.

As far as temperate water clawed lobsters and crabs are concerned, the carrying capacity of new artificial habitat (wrecks, artificial reefs and islands, coastal protection schemes) seems more likely to be limited by food availability than by the number of crevices. Early seeding of new structures with, for example, mussels is also being considered in Britain (Wickins, unpublished) and if resident populations could be established, could offer a number of advantages:

(1) Seeding would reduce the opportunities for colonisation by organisms less nutritious for lobsters.
(2) Mussels would provide a valuable food resource more readily available to lobsters and crabs than to fish.
(3) Lobsters in particular cannot forage in high current speeds and a close supply of food might reduce the need to forage away from the habitat and, at the same time, might extend the period of tide over which they can safely feed.
(4) An overall increase in seabed productivity might occur provided the mussels did not attract undue numbers of predators (e.g. starfish).

5.7 References

Alderman D.J. & Wickins J.F. (1990) *Crayfish Culture*. Lab. Leafl. (62). MAFF Direct. Fish. Res., Lowestoft.

Anon. (1988) UK power plant farm closes: pioneer project hit by royalties demand. *Fish Farming International*, **15** (6) 1, 5.

Anon. (1989) Thermal power for N Z *Macrobrachium* farm. *Austasia Aquaculture Magazine*, **3** (6) 5.

AQUACOP (1984) Review of ten years of experimental penaeid shrimp culture in Tahiti and New Caledonia (South Pacific). *J. World Maricult. Soc.*, **15**, 73–91.

AQUACOP (1989) Production results and operating costs of the first super-intensive shrimp farm in Tahiti. *Aquaculture '89 Abstracts*, from World Aquaculture Society Meeting 1989, Los Angeles.

Arnason R. (1989) *On the external economies of ocean ranching*. ICES 1989 EMEM/ No. 57, (mimeo), Int. Counc. Explor. Sea.

Arrignon J. (1981) *L'écrivisse et son elevage*. Gauthier-Villars, Paris.

Arrobas I. (1989) *Culture of kuruma prawn (*Penaeus japonicus*) in Portugal*. Spec. Publ. No. 10, pp. 15–16, European Aquacult. Soc.

Aston R.J. (1981) The availability and quality of power station cooling water for aquaculture. In *Aquaculture in heated effluents and recirculation systems* (Ed. by K. Tiews), **1**, 39–58. Heenemann Verlagsgesellschaft, Berlin.

Bannister R.C.A. & Wickins J.F. (1989) A new perspective on lobster stock enhancement. *Proc. 20th Ann. Shellfish Conf.*, 16–17 May 1989, pp. 38–47. Shellfish Association of Great Britain, London.

Bannister R.C.A., Howard A.E., Wickins J.F., Beard T.W., Burton C.A. & Cook W. (1989) *A brief progress report on experiments to evaluate the potential of enhancing stocks of the lobster (*Homarus gammarus L.*) in the United Kingdom*. ICES Shellfish Committee CM 1989/K:30 (mimeo).

Briggs M. (1989) Shrimp farming developments in Thailand. *Aquaculture News*, (7) 12–13. Institute of Aquaculture, University of Stirling, Scotland.

Campbell M. (1989) Prince Edward Island. Lobster culture technology on hold. *Atlantic fish farming*, 20 Aug. 1989.

Cañavate J.P. & Sanchez M.P. (1989) *Effect of nursery stocking density on further growing of* Penaeus japonicus *cultured in earthen ponds*. Spec. Publ. No. 10, pp. 51–2. European Aquacult. Soc.

Chakraborti R.K., Halder D.D., Das N.K., Mandal S.K. & Bhowmik M.L. (1986) Growth of *Penaeus monodon* Fabricius under different environmental conditions. *Aquaculture*, **51** (3/4) 189–94.

Chamberlain G.W. (1989) *Status of shrimp farming in Texas*. Presented at 20th WAS meeting, 12–16 Feb. 1989, Los Angeles.

Chen J-C., Liu P-C. & Lin Y.T. (1989) Culture of *Penaeus monodon* in an intensified system in Taiwan. *Aquaculture*, **77** (4) 319–28.

Chen L.-C. (1990) *Aquaculture in Taiwan*. Fishing News Books, Blackwell Scientific Publications, Oxford.

Chiang P. & Liao I.C. (1985) The practice of grass prawn (*Penaeus monodon*) culture in Taiwan from 1968 to 1984. *J. World Maricult. Soc.*, **16**, 297–315.

Cogan P. (1987) Marron battery study proves encouraging. *FINS* **20** (3) 5–6, Fishing Industry News Service, Perth, West Australia.

Cohen D., Ra'anan Z. & Barnes A. (1983) Production of the freshwater prawn *Macrobrachium rosenbergii* in Israel. 1. Integration into fish polyculture systems. *Aquaculture*, **31** (1) 67–76.
Collins K.J. & Jenson A.C. & Lockwood A.P.M. (1990) Fishery enhancement reef building exercise. *Chemistry and Ecology*, **4**, 179–87.
Cowan L. (1983a) Hatchery production of crabs in Japan. In *Proc. 1st Int. Conf. on Warm Water Aquaculture–Crustacea*, 9–11 Feb. 1983 (Ed. by G.L. Rogers, R. Day & A. Lim), pp. 215–20. Brigham Young University, Hawaii.
Cowan L. (1983b) *Crab farming in Japan, Taiwan and the Philippines*. Information series Q184009, Queensland Dept. of Primary Industries.
CPC (1989) *Libro blanco del camarón*. Cámara de Productores de Camarón, May 1989, Guayaquil, Ecuador.
Culley D.D. & Duobinis-Gray L. (1989) Soft-shell crawfish production technology. *J. Shellfish Res.*, **8** (1) 287–91.
D'Abramo L.R. & Conklin D.E. (1985) Lobster Aquaculture. In *Crustacean and Mollusk Aquaculture in the United States* (Ed. by J.V. Huner & E.E. Brown), pp. 159–201. AVI Inc. Westport, USA.
de la Bretonne L.W. Jr. & Romaire R.P. (1989) Commercial crawfish cultivation practices: a review. *J. Shellfish Res.*, **8** (1) 267–75.
Dueñas J., Harmsen A. & Emberson C. (1983) Penaeid shrimp pond culture in Ecuador. In *Proc. 1st Int. Conf. on Warm Water Aquaculture – Crustacea*, 9–11 Feb. 1983 (Ed. by G.L. Rogers, R. Day & A. Lim), pp. 99–108. Brigham Young University, Hawaii.
Eggleston D.B., Lipscius R.N. & Miller D.L. (1990) Stock enhancement of Caribbean spiny lobster. *The Lobster Newsletter*, **3** (1) 10–11.
Fee R. (1986) Artificial habitats could hike crab and lobster catches. *National Fisherman*, **67** (8) 10–12, 64.
Furst M. (1989) Methods of stocking and management of freshwater crayfish. In *Crayfish Culture in Europe* (Ed. by J. Skurdal, K. Westman & P.I. Bergen), pp. 152–63. Norwegian Directorate for Nature Management, Trondheim.
Henocque Y. (1983) Lobster aquaculture and restocking in France. In *Proc. 1st Int. Conf. on Warm Water Aquaculture – Crustacea*, 9–11 Feb. 1983 (Ed. by G.L. Rogers, R. Day & A. Lim), pp. 235–27. Brigham Young University, Hawaii.
Herdman A. (1988) Heating of hatchery water supplies. In *Aquaculture engineering technologies for the future*, pp. 343–56. Institution of Chemical Engineers Symposium series No. 111, EFCE Publication series No. 66, Hemisphere, London.
Holdich D.M. & Lowery R.S. (Eds.) (1988) *Freshwater Crayfish, biology, management and exploitation*. Croom Helm, London.
IFC (1987) *Marine shrimp farming: A guide to feasibility study preparation*. Aquafood Business Associates, International Finance Corporation, P.O. Box 16190, Charleston, SC 29412, USA.
Issar G., Seidman E.R. & Samocha Z. (1987) Preliminary results of nursery and pond culture of *Penaeus semisulcatus* in Israel. *Bamidgeh*, **39** (3) 63–74.
Johnston W.E. & Botsford L.W. (1981) Systems analysis for lobster aquaculture. In *Aquaculture in heated effluents and recirculation systems* (Ed. by K. Tiews), **2**, 455–64. Heenemann Verlagsgesellschaft, Berlin.
Jones C.M. (1990) *The biology and aquaculture potential of the tropical freshwater*

crayfish Cherax quadricarinatus. Information Series QI 90028, Queensland Department of Primary Industries.

Juario J.V. & Benitez L.V. (Eds.) (1988) *Perspectives in aquaculture development in south-east Asia and Japan.* South-east Asian Fisheries Development Centre, Aquaculture Department, Iloilo, Philippines.

Keller M. (1988) Finding a profitable population density in rearing summerlings of European crayfish *Astacus astacus* L. In *Freshwater crayfish* 7 (Ed. by P. Goeldlin de Tiefenau), pp. 259–66. Musée Zoologique Cantonal, Lausanne, Switzerland.

Khmeleva N.N., Kulesh V.F. & Guiguiniak Y.G. (1989) Growth potentialities of the giant tropical prawn, *Macrobrachium rosenbergii* (de Man), in waste-heat discharge waters of a thermoelectric power station. *Aquaculture,* **81** (2) 111–17.

Kim J.H. (1990) The culture of *Penaeus chinensis* and *P. japonicus* in Korea. In *The culture of cold-tolerant shrimp* (Ed. by K.L. Main & W. Fulks), pp. 64–9. Oceanic Institute, Honolulu, Hawaii.

Kittaka J., Henocque Y., Yamada K. & Tabata N. (1983) Experimental release of juvenile lobster at Koshiki Islands in south Japan. *Bull. Jap. Soc. Sci. Fish.,* **49** (9) 1347–54.

Kuthalingam M.D.K. (1990) Studies on the culture of spiny lobsters, *Panulirus homarus* and *P. versicolor* in marine floating cages with an account of lobster fishery resources of India. *Abstract from II Congreso de Ciencias del Mar,* p. 170, 18–22 Junio 1990, La Habana, Cuba.

Kwei Lin C. & Boonyaratpalin M. (1988) An analysis of biological characteristics of *Macrobrachium rosenbergii* (de Man) in relation to pond production and marketing in Thailand. *Aquaculture,* **74** (3/4) 205–15.

Liao I.C. & Chien Y.H. (1990) Evaluation and comparison of culture practices for *Penaeus japonicus, P. penicillatus* and *P. chinensis* in Taiwan. In *The culture of cold-tolerant shrimp* (Ed. by K.L. Main & W. Fulks), pp. 49–63. Oceanic Institute, Honolulu, Hawaii.

Lorena A.S.H. (1986) The status of the *Procambarus clarkii* population in Spain. In *Freshwater crayfish 6* (Ed. by P. Brinck), pp. 131–3. Int. Assoc. Astacology, Lund, Sweden.

Lovatelli A. (1990) *Regional seafarming resources atlas.* FAO/UNDP Regional seafarming development and demonstration project, RAS/86/024 Jan. 1990.

Lumare F., Amerio M., Arata P., Guglielmo L., Casolino L., Marolla V., Serra A., Schiavone R. & Ziino M. (1989). Semi-intensive culture of the kuruma shrimp *Penaeus japonicus* by fertilizer and feed applications in Italy. In *Aquaculture – a biotechnology in progress* (Ed. N. De Pauw, E. Jaspers, H. Ackefors & N. Wilkins), pp. 401–7. European Aquaculture Society, Bredene, Belgium.

Lutz C.G., Caffey R.H. & Avault J.W.Jr. (1990) Production of soft-shell prawns *Macrobrachium rosenbergii* in commercial crawfish shedding facilities. *Abstracts from World Aquaculture 90,* p. 85. 10–14 June 1990, Halifax, Nova Scotia.

Macintosh D.J. (1982) Fisheries and aquaculture significance of mangrove swamps, with special reference to the Indo-West Pacific region. In *Recent advances in aquaculture* (Ed. by J.F. Muir & R.J. Roberts), pp. 3–85. Croom Helm, London.

Malecha S.R. (1983) Commercial pond production of the freshwater prawn *Macrobrachium rosenbergii* in Hawaii. In *Handbook of Mariculture, Vol. 1 Crustacean aquaculture* (Ed. by J.P. McVey), pp. 231–60. CRC Press, Boca Raton, Florida.

McCoy H.D. II. (1986) Intensive culture systems past, present and future, Pt 1. *Aquaculture Magazine*, **12** (6) 32–5.

Miller D.L. (1983) Shallow water mariculture of spiny lobster (*Panulirus argus*) in the Western Atlantic. In *Proc. 1st Int. Conf. on Warm Water Aquaculture – Crustacea*, 9–11 Feb. 1983 (Ed. by G.L. Rogers, R. Day & A. Lim), Brigham Young University, Hawaii.

NACA (1986) *Giant freshwater prawn breeding and farming in Thailand: an introduction*. NACA TV Video production, FAO network of aquaculture centres in Asia.

Nagaraj M. & Neelakantan B. (1982) Prawn culture in the paddy fields of Uttara Kannada District, Karnataka. *Proc. Symp. Coastal Aquacult. (1)*, pp. 392–3. Mar. Biol. Assn. India, Cochin, India.

New M.B. (1988a) Shrimp farming in other areas. In *Shrimp '88 Conference proceedings*, 26–28 Jan. 1988, pp. 102–22. Bangkok, Thailand. Infofish, Kuala Lumpur, Malaysia.

New. M.B. (1988b) *Freshwater prawns: status of global aquaculture, 1987*. NACA Technical manual No. 6, network of aquaculture centres in Asia.

Oesterling M.J. & Provenzano A.J. (1985) Other crustacean species. In *Crustacean and Mollusk Aquaculture in the United States* (Ed. by J.V. Huner, & E. Evan Brown), pp. 203–34. AVI Inc. Westport, Connecticut.

Oshima Y. (1984) Status of 'fish farming' and related technological development in the cultivation of aquatic resources in Japan. *Proc. ROC-Japan Symp. Maricult.* (1) 1–11.

O'Sullivan D. (1990) Intensive freshwater crayfish system tested. *Austasia Aquaculture Magazine*, **5** (4) 3–5.

Pai M.V., Somvanshi V.S. & Telang K.Y. (1982) A case study of the economics of a traditional prawn culture farm in the North Kanara District, Karnataka, India. *Proc. Symp. Coastal Aquacult. 1982, Pt 1*, pp. 392–3. Mar. Biol. Assn., Cochin, India.

Palmegiano G. & Saroglia M.G. (1981) Winter shrimp culture in thermal effluents. In *Aquaculture in heated effluents and recirculation systems* (Ed. by K. Tiews), **2**, 297–302. Heenemann Verlagsgesellschaft, Berlin.

Park B-H. (1990) The status of shrimp culture in Korea. In *The culture of cold-tolerant shrimp* (Ed. by K.L. Main & W. Fulks), pp. 3–15. Oceanic Institute, Honolulu, Hawaii.

Ponticelli A., Corbari L., Caggiano M. & Aliberti A. (1989) Growth experiments with *Penaeus monodon* in southern Italy in semi-intensive culture. In *Aquaculture – a biotechnology in progress* (Ed. by N. De Pauw, E. Jaspers, H. Ackefors & N. Wilkins), pp. 415–19. European Aquaculture Society, Bredene.

Pretto R.M. (1983) Penaeus shrimp pond grow-out in Panama. In *Handbook of Mariculture, Vol. 1, Crustacean aquaculture* (Ed. by J.P. McVey), pp. 169–78. CRC Press, Boca Raton, Florida.

Rensel J.E. & Prentice E.F. (1979) Growth of juvenile spot prawn, *Pandalus platyceros*, in the laboratory and in net pens using different diets. *US Fish Wildl. Serv. Fish. Bull.*, **76** (4) 886–90.

Rosenberry R. (Ed.) (1982) Shrimp Systems International Ltd., *Aquaculture Digest*, **7** (12) 1–5.

Rosenberry R. (ed.) (1989a) Status and history of shrimp farming in Texas, USA. *World Shrimp Farming*, **14** (7) 12.

Rosenberry R. (Ed.) (1989b) Shrimp farming in China. Aquaculture Digest, **14** (9) 3.

Rosenberry R. (1990) Shrimp farming in the United States. *World Shrimp Farming*, **15** (11) 11.

Scott G., Perry W.G. Jr. & Avault J.W. Jr. (1988) Polyculture of the giant Malaysian prawn and the golden shiner in south-western Louisiana. *J. World Aquacult. Soc.*, **19** (3) 118–26.

Shang (1989) Marine shrimp farming in P.R. China. *Infofish International*, (2) 16–17.

Sheehy D.J. (1982) The use of designed and prefabricated artificial reefs in the United States. *Mar. Fish. Rev.*, **44** (6–7) 4–15.

Spotts D. (1984) The development of the Shigueno-style shrimp culture in southern Japan. *Aquaculture Magazine*, **10** (4) 26–31.

Staniford A.J. (1989) The effect of yield and price variability on the economic feasibility of freshwater crayfish *Cherax destructor* Clark (Decapoda: Parastacidae) production in Australia. *Aquaculture*, **81** (3/4) 225–35.

Sundararajan D., Chandra Bose V.S. & Venkatisan V. (1979) Monoculture of tiger prawn, *Penaeus monodon* Fabricius, in a brackishwater pond at Madras, India. *Aquaculture*, **16** (1) 73–5.

Syslo M. (1986) Getting the 'bugs' out. *Massachusetts Wildlife*, **36** (3) 4–10.

Tiews K. (Ed.) (1981) *Aquaculture in heated effluents and recirculation systems*, **1**, 1–513; **2**, 1–666. Heenemann Verlagsgesellschaft, Berlin.

Van Olst J.C., Carlberg J.M. & Ford R.F. (1977) A description of intensive culture systems for the American lobster (*Homarus gammarus*) and other cannibalistic crustaceans. *Proc. 8th Ann. Wkshop World Maricult. Soc.*, 9–13 January, pp. 271–92. San Jose, Costa Rica.

Van Olst J.C., Carlberg J.M. & Hughes J.T. (1980) Aquaculture. In *The biology and management of lobsters, Vol. 2, Ecology and management* (Ed. by J.S. Cobb & B.R. Phillips), pp. 333–84.

Villarreal H. (1988) Culture of the Australian freshwater crayfish *Cherax tenuimanus* (marron) in Eastern Australia. In *Freshwater Crayfish 7* (Ed. by P. Goeldin de Tiefenau), pp. 401–8. Musée Zoologique Cantonal, Lausanne, Switzerland.

Waddy S.L. (1988) Farming the Homarid lobsters: state of the art. *World Aquaculture*, **19** (4) 63–71.

Westman K., Pursiainen M., Jarvenpåa T. & Westman P. (1989) Situation of the crayfish stocks and crayfish culture in Europe. In *Crayfish Culture in Europe* (Ed. by J. Skurdal, K. Westman & P.I. Bergen), pp. 175–92. Norwegian Directorate for Nature Management. Trondheim, Norway.

Wickins J.F. (1976) Prawn biology and culture. In *Oceanogr. Mar. Biol. Ann. Rev.* (Ed. by H. Barnes), **14**: 435–507.

Wickins J.F. (1982) Opportunities for farming crustaceans in western temperate regions. In *Recent advances in aquaculture* (Ed. by J.F. Muir & R.J. Roberts), pp. 87–177. Croom Helm, London.

Wickins J.F. (1986) Prawn farming today: opportunities, techniques and developments. *Outlook on agriculture*, **15** (2) 52–60.

Wyban J.A. & Sweeney J.N. (1989) Intensive shrimp growout trials in a round pond. *Aquaculture*, **76** (3/4) 215–25.

Wyban J.A., Sweeney J.N., Kanna R.A., Kalagayan G., Godin D., Hernandez H. & Hagino G. (1989) Intensive shrimp culture management in round ponds. In *Proceedings of the Southeast Asia shrimp farm management workshop*, 26 Jul.–

11 Aug. 1989 (Ed. by D.M. Akiyama). Philippines, Indonesia, Thailand. American Soybean Association, Singapore.
York R.H. (1983) Economics of pond polyculture of clams (*Mercenaria mercenaria*) and shrimp (*Penaeus* spp.) in Hawaii. In *Proc. 1st Int. Conf. on Warm Water Aquaculture – Crustacea*, 9–11 Feb. 1983 (Ed. by G.L. Rogers, R. Day & A. Lim), pp. 43–5. Brigham Young University, Hawaii.
Zhang N. (1990) On the growth of cultured *Penaeus chinensis* (Osbeck). In *The culture of cold-tolerant shrimp* (Ed. by K.L. Main & W. Fulks), pp. 97–102. Oceanic Institute, Honolulu, Hawaii.

Chapter 6
Site selection

6.1 Introduction

Siting decisions have a critical influence on the success of a crustacean farming project because they play a large part in determining potential yield levels and greatly affect the costs of construction and operation. Evidence of their importance is provided by the many abandoned projects in which poor siting has been identified as either the principal reason or a major contributory factor for failure. Of course no site can be optimal in all respects, but it is important to realise that some siting requirements are fundamental to success, and serious deficiencies in these cannot be compensated by any amount of other favourable site characteristics (Muir & Kapetsky 1988).

In this chapter the factors that govern site selection have been arranged firstly into those which operate at the level of countries or regions, and secondly into those which have much more to do with the specific locality and proposed ground site for the operation. Though both types of factor are equally critical to successful site selection, this distinction is useful because if a country or region is unsuitable it can be omitted from the selection process without the need to identify possible sites within it.

Ideally, in the pre-investment stages of a crustacean project, the process of site selection will follow an appraisal of which species to produce, which techniques and phases of culture to perform, and what represents an economically viable scale of operation (Chapter 9). Although this is the most logical sequence for planning purposes, in some cases interest in the potential of crustacean aquaculture arises from possession of a particular piece of land and the wish to use it more productively. In these cases no site selection is performed, rather a species and the approach to its culture are chosen to fit the site's attributes. Crustacean production may be just one of several possibilities under consideration which, apart from other forms of aquaculture, may include projects involving agriculture or leisure amenities. In any circumstance it is certainly worth considering what value a site may have for alternative uses should crustacean farming become unprofitable.

Some of the most critical aspects of site selection relate to climate, water supply, soil type, the availability of broodstock and seedstock, and the proximity of markets. In this chapter these are considered along with a whole range of other factors which, depending on the project being planned, are likely to have an impact on overall profitability.

6.2 Country or region

Site selection factors that can be conveniently taken into account at the level of countries and regions include: seasonal temperature range, the availability of essential inputs (labour, materials and services), the size and type of market to be exploited, and various political, institutional and legal considerations.

6.2.1 Climate

Knowledge of the usual weather patterns of a country and the possible diurnal, seasonal and annual extremes is essential to the site selection process. Armed with accurate and detailed information, it is possible to assess if conditions are suitable for continuous year-round operations, if some phases of culture will have to be restricted to certain months of the year, or if some environmental control will be required.

The climate will not only influence production but also the timing and duration of construction work, since at some sites earthworks can only be performed during the dry season. Any delays in generating revenue from production can strongly influence the financial viability of a project (Section 10.7).

6.2.1.1 Temperature

Temperature is a fundamental determinant of crustacean growth rates and can only be cost-effectively managed through site selection (Malecha 1983). If it falls outside optimum ranges (Table 8.3; Appendix I) for significant periods, production potential will be impaired.

Both mean temperatures and likely extremes must be considered. For example, although the mean annual air temperature in Kuwait is around 25.7°C, within the optimal range for the culture of most penaeid shrimp, it varies between a mean of 37.4°C in July and 12.7°C in January, both well outside ideal limits. In addition, the difference between night and day temperatures averages 14°C, greatly reducing the potential viability of outdoor culture operations (Farmer 1981). As a general rule, inland regions and those with little maritime influence on their climate will experience the greatest seasonal and diurnal fluctuations in temperatures. Water does not heat up or cool down immediately in response to changes in air temperatures and this provides some protection against daily extremes and short-lived spells of unusually hot or cold weather. This buffering effect is only significant, however, in embayments, deep ponds, lakes and some lagoons.

6.2.1.2 Rainfall

The quantity and seasonal patterns of rainfall should be studied for proposed aquaculture sites. Heavy rainfall, for example during monsoons, can cause severe flooding and embankment erosion and hinder farm access by turning earthen tracks to mud and by inundating waterways. In 1988 in southern Thailand, 1000 shrimp farms suffered stock losses and severe damage when five days of heavy rain caused serious flooding.

The salinity levels in marine and brackish water ponds will also be strongly influenced by rainfall, or a lack of it (Section 6.3.1.4).

6.2.1.3 Wind

Air movement over outdoor ponds promotes gaseous exchange between the atmosphere and the water and generates surface currents which destratify the water column. Although this effect is desirable, wind, particularly in combination with low humidity and high temperature, can cause significant evaporation in ponds, which raises the salinity of brackish water and so increases requirements for water renewal.

On the other hand, calm periods accompanied by high temperatures or heavy rainfall can induce thermal or salinity stratification of the water, and prevent oxygen from reaching the pond depths. If these periods are frequent in a particular region, then requirements for aeration or water exchange will be increased. Usually, at coastal sites on sunny days, sea breezes are generated as a result of the differential heating and cooling rates of land and water. These breezes can ameliorate the sun's heating effects. In regions with very strong winds, wave damage to pond embankments can be severe unless sheltered sites can be located (Section 6.3.2.3).

Wind blown dust can also present problems for aquaculture in some desert regions. Outdoor tanks in Kuwait trapped on average 11.6 g of dust m^{-2} per month and as much as 2.43 g m^{-2} day^{-1} during a particularly severe storm (Farmer 1981). Even low levels of dust (0.02 g ℓ^{-1}) suspended in larvae cultures have been found to cause heavy mortalities (Al-Hajj *et al.* 1981).

6.2.1.4 *Evaporation and humidity*

Evaporation is most significant when low humidity levels are combined with wind and elevated temperatures. In arid regions at exposed sites mean evaporation rates can exceed 15 mm per day.

The water supply for an aquaculture operation must replace evaporative losses, and for marine and brackish water operations must permit some control over salinity levels. For the latter, particular problems may arise during hot, dry seasons when the salinity levels rise both in ponds and in the incoming water supply, hampering attempts at salinity control through water renewal. Salinities in shrimp ponds in Sri Lanka have reached 50 ppt. as a result of monsoon failure (Chamberlain 1988) (Section 6.3.1.4).

Very humid conditions result in reduced pond evaporation but make indoor work environments very unpleasant. They may necessitate air-conditioning in offices, laboratories and staff accommodation, which increases energy costs.

6.2.1.5 *Insolation (sunshine)*

Insolation levels are determined largely by cloud cover and latitude. At high latitudes intensities will be less than at tropical sites, but total insolation will greatly increase in spring and summer with the longer days.

Insolation is a very important factor in the calculation of energy budgets in controlled-environment and greenhouse systems. In unheated structures it should be sufficient to raise temperatures and promote growth in early spring. However, in all greenhouses in the summer, insolation can cause over-heating unless adequate ventilation and shading is provided.

For outdoor algae cultures made in support of hatcheries, very strong sunlight may inhibit growth and raise temperatures excessively, necessitating the use of shade cloth. Shade cloth is also required in some nursery operations and for transit/holding tanks, where it prevents algal growth as well as reducing temperatures.

6.2.1.6 *Climate change*

It is worth considering the likely impacts of climate change on an aquaculture venture (Section 11.5.4). If sea level rises it will expose coastal and estuarine sites to increased wave damage and flooding, particularly on extreme tides associated with storm surges. Also, if rainfall increases in the tropics, it will cause more flooding of rivers and surrounding areas and thereby lead to increased embankment erosion and greater levels of suspended solids in water. Silt which accumulates in ponds and canals demands periodic removal.

6.2.2 Availability and costs of essential inputs

The costs and availability of essential inputs will have a direct bearing on the economics and ease of operating a crustacean farm or hatchery. Although it is possible to transport men and materials almost anywhere in the world, it makes economic sense to set up operations where essential resources are readily available and inexpensive, or at least competitively priced.

6.2.2.1 Broodstock and seedstock

Broodstock and/or seedstock are essential to crustacean aquaculture and great attention should be given to locating reliable sources. Although techniques exist for shipping live material over long distances, there are several advantages to having a local supply. Not least is the simple fact that local supplies are usually cheaper, and because broodstock and seedstock costs are often significant components of operating budgets, even modest savings can influence profitability. Although natural stocks may only be available for certain periods each year, when they do occur they may be abundant and therefore low-priced. Production schedules can be arranged to capitalise on patterns of seasonal abundance, as has been the case with the culture of *P. chinensis* in China.

A conveniently located wild stock can support a culture industry at more than one level of operation by providing both wild caught juveniles for farms, and broodstock for hatcheries. The bulk of the Ecuadorian shrimp farming industry has been based on supplies of wild *P. vannamei* seed. When these supplies became limiting to production in the mid 1980s, local supplies of broodstock were employed for the production of juveniles in hatcheries. This situation can be contrasted with the experience of one farm in Gambia, which, faced with the absence of a locally available fast growing penaeid species, was forced to go to the expense of importing *P. monodon* broodstock from Asia (Skabo 1988).

Shrimp production in countries that rely on introduced species has remained small by comparison to production from countries where suitable cultivable species occur in the wild. Output in Brazil, for example, relies partly on the slow growing indigenous species *P. schmitti* and *P. brasiliensis*, but the desire to obtain better yields led to the introduction of *P. vannamei*, *P. monodon* and *P. japonicus*. Despite these measures and the presence of good sites and a favourable climate, shrimp production in Brazil at 2 000 mt yr^{-1} has grown very slowly compared to Ecuador where *P. vannamei* occurs in the wild and supports an annual output of around 73 000 mt.

If a particular crustacean farming operation is to rely on the shipping of broodstock or seedstock over long distances, several disadvantages should be borne in mind. Firstly, live transport necessitates crowding for extended periods, and stress and mortality of the animals is likely to result. Secondly, obtaining supplies from distant sources often allows fewer opportunities for the buyer to inspect the quality of the

Plate 6.1 Semilleros of Ecuador using scissor push nets to catch wild post-larval shrimp.

purchase. This can be especially problematic if prepayment is demanded and a shipment subsequently arriving in poor condition cannot be returned to the supplier. Thirdly, imported stocks can have a significant ecological impact (Section 11.4.2) and may be subject to complex regulations regarding movements of live animals (causing delays at airports). Sometimes shrimp broodstock and seedstock are smuggled out of countries where legislation exists specifically to prevent the export of resources needed for domestic farming activities. Such supplies are usually sold at inflated prices and would become vulnerable if the relevant anti-contraband laws were strictly enforced. Distance is certainly a disadvantage in the face of competition for declining stocks.

Any factor likely to jeopardize sources of broodstock or seedstock should be investigated wherever large-scale projects are proposed. Sometimes fishing regulations specify periods each year when wild stocks cannot be fished, and sometimes complete bans operate to protect wild stocks. In the case of shrimp farming, if a fishery for wild post-larvae does not exist it can take time to establish one. It may require the training of seed fishermen in capture and handling techniques, and some research to establish which species are available and what is the pattern of seasonal abundance (Section 7.2.6.1). Ideally, investigations into the availability of wild seed should be supported by governments who wish to promote shrimp farming. However, resistance to the fishing of both post-larvae and broodstock may come from established fishing interests concerned about real or perceived detrimental impacts on wild stocks and competition from aquaculture (Section 11.4.1). This has occurred recently in Mexico following the capture of increasing numbers of shrimp post-larvae for aquaculture (Wood P., 1989, pers. comm.).

When the best places to catch mature adults of the preferred crustacean species do not coincide with the most productive fishing grounds, special and often costly arrangements may be necessary. For example, where fishermen are not already engaged in supplying a live trade or broodstock for shrimp hatcheries, it may be necessary to go to the expense of periodically chartering a trawler.

The situation for freshwater prawns and crayfish is somewhat different from that for the most important penaeid species: when they become established in farms, ample supplies of egg-bearing broodstock can usually be obtained from production or holding ponds. This provides independence from wild stocks and has given great impetus to the spread of *Macrobrachium rosenbergii* farming throughout the world; it has also led to the establishment of populations of several crayfish species outside their natural range.

6.2.2.2 Feeds and feed raw materials

Large quantities of feed are needed for semi-intensive and intensive culture operations, and they represent a major production expense. Feed of suitable quality must be readily available at a price which does not significantly narrow profit margins. Specialised crustacean diets are manufactured in countries where significant farming operations are established, but outside these locations suitable diets must be imported. In West Africa, for example, the absence of a shrimp feed manufacturer led one operation to seek alternative sources in Europe, the USA and Asia (Skabo 1988). Extra costs for shipping and storage can be expected with all bulk imports of feed.

A feed mill to manufacture diets, at or close to the farm, may be set up either to support the requirements of the particular project or as a larger, independently-viable, commercial operation serving many customers. In either case, raw materials such as fishmeal, wheat and soybeans will be required, possibly together with binding agents and vitamin and mineral mixes. Some or all of these materials may be unavailable locally, creating dependence on imports once more. By-products from animal processing operations can be used, including shrimp head meal, and if trash fish (especially marine species) is available this can be a valuable ingredient in simply-prepared moist diets. Such diets however, cannot be stored easily, requiring new batches to be prepared every few days, and they may only be suitable for small farms. For freshwater prawns it is possible to use diets made for pigs or poultry, though these are not designed to be stable in water and they disintegrate rapidly.

Diets for use in hatcheries are mostly produced in Japan, the USA and Europe, but because they are used in relatively small volumes they can be easily shipped to other parts of the world at little extra expense. Some companies even offer their products at a fixed basic price all over the world. *Artemia* cysts, mostly produced or processed and packed in the USA, are also easy to transport and have long storage lives (up to five years). Many countries harvest *Artemia* but care should be taken with cysts from unrecognized sources because hatch rates, nutritional value and levels of contamination with pesticides vary greatly between batches and sources (Sorgeloos *et al.* 1983).

6.2.2.3 Fertilizer and lime

Supplies and prices of inorganic fertilizers (Section 8.3.5.3) should be easy to find because these products are widely used in agriculture. Organic fertilizers, which are mostly by-products of poultry and livestock production, are bulky, and if sources are distant transport costs may be prohibitive. Liming materials are often widely available because of their use in farming and the manufacture of cement (Section 8.3.5.1).

6.2.2.4 Energy

Many important factors relating to energy supplies are dependent on the locality chosen (Section 6.3.5). At the level of a country or region it may be important to investigate the overall national energy policy. Sometimes electricity prices for industry are subsidised to promote regional development. Some oil exporting countries provide fuel at subsidized prices, which can make the installation and operation of generators a competitive alternative to reliance on mains electricity. Countries with a well-developed electrical network can provide a more reliable supply and, of course, the national voltage must be considered when electrical equipment is ordered.

6.2.2.5 Staff

Total labour costs are greater than simply the sum of wages and salaries. In most countries detailed legislation covers subjects such as overtime pay, national holidays, social benefits, health care benefits, seasonal bonuses and company liabilities if permanent staff are dismissed. The full impact of these regulations on labour costs should be investigated and understood. Labour rates for construction workers will have a big impact on the overall cost of setting up a project.

Customs relating to work and the differing roles for men and women may be significant. For instance, in Moslem and Christian communities the workload may need to be reduced during the periods of Ramadan or Christmas. In processing plants women often make up a large part of the workforce, but this may not be considered acceptable in certain societies. Some other aspects of labour in relation to locality are discussed in Section 6.3.6.

Suitably qualified technical staff can usually be obtained more easily in regions where crustacean aquaculture is established. People without suitable 'hands-on' experience, but with some technical background and an education in science, can be found in most developed and some developing countries and can be trained accordingly. Requirements for well-trained and experienced technical staff are at their greatest in more intensive projects.

People with good managerial skills, biological knowledge and experience immediately relevant to crustacean farming are often needed to run farms and hatcheries, but are in short supply worldwide. Although this problem is somewhat independent of location, it is worth considering that very remote regions can result in demands for higher salaries and can deter the long-term involvement of expatriate staff (Section 11.2.6).

6.2.2.6 Construction materials and engineering services

The availability of essential building materials should be investigated and unit costs calculated to include allowances for delivery. Materials needed may include timber, cement, sand, stones, aggregate, bricks, building blocks, steel reinforcement bars, steel roof supports, and roofing materials such as asbestos and transparent corrugated sheets. For materials such as sand and aggregate, the main part of the cost is usually transport, so full costs cannot be established until possible sites are located.

Earth moving is the main expense in the construction of ponds, so the likely cost per cubic metre should be investigated. Unless ponds are to be built by hand, bulldozers, graders, compactors and excavators may be needed. These can be located easily in most countries because of their use in road building.

A considerable quantity of plumbing materials, including plastic valves and PVC piping, will be needed for most hatcheries. Decisions on what to use for making hatchery tanks should be influenced by the availability of local materials and the possibility of using prefabricated tanks of plastic or GRP. Some plastics may need to be imported and can be very expensive. The suitability of different materials is discussed in Section 8.1. In one alternative building technique, inexpensive tanks can be created using concrete reinforced with chicken wire.

The availability of professional services, for example for quantity surveying, site engineering and contracting, well drilling and electrical installation, will probably need to be checked (Section 9.5.2).

6.2.2.7 Equipment

Reliance on imported equipment can be high in some developing nations, especially if a 'high-tech' turnkey package is acquired. This can create a dependence on outside suppliers of spare parts, which may result in high maintenance costs and extended periods during which equipment is under repair. It is therefore often worth investigating locally-made alternatives before resorting to imports. Often development aid agencies specify equipment sources to ensure quality, even though this results in increased dependence on imports.

Examples of the kind of equipment needed to establish a shrimp hatchery and a shrimp or prawn farm are given in Tables 10.8 and 10.11 respectively.

In some countries imports of various chemicals and therapeutants may be essential, particularly in order to maintain a hatchery operation. If algae are to be grown from stock cultures a series of high grade nutrients will be required, and sometimes compressed carbon dioxide. For live transport compressed oxygen may be needed, though this is generally widely available because of its use in hospitals and for welding torches.

6.2.2.8 Technical services and support

In addition to those professional and engineering services mentioned above, others that may be needed include legal assistance, specialist plumbing, pond cleaning and harvesting services. Obviously the requirement for and availability of each of these will vary from one location to the next. Especially difficult to obtain in some countries is technical support, which can be very important when problems arise during production (Section 9.5.2). Technical support is often provided by extension services, disease specialists and laboratory services which can perform water quality and nutritional analyses, albeit at a cost. Some feed companies take on the role of extension agents in return for purchase of their feeds (Section 11.6.2).

6.2.3 Markets

The identification of the market to be exploited (Chapter 3) is very important to site selection. Although much crustacean farming relies on distant bulk frozen markets, in which proximity is not essential, when live and fresh crustaceans are aimed at local specialist markets there are usually cost and quality benefits associated with basing production near the consumer.

Frozen crustacean products like whole and headless shrimp are traded worldwide and the best locations to farm them are not simply determined by the fact that most consumption is centred in Japan, the USA and Europe. For example, some shrimp farming businesses who are targeting the North American market originally started their operations within the USA but have now relocated to Central or South America to take advantage of more favourable climates and cheaper production costs. The savings easily outweigh any extra shipping costs. Apart from penaeids, other farmed crustaceans which are shipped over long distances include freshwater prawns from Thailand and crayfish from the USA, both sold in Europe.

If live or fresh products are to reach the consumer before their condition and value deteriorates, the duration of transport should be just a matter of hours. Proximity to the market, or at least very efficient transport links, are essential. When limited small and local markets are targeted – for example for live crabs in Asia, for live shrimps in Japan or for sales of *Macrobrachium* and crayfish either at the farm gate or direct to the catering trade – attention should be given to finding sites within reach of population centres or tourist areas. In southern Spain many crayfish farmers have been able to take advantage of the large numbers of holidaymakers arriving each summer.

Knowledge of regional and local market conditions can also be significant to the site selection process. Often consumption rates vary greatly with season and sometimes religious taboos are significant and can severely limit local marketing opportunities. In Israel, shrimps and prawns are not considered 'kosher' and are either exported or consumed by tourists.

6.2.4 Processing facilities

Processing requirements may be minimal if live or fresh products are sold on local markets, but processing facilities for washing, deheading, freezing, packing or canning will usually be required for bulk and export markets.

Farming operations should be located so that only short journeys are required for their perishable products to reach the processor, and the availability of processors and the quality of their installations should be investigated to ensure they meet acceptable standards of hygiene and quality control. Reputations for quality vary between plants and countries and will influence the prices that can be obtained. Not all processing facilities are capable of producing the range of high quality and value-added products increasingly demanded by consumers, and this shortcoming has been identified as a likely constraint for some proposed shrimp farming operations.

Processing plants often exist to handle wild-caught products and they can easily deal with additional yields from aquaculture. As part of their service some supply trucks and ice during a harvest, provided that the farm is not too distant and has good road access. In other situations a source of ice may be needed. Ice suppliers are often easy to find in coastal areas where fishing is an important activity, but they may be rare inland.

6.2.5 Political, institutional and legal factors

6.2.5.1 *Civil stability*

Some countries or regions are troubled by civil strife due to guerilla or terrorist activity, military coups, separatist conflicts and internal security problems related to drug production, drug smuggling and resulting violent crime. The impact of such problems extends to a country's overall economic and investment climate. Some conflicts have impinged directly on crustacean farming enterprises. For example, a massacre of Tamil workers by military personnel took place on one Sri Lankan shrimp farm and at least one shrimp farm in the Philippines has relied on access by helicopter because of guerilla activity in surrounding areas. Remote farms producing valuable crops in countries where security cannot be guaranteed are certainly easy targets for extortion or payroll robberies by armed gangs.

While potential dangers should not be underestimated, knowledge of the actual conditions in a country cannot be gained simply from the coverage of events on international news media. These often give impressions of widespread turbulence in countries which are for the most part stable and peaceful except for occasional or localised violent incidents which attract disproportionate attention. A more complete picture of the security situation of a country can be obtained by speaking to knowledgeable residents or experienced travellers who can make judgements based on first-hand experience.

Also of great importance to any proposed business is a measure of political and economic stability. Commenting on the situation in some African states, Skabo (1988) noted that 'high inflation rates, depreciation of the local currencies, coupled with monetary and political instability make long-term investments unattractive, in spite of the projected returns.' For any project that operates in an unpredictable economic environment, financial planning can be a very complicated process.

6.2.5.2 *Taxes and duties*

Tax regulations differ widely between countries and always have an important impact on profitability. Of particular interest to investors will be levels of corporation tax, the availability of tax holidays and the possibility of tax concessions or tax write-offs on research and development projects.

Levels of import duties need to be known in order to calculate the true costs of imports. In many countries the relevant legislation is complex, with differing levels of duty applying to materials destined for different purposes. For instance, while medical supplies may be duty-free, some items of equipment destined for industry may attract duties of up to 50% or more. Sometimes very high duty is applied to discourage the importation of goods such as motor vehicles that are assembled or produced by a home-based industry. Often high rates extend to materials for which no real equivalent product is locally available. Nevertheless, when a strong case is made for the value of crustacean farming to a national economy, special rates of import duty may be applied. In Ecuador, when shrimp farms were faced with acute shortages of wild shrimp post-larvae, all import duties on equipment destined for the construction of shrimp hatcheries were waived. The importance of positive government attitudes in general is discussed later (Section 11.6.3.3).

6.2.5.3 *Exchange controls*

In an attempt to regulate the flow of money in and out of some countries, currency exchange rates are maintained at artificial levels. These can delay and complicate the processes of importing and exporting and moving capital in and out of a country. Sometimes special regulations exist specifically to limit the repatriation of profits. One result of such measures is that two rates of exchange exist, one for free (or in some cases 'black') market exchange and another based on control by a central bank. Unpredictable movements in exchange rates can greatly confuse financial planning, especially for an export-based or foreign-funded operation.

6.2.5.4 *Land costs and concessions*

Many questions may need to be answered regarding the availability and cost of land. For example: can low cost leases be obtained from government or regional authorities, or will land need to be purchased? In some countries the ability to buy land is restricted to nationals. Sometimes a government will provide land for a project and take a portion of the equity in return. Obtaining large tracts of contiguous land can be a particular problem if areas are divided into many small parcels, each with a different owner.

Often regulations exist which specify that certain zones be used preferentially for industry or tourism, or be reserved for wildlife. Even if sites are specifically designated as suitable for aquaculture development, the full site selection process should not be by-passed because 'suitable for aquaculture' can often be translated merely as meaning 'unsuitable for anything else so why not try aquaculture?' (Section 11.5.1).

6.2.5.5 *Availability of loans and grants*

The sources and nature of finance need to be investigated, together with the availability of investment incentives (Section 10.3). Projects likely to meet identified goals for economic development may qualify for grants or loans on especially favourable terms. The form of financial assistance can vary with respect to interest rates, the duration of grace periods before repayments are due, the proportion of total project cost that can be met by the loan or grant, and whether money is available to cover all project expenses or just capital investment in facilities and infrastructure. There may be a minimum amount of capital assistance that can be provided, and unfortunately when this level is set too high it renders the credit unsuitable for small-scale operations and only benefits large schemes. For smaller operations credit extension schemes operating at an appropriate scale may need to be located (Section 11.6.1).

The interest rates payable on bank overdrafts should be taken into account, particularly for small operations in which cash may be needed at short notice to cover operating and pre-operating expenses.

6.2.5.6 *Traditions*

It is more problematic to start an aquaculture operation in regions like Africa, where the activity is largely unknown, than it is in areas such as south-east Asia where aquaculture has been practised for centuries. A lack of tradition and technology in some countries can lead to strong feelings of isolation for aquaculture pioneers. When crustacean farming does become established trade organisations are often formed which can promote the exchange of information and serve as a lobbying force.

Local customs can affect aquaculture projects in many ways and need to be investigated during the site selection process. Strict vegetarian communities, anti-aquaculture lobbies and traditional multiple land ownership can all exert significant influence (Section 11.2.5).

6.2.5.7 Legal requirements

Legal requirements affecting aquaculture are often complex and stringent and in some regions the bureaucratic 'red-tape' they generate stifles interest in new projects. Some laws are applied to aquaculture even though they were originally written for industry, fisheries or agriculture (Section 11.6.3). Areas where legal requirements are likely to affect crustacean farming projects include:

- *Land ownership* Laws may differ for foreigners and for land with sea frontage.
- *Water usage* Laws are often particularly strict with respect to ground water abstraction, permits may be needed.
- *Water discharge* In some countries farm effluents must comply to set standards.
- *Introduction of new species* See ecological impacts (Section 11.4.2).
- *Joint ventures* Equity limits may operate for foreigners (Section 10.4).
- *Construction* Permits will be needed and many planning regulations usually apply.
- *Visa requirements* For foreign investors and technical staff and the need for foreign experts to train local staff.
- *Import/export* Restrictions relating to materials, broodstock, seedstock, and product.

6.3 Locality

The most obvious requirement for a specific site is that it should provide enough land and water of sufficient quality for the proposed project. Soil characteristics are especially important if earthen ponds are to be constructed and attention must also be paid to the suitability of the local infrastructure, communications and labour force, as well as to a number of social and environmental factors.

6.3.1 Water

6.3.1.1 Quantity

Water at aquaculture sites may be obtained from surface waters or from underground reserves. Outdoor on-growing operations have the greatest requirements (Section 8.3.5.5). If surface water is to be taken from estuaries or rivers, seasonal variations in flow rates can be expected. During the dry season in some tropical river estuaries, it is possible for negative flow to occur with seawater moving inland to replace freshwater lost by evaporation. Under these conditions salinity levels can rise sharply.

Ideally, streams for freshwater farms should not be allowed to flow directly through ponds because of the risk of siltation and flooding. Ponds should be built to one side of the stream or the stream diverted to one side of the ponds. For simplicity, however, some farm dams in Australia which have been stocked with crayfish consist of a straightforward transverse embankment across the path of a stream. Sites are usually selected so that water run-off each year will at least be sufficient to fill the dam.

Groundwater is more consistent year-round than surface water but it is still important to know how much is available and from what depth it needs to be pumped. The latter will influence pumping costs. Where underground freshwater is already being tapped for agriculture or to provide a supply of municipal water, the reserve may already be known. If not, a series of test boreholes will be needed. For all sites requiring freshwater, the needs of any nearby irrigation schemes should be reviewed in relation to the amount of freshwater available (Section 11.5.2). If a water shortage ever arises, priority will usually be given first to domestic needs and then to the irrigation of agricultural crops rather than crustacean farms. In streams and rivers, traditional usage for such things as washing clothes and watering livestock should be investigated.

Ideally, hatcheries should be located where an unlimited supply of high quality water is available. Marine seawater hatcheries located adjacent to an ocean, and most freshwater hatcheries with a reliable borehole supply, are more likely to have good quality water than estuarine or lagoon-side hatcheries. Borehole water and seawater taken from beach wells contain only very small amounts of suspended solids, thereby greatly reducing or even eliminating the need for further filtration (Section 8.4). The ideal site for a *Macrobrachium* hatchery is one where wells sunk at different depths provide both fresh and saltwater which can then be mixed to achieve the optimum salinity of 12 ppt. (New & Singholka 1985).

A reliable supply of freshwater is also needed for washing operations in processing plants. If no mains supply exists a well may need to be installed or surface water pumped and treated.

6.3.1.2 Distance from source

Water supplies should be as close as possible to a farm or hatchery to avoid the expense, ownership problems and maintenance costs of long supply channels or piping installations. A notable exception to this general rule is found in some *Macrobrachium* hatcheries in Thailand, which are located at inland sites and rely on road tankers for a supply of concentrated brine from salt pans. The brine is diluted with freshwater for larvae culture.

6.3.1.3 Tides

If tides are to be used for water exchange in brackish water ponds, the tidal range must be sufficient for flushing but not so great as to require massive embankments and embankment protection. Tide tables should be studied to establish tidal ranges and variations between spring and neap tides. Spring tide ranges of around 2 m are about the minimum necessary, while anything greater than about 4 m will require excessive embankment protection (ASEAN 1978). Little or no water exchange will be possible during neap tides. Embankments should be able to cope with unusually high tides caused by storm surges. Where land is scarce in Japan, some farmers have built

Plate 6.2 An inland backyard hatchery for freshwater prawns in Thailand. The owner and her family are seen discussing alternative seawater supply prospects with an extension worker from the Department of Fisheries.

tidal impoundments in which the embankments become submerged at high tides (Section 7.2.6.4).

6.3.1.4 *Quality*

Details of water quality and likely fluctuations at a site can never be known with precision. An estimate has to be made of the risk of water quality falling at times to unacceptable levels. An initial site survey will usually only be done on one or two days and even a full survey will usually draw heavily on previously gathered information which may not be completely suitable. Data may have been gathered by navies or prospectors for purposes unrelated to aquaculture.

Aspects of water quality that are of particular significance at the site selection stage are covered here while the water quality tolerance of crustaceans is discussed later (Section 8.5).

Although inland groundwaters can be saline, salinity levels are usually only of relevance to marine and brackishwater farms where seasonal fluctuations in pond salinities can be expected as a result of precipitation and evaporation. Water at some estuarine sites can become virtually fresh during rainy seasons and super-saline following hot dry periods. During the latter, an alternative source of abundant fresh or low salinity water is at a premium.

Sub-optimal levels for more than one water quality factor can have deleterious synergistic interactions and although crustaceans can sometimes display remarkable resilience to sub-optimal conditions, site selection should pay close attention to meeting water quality requirements if reliable survival and growth rates are to be obtained. For example, reports of respectable survival rates for certain stocks of *P. monodon* at salinities up to and in excess of 50 ppt. (Chamberlain 1988) should not be taken as a reason to deviate from the usual preference for low salinity sites for the culture of this species. Some semi-intensive *P. monodon* farmers raise salinity as the shrimp grow older to increase growth. At these sites access to supplies of both brackish water and freshwater are essential.

The temperature of a water supply can differ significantly from ambient air temperatures. In some temperate crayfish farms well water may be particularly cool, in which case shallow ponds would usually be employed to allow this water to warm up before use. On the other hand, excessively hot water from geothermal sources may need to be left to cool before use, or blended with cooler water, or used indirectly as a heat source for controlled-environment cultures (Sections 5.2, 5.3 and 8.4.4). Sometimes geothermal water may need to be 'degassed' by exposure to the atmosphere, aeration and agitation to help eliminate hydrogen sulphide and ammonia and remove iron which may otherwise precipitate on the gills of crayfish. Many geothermal waters are acceptable for use in culture operations but a few may contain unsuitable ionic compositions, for example excessive hardness.

The concentrations of phytoplankton and zooplankton in surface water can give an indication of potential productivity. For outdoor on-growing operations, clear waters may require extra fertilizer or higher quality feeds to compensate for their lack of nutrients and natural productivity. On the other hand, an excess of nutrients, particularly nitrates and phosphates, can lead to excessive eutrophication (see Glossary). Water contaminated with agricultural fertilizers or sewage should be avoided. Of particular significance to site selection are the likely seasonal changes in the levels of chemical nutrients and the risk of toxic algae blooms occurring. Neither of these can be assessed simply with spot checks. Seasonal storms can increase the sediment load in some coastal waters and some regions are known to have waters with different characteristic levels of nutrients. For example, on the Caribbean coast of Mexico waters are known to contain less nitrogen and phosphorus than on the Pacific coast. These nutrients have also been found to be deficient in oxygen-rich surface waters off Oman.

Pollution can arise from many different sources and adversely affect crustacean farms and hatcheries. A thorough assessment must be made of the proximity of the proposed site to agricultural land where fertilizers and pesticides are used, to any industrial activity where chemicals are used or discharged, and to any other human activity resulting in effluent or run-off that could contaminate water supplies (e.g. sewage or mine tailings). Contamination can be direct (into streams etc.) or indirect via soil, wind-blown dust, birds or insects. However, the most significant sources of pollution at many aquaculture sites are existing aquaculture operations. The best known examples of this can be found in southern Taiwan and in the Bay of Bangkok. To avoid the worst effects, prevailing current directions and their seasonal changes should be considered. Hatchery discharges can contain pathogens, including bacteria with resistance to antibiotics which can severely limit the impact of antibiotic therapy (Sections 11.5.3 and 11.4.4).

6.3.2 Topography

The principal factors to be considered in selecting a farm site with suitable topography concern elevation, gradient and exposure to winds.

6.3.2.1 Elevation

The elevation of marine and brackishwater ponds is a compromise between good tidal exchange and (where pumped) low pumping costs, and the need to drain completely. Since tidally flushed ponds require land with intertidal elevations this often involves occupying areas covered by mangroves, salt marshes or mud flats. Tidal ponds typically permit water exchange only during spring tides and allow complete drainage only at low water during these periods. Suitable areas for tidal ponds thus need to be selected according to tidal levels and ranges.

Pumped ponds can be built above the level of high water (spring tides) to allow for free drainage at nearly all times. Ground with an elevation of up to 5 m is often used for large shrimp farms with pumping costs closely related to the height that the water must be raised. In some very large operations the water supply for high level ponds (>5 m) is provided by two separate sets of pumps via an intermediate reservoir.

6.3.2.2 Gradient

Gradients have important implications for the construction and drainage of ponds. Sites with steep slopes necessitate either small ponds, deep water or excessive soil movement during construction (Wheaton 1977) and are thus inappropriate for large extensive and semi-intensive ponds. Land gradients for these should be between 0.1–0.5% whereas for ponds of less than 1 ha gradients of 0.5–5% are suitable.

For freshwater ponds relying on streams or rivers, valleys with a V-shaped cross-section may be acceptable for small and raceway type ponds, if the slopes are not too acute (Huet 1972).

In the design of some hatcheries a raised reservoir is included to allow for gravity flow down to hatchery level. If this approach to water supply is planned then a raised area of land in the vicinity of the hatchery is needed to avoid the cost of building a large base for the reservoir. Some hatcheries make do with a metal or fibrocement header tank.

6.3.2.3 Exposure

Although steady light breezes are beneficial for ponds because they promote gaseous exchange, high winds at very exposed sites can result in severe wave damage to embankments and excessive evaporation, unless measures are taken to protect downwind embankments and install wind breaks. In one shrimp farm built at an exposed site on a windy section of the Spanish coast, severe wave damage necessitated the remodelling of ponds. At considerable expense, long, narrow ponds were constructed with the long embankments at right angles to the prevailing wind.

The level of exposure and amount of vegetation at a site can influence evaporation rates. For example, while an evaporation rate of 6016 mm yr^{-1} has been recorded at Kuwait International Airport, in the same country at another site with plant and tree cover the equivalent rate was only 1598 mm yr^{-1} (Farmer 1981). Vegetation serves to cut evaporation from ponds by reducing temperatures, raising humidity levels and reducing wind speeds.

6.3.3 Soil

The suitability of a site for pond construction is greatly influenced by the physical and chemical characteristics of its soil. Physical properties must provide for reliable and sufficiently impermeable embankments and pond beds, and soil chemistry should ideally contribute to, or at least not detract from, the fertility of the pond water.

An initial reconnaissance survey can be carried out on a proposed site in order to make a simple map and to collect soil samples. Later on, when the most suitable areas of a site have been located, a more detailed survey is performed by collecting around 0.5–2 soil core samples per hectare. Soil profiles are examined in their natural state by excavating an open pit (0.8 × 1.5 × 2 m deep, or down to parent rock or water table). Soil samples can be taken from within the pits or collected using an auger and should be bagged and labelled in preparation for analysis. A thin walled steel tube (30–60 cm long and 4–7 cm in diameter) is inserted into the ground to collect the undisturbed sample cores needed for certain laboratory tests. Soil quality usually needs to be investigated to a depth of at least 1 m below the intended base of ponds.

To provide a detailed soil chemical analysis, samples weighing around 1 kg each can be sent to a soil testing laboratory. However, one factor of particular significance to aquaculture – pH – is best measured in the field with a pH meter (test 1 part of soil mixed with two parts of distilled water). Soils with pH values of between 5.5 and 9.5 are preferable and the range 6.5–8.5 is optimal for pond productivity.

For pond construction the most important physical properties of soil are texture, consistency, structure, permeability, and colour.

6.3.3.1 Texture

Texture is largely determined by the proportions of different sized soil particles. To establish the proportions with precision it is necessary to dry and sieve a sample. Fine particles (less than 2 mm) are classified as sand, silt and clay in decreasing order of size. For low porosity in embankments a minimum clay content of around 25% is recommended, although as little as 10% may be acceptable for well compacted soils (O'Sullivan 1988). For pond construction in general a minimum of 50% silt and clay particles is recommended (Coche 1985). However, soils with a clay content in excess of 60% are not recommended because they are subject to excessive cracking when ponds are dried (New & Singholka 1985). Very heavy clay soils also make earth-working hard.

Detailed particle analysis of soils can be performed in a specialised laboratory but a number of simple tests can be performed in the field which may suffice for small projects. A simple indication of texture can be gained by taking a handful of moist soil, squeezing it into a ball, throwing it 50 cm into the air and catching it. If the ball falls apart it is probably too sandy and unsuitable for ponds, whereas if it sticks together the clay content is probably adequate.

Further simple and quick field tests to indicate basic soil texture are detailed by Coche (1985). The bottle test involves putting 5 cm of soil in a glass bottle, filling the bottle with water, suspending the soil by vigorous mixing and then leaving the sample for an hour. The proportions of sediment in sand, silt and clay can be estimated from the layers which settle in the bottom of the bottle. Another method that gives a better indication of soil texture is the manipulative test, which involves making a ball of moist soil in the hand and rolling it into a thin sausage 15–16 cm long. If the sausage cannot be formed the soil is probably too sandy for ponds. If it can be formed but cannot be bent into a semicircle the soil is a loam which is a fair material for building embankments. If the semicircle can be formed and then bent into a ring without cracks forming, then the soil is clay which is excellent for embankment construction. If cracks do form, the soil is a light clay which is also suitable for embankment building. Another simple test of the suitability of soil for embankments involves kneading moist soil to make several 10 cm diameter balls and leaving them submerged in water for a day. Good material will remain intact while balls of unsuitable soil will disintegrate within a few hours.

6.3.3.2 Consistency

Consistency is a measure of the strength with which soil is held together when dry, moist or wet and has important implications for pond construction. It can be gauged roughly in the field by using simple tests, or assessed with precision in the laboratory by using standard methods to determine the Atterberg limits. An Atterberg limit corresponds to the moisture content at which a soil sample changes from one consistency to another. Of particular interest to pond construction are the liquid limit and the plastic limit. They represent the percentage moisture content at which a soil changes (with decreasing wetness) from liquid to plastic consistency and from plastic to semi-solid consistency. For normal pond embankments best compaction results are obtained with a liquid limit of 35%. Clay which is to be used for an impervious core within an embankment, or as a layer lining an embankment (Section 8.2.2), should have a liquid limit below 60% and a plastic limit below 20%.

The range of moisture contents at which soil remains plastic is measured by the difference between the plastic and liquid limits. The resulting value is known as the plasticity index (PI). In general it depends only on the amount of clay present and indicates the fineness of a soil sample and its capacity to change shape without altering its volume. For normal pond embankments PI values between 8 and 20% are acceptable, with 16% ideal for compaction. For clay used for impervious cores the PI should be greater than 30%.

6.3.3.3 Permeability

Soil permeability has obvious significance for aquaculture ponds and depends largely on the number and size of pores present in the soil – two factors which are principally determined by soil texture and structure (the way individual soil particles are assembled into larger particles known as aggregates). Permeability can be measured as a seepage rate in cm day^{-1}. Ponds with daily seepage losses of more than 1–2 cm can be difficult to manage. Compaction can be used to alter soil structure and reduce pore size and permeability.

One procedure for testing permeability is: dig a hole about 0.8 m deep, line the walls with clay (to prevent horizontal seepage), fill with water in the morning, check the level in the evening, top up and cover the hole with vegetation. If most of the water is still present the next morning, the soil permeability is acceptable for pond construction.

To test the permeability of soil material for embankments the bottleneck test can be performed: cut the bottom off a bottle and fill the neck with a slightly compacted soil sample. Fix the bottle upside down and fill with water. The soil will be suitable for dams if no water has leaked out after 24 hours (O'Sullivan 1988).

Precise measurements of permeability can be made in a laboratory using an 'undisturbed' soil sample carefully collected with a thin-walled tube. Normal embankments require soil with a coefficient of permeability less than 1×10^{-4} m s^{-1}. Pond bottoms require less than 5×10^{-6} m s^{-1}.

If the soil proves to be light and porous, it may be necessary for clay to be brought in from an outside source, or for pond sealants or pond liners to be used to provide impermeability (Section 8.2.2). Such measures will increase construction costs and it will be necessary to investigate the availability of the relevant materials. The porosity of earthen ponds can be a particular problem in desert areas (Farmer 1981). Perhaps the only advantage of light soils is that they are easy to move and can thus reduce the costs of earthworking. Ponds should not be established on seepage areas with gravel or sand seams or aquifers, and should not be constructed below the water table. It is obviously difficult to make ponds on a rock substrate but O'Sullivan (1988) notes that it is possible if the topography is suitable and clay embankments are used on impervious, fracture-free rock.

A site close to a beach with deep sand is ideal for a coastal hatchery because it allows a sub-sand filter to be installed for the water supply. If, however, a hatchery is planned with broodstock and nursery ponds, the need for impermeable soils or pond liners must also be taken into account. Some sites which have been previously used as salinas (salt pans) may present problems for aquaculture because of brine upwelling.

6.3.3.4 Colour

Soil colour may be a useful indicator of organic content and drainage conditions. Dark shades can indicate high organic content and/or poor soil drainage. Abundant pale yellow mottles coupled with a low pH characterise acid sulphate soils (see next section). Although layers of topsoil rich in organic matter can and must be removed, a site with deep peaty soil is unsuitable for pond construction.

6.3.3.5 Acid sulphate soils

One particular type of soil that needs to be identified during site selection is characterised by its potential to become highly acidic when drained and exposed to air. Known as potential acid sulphate soil, it can reach a pH of 4 or less following earthworking and in general should be avoided for pond building because it necessitates special construction and management procedures which can delay the start-up of production and greatly add to overall project costs.

Potential acid sulphate soils are found in saline areas such as coastal mangroves and in freshwater areas such as river plains, where sediments with a high organic content accumulate. They contain iron sulphides precipitated as a result of sulphide excretion by sulphur-reducing bacteria which decompose organic matter under anaerobic waterlogged conditions. While these soils remain waterlogged and anaerobic they undergo little change, but in contact with air bacterial oxidation produces sulphuric acid and precipitates of ferric hydroxide. The latter often imparts a red colour to the soil surface and can accumulate in the gills of crustaceans (Nash *et al.* 1988), but more critically the sulphuric acid can increase acidity by as much as 3 pH units. Potential acid sulphate soils can be identified by measuring pH before and after exposure to air. The test procedure involves taking a handful of soil (moisten if dry), working it into a cake 1 cm thick and placing it in a thin plastic bag. The bag is sealed to retain moisture and encourage bacterial activity. If after 1 month the pH measures four or less the soil is acid sulphate.

Actual acid sulphate soils (potential acid sulphate soils that have already been oxidised) are relatively rare but are easily recognised by their low pH values and often by mottles of the pale yellow mineral jarosite.

When ponds constructed with acid sulphate soils are filled, sulphuric acid leaches into the pond water. Embankments generally produce more acid than pond bottoms because, for the most part, the latter remain waterlogged and are less exposed to the air. Particularly heavy leaching of acid can occur from embankments following rainstorms. Acidic conditions are detrimental to pond productivity, adversely affecting bacteria and phytoplankton as well as the crop. Acid embankments usually restrict the growth of vegetation and are vulnerable to erosion.

Where crustacean farms have been constructed at sites with acid sulphate soils, a range of measures can be taken to limit pond acidity both during pond preparation and operation (Sections 8.2.2.3 and 8.3.5.8). Useful accounts of ways of coping with acid sulphate soils include Webber & Webber (1978) and Simpson *et al.* (1983), and the subject is also covered in ASEAN (1978).

6.3.4 Vegetation

A site with dense, deep-rooted vegetation will result in greatly increased land clearance costs. Former agricultural land often has good soils but care must be taken to check for residual pesticides and other contaminants. Some tea plantations, for example, are sprayed with zinc.

6.3.5 Communications and infrastructure

Good approach routes either via land or water are essential to gain access to an operation all year round. For sites without access from an existing road network, the cost of building a road capable of withstanding the usual weather conditions must be taken into account.

The proximity of a site to a source of ice and to processing facilities will need to be considered. One possibility for remote sites is to construct a processing plant alongside an on-growing operation. However, even if this strategy is followed, good transport links will still be needed to reach markets or market channels. For exports, international air or seaports will be needed with appropriate facilities for storage and handling cargo and containers. Freight rates will need to be established for economic evaluation.

Hatcheries are most conveniently placed at the site of the farm, but for shrimps and prawns the different water quality requirements of the larvae rearing and on-growing phases of culture often make this an impossibility. Nevertheless, this does not often represent a major constraint since post-larvae can be transported quite easily (Sections 7.2.4 and 7.3.2).

Good telephone, telex and facsimile links are very useful for the efficient running of any business. In the absence of these, two-way radios can be used provided that the relevant authorities will allocate transmission frequencies for private use.

If mains electricity is to be used the extent of the existing electricity network should be taken into account, since the cost of extending powerlines can be prohibitive. The quality of the supply with regard to interruptions, overall capacity (some limited networks can only cope with extra demand on a domestic scale), voltage consistency and the number of phases available, should be investigated. This information is needed to properly compare the different options and costs for energy. Inadequate or non-existent electricity supplies at many sites necessitate the use of diesel powered generators either as a back-up or to provide for all electrical needs.

The location of other utility networks for freshwater and gas sometimes plays a part in the site selection process. If there is no mains supply or underground source of freshwater, a supply from water tankers may be needed and this may have significant cost implications. Municipal freshwater is likely to be chlorinated so it is not suitable for direct use in culture operations.

6.3.6 Labour force

The remoteness of a site can have implications for the availability and well-being of a project's workforce. An isolated site away from towns or villages may reduce the risks of poaching but can mean considerable expenditure on transport or the construction of housing. Distant community facilities, such as schools and health centres, can present problems for employees (Section 11.2)

The availability of a casual and seasonal labour force should be investigated if it will be needed for construction work or for harvesting and processing operations. Sometimes there are other seasonal activities which will compete for a local workforce. The limited availability of experienced construction workers in some rural areas can mean bringing in labourers from urban centres. Other considerations concern the reliability of the available labour, local attitudes to work, and possible liaison difficulties.

6.3.7 Social, environmental and ecological impacts

The social, ecological and environmental impacts of crustacean farming should be taken into account from the early planning stages of all projects (see Chapter 11). Good site selection can help to minimise the negative impacts of crustacean farming while maximising the potential benefits. The kinds of issues which should be given attention include possible conflicts of land use with rights of way, recreation, and existing grazing or agriculture, and possible conflicting use of water resources including irrigation and navigation.

Dutrieux and Guélorget (1988) stress the importance of an ecological approach to the planning of aquaculture and note that very few sites on the French Mediterranean have been chosen simply for their productivity potential. Among 15 studied, nine were chosen for socio-economic reasons such as availability of the plot or planning permission from the authorities.

6.4 Modifications to an existing facility

The first and most important task when acquiring a facility is to establish why it is for sale or why it has been abandoned. The previous owners should be closely questioned, possibly with the guidance of an independent expert. Sites and facilities should be carefully inspected. Reasons for sale may include specific problems such as disease and poor water quality or wider problems relating to poor siting, inadequate management, lack of finance or changes in market conditions.

The roots of all general and site specific problems must be exposed. Management problems can be related to the remoteness of a site and its lack of appeal to highly qualified staff. Disease may be due to increasing pollution from a whole range of sources including neighbouring aquaculture operations. Failure to operate profitably can sometimes relate to the use of extensive techniques which do not provide sufficient yields per hectare. Equally, failure may be attributed to the use of high cost intensive or super-intensive methods that cannot be adjusted in line with falling market prices. Some shrimp hatcheries have been forced out of business because of low prices obtained for post-larvae; in Ecuador this has been associated with gluts of wild-caught post-larvae, and in Taiwan with the changing fortunes of the shrimp farming business as a whole. New owners who acquire an existing facility may be able, initially at least, to improve financial and managerial inputs, but they may be powerless to rectify many underlying problems.

6.4.1 Hatchery

When acquiring a used hatchery, special attention should be given to the quality of its water supply since larval crustaceans are more sensitive to water conditions than adults or juveniles (Section 8.5). If the site appears to be favourable the hatchery can be inspected to assess the quality and value of its installations. Abandoned hatcheries may seem to be complete from a visual inspection, but the process of bringing them back into production can be lengthy and may reveal many hidden system problems and design deficiencies. The purchase of a working hatchery may be preferable, although this is no guarantee that smooth running can be maintained with the existing facilities. Hatchery designs usually closely match the intended operating methods so if new techniques are to be applied, remodelling may be required (e.g. to permit more intensive management, the size of larval rearing tanks may need to be reduced).

6.4.2 Farm

In the acquisition of existing farms, the quality and value of the installations should be assessed, together with the attributes of the site. Investment in renovation will only be worthwhile if the site meets all the important selection criteria. If extensive farms are to be up-graded, the kinds of modifications that can be considered include: deepening ponds and increasing the bottom slope, strengthening bunds, separating inlet and outlet gates and incorporating nursery ponds.

6.5 References

Al-Hajj A.B., Farmer A.S.D., Al-Hassan K.E. & Saif M.A. (1981) *The effect of airborne dust and suspended marine sediments on the survival and larval stages of* Penaeus semisulcatus *de Haan*. KISR 201, Mariculture and Fisheries Department, Kuwait Institute for Scientific Research.

ASEAN (1978) *Manual on pond culture of penaeid shrimp*. ASEAN National Coordinating Agency of the Philippines, Ministry of Foreign Affairs, Manila.

Chamberlain G.W. (1988) Shrimp culture news from around the world. *Coastal aquaculture*, **5** (1) 13–15.

Coche A.G. (1985) Soil and freshwater fish culture. *Simple methods for aquaculture*, FAO training series 6, FAO, UN, Rome.

Dutrieux E. & Guélorget O. (1988) Ecological planning: A possible method for choice of aquacultural sites. *Ocean and Shoreline Management*, **11** (6) 427–47.

Farmer A.S.D. (1981) Prospects for penaeid shrimp culture in arid lands. In *Advances in food producing systems for arid and semi-arid lands*. Academic Press Inc., London.

Huet M. (1972) *Textbook of fish culture*. Fishing News Books, Blackwell Scientific Publications, Oxford.

Malecha S.R. (1983) Commercial pond production of the freshwater prawn, *Macrobrachium rosenbergii*, in Hawaii. In *Handbook of Mariculture, Vol. 1, Crustacean aquaculture* (Ed. by J.P. McVey), pp. 231–60. CRC Press, Boca Raton, Florida.

Muir J.F. & Kapetsky J.M. (1988) Site selection decisions and project cost: The case of brackish water pond systems. In *Aquaculture engineering technologies for the future*. Institution of Chemical Engineers Symposium series No. 111, pp. 45–63. EFCE Publication series No. 66, Hemisphere, London.

Nash G., Anderson I.G. & Shariff M. (1988) Pathological changes in the tiger prawn, *Penaeus monodon* Fabricius, associated with culture in brackishwater ponds developed from potentially acid sulphate soils. *J. Fish Diseases*, **11** (2) 113–23.

New M.B. & Singholka S. (1985) *Freshwater prawn farming*. A manual for the culture of *Macrobrachium rosenbergii*. FAO Fish. Tech. Pap. 225, Rev. 1.

O'Sullivan D. (1988) Backvard and farm dam culture. *Austasia Aquaculture Magazine*, **3** (2) 17–18.

Simpson H.J., Ducklow H.W., Deck B. & Cook H.L. (1983) Brackish-water aquaculture in pyrite-bearing tropical soils. *Aquaculture*, **34** (3/4) 333–50.

Skabo H. (1988) Shrimp farming developments in West Africa. In *Shrimp '88, Conference proceedings*, 26–28 Jan. 1988, pp. 95–102. Bangkok, Thailand. Infofish, Kuala Lumpur, Malaysia.

Sorgeloos P., Bossyut E., Lavens P., Léger P., Vanhaecke P. & Versichele D. (1983) The use of brine shrimp *Artemia* in crustacean hatcheries and nurseries. In *Handbook of Mariculture, Vol. 1, Crustacean aquaculture*, pp. 71–96. CRC Press, Boca Raton, Florida.

Webber R.J. & Webber H.H. (1978) Management of acidity in mangrove sited aquaculture. *Rev. Biol. Trop.*, 26 (Supl. 1) 45–51.

Wheaton F. W. (1977) *Aquacultural engineering*. Wiley-Interscience, New York.

Chapter 7
Techniques: species/groups

7.1 Introduction

To assess or evaluate an aquaculture proposal, a reasonable knowledge is needed of the techniques and management practices appropriate for each stage of the operation. It is equally important to know, at least in broad terms, the water quality and food requirements and the environmental tolerances of the animals being farmed. In this chapter the main culture methodologies are outlined for each species or group in accordance with the scheme shown in Table 7.1. However, to avoid undue repetition, fine differences in methods for each species within a group have been omitted. References to the literature cited for each species/group have been deliberately kept separate and are listed at the end of each section in order to facilitate the selection of material for further study.

While the majority of culture techniques are specific to species, some aspects relate to nearly all the cultivable species, for example materials toxicity, diet, tolerances to metabolites and the general principles of animal husbandry. These are discussed in Chapter 8.

7.2 Penaeid shrimp

7.2.1 Species of interest

Giant tiger (*Penaeus monodon*), fleshy (*P. chinensis*), whiteleg (*P. vannamei*), banana (*P. merguiensis*), Indian white (*P. indicus*), blue (*P. stylirostris*), kuruma (*P. japonicus*), redtail (*P. penicillatus*), green tiger (*P. semisulcatus*), southern white (*P. schmitti*), red spotted (*P. brasiliensis*), southern pink (*P. notialis*), western king (*P. latisulcatus*), greasyback (*Metapenaeus ensis*), speckled (*M. monoceros*).

The techniques for culturing penaeid shrimp vary greatly in complexity during the on-growing phase, and in dependence on hatcheries for the supply of juveniles for stocking. Traditional extensive methods, for example, simply trap and grow wild juveniles in shallow tidal impoundments (Section 5.1.1) while modern super-intensive systems rely on a regular supply of hatchery-reared juveniles for culture at high densities in specially designed concrete tanks or covered raceways (Section 5.1.3). This section attempts to cover the wide range of practices commonly applied both in the Far East, where aquaculture has a long tradition, and in the Americas, where shrimp farming is a relatively recent activity. For more detailed accounts of shrimp farming techniques the reader is referred to the various authoritative manuals and technical publications published since 1980: McVey 1983; Kungvankij & Chua 1986; Tseng 1987; Lucien-Brun 1988; SEAFDEC 1988; Akiyama 1989; Chavez 1990; Main & Fulks 1990; Tacon 1990; Villalon & Treece (in press).

Table 7.1 Steps in the farming of crustaceans.

	Phase	Inputs	Facilities/operation
	Acquisition of broodstock		
	Maturation	Natural feeds	Ponds/tanks/pens
┌	Spawning	Minimal disturbance	
Penaeid shrimp	Incubation	Minimal disturbance	Tanks/bins/kreisels
└→ ┌	Hatching	Clean, filtered water	
Crayfish	Larval rearing	Hatching and culture of live food	Tanks/bins/kreisels
└→	Nursery	Hatchery reared or wild-caught juveniles; live and compounded feeds	Ponds/tanks/raceways/trays
	On-growing	Natural or compounded feeds	Ponds/tanks/raceways/ trays/cages/artificial reefs/ protected fishery areas
┌	Harvesting	Bait, ice, seasonal labour	Seine/electro-netting/ draining/trapping/direct collection
Live	Processing	Ice, preservatives, seasonal labour	Chilling/washing/sorting/ value added treatment/ freezing
└→	Marketing	Packaging	Distribution/sales

7.2.2 Broodstock

7.2.2.1 Acquisition

Although the option of producing shrimp broodstock in captivity is sometimes adopted where non-native species are farmed, supplies of broodstock shrimp are usually obtained from the wild. Capture techniques are basically the same as for normal shrimp fishing, although when trawling is employed the duration of the tows must be reduced to around 20 minutes or less to avoid crushing and stress which can lead to ovary resorption. Tangle nets and traps are particularly suited to catching live shrimp for use as broodstock, since these methods generally provide animals in better condition than do trawls.

Most hatcheries obtain their broodstock from middlemen or directly from fishermen. Sometimes, however, supplies can be interrupted by close seasons and special dispensation may be needed to catch shrimp during these periods. In Japan, the market for live *P. japonicus* for human consumption provides a ready source of broodstock.

7.2.2.2 Transport

Three methods are commonly used for the transport of broodstock from the point of capture to the hatchery:

(1) En masse in tanks or bins;
(2) In plastic bags with water and oxygen;
(3) In chilled sawdust.

Table 7.2 shows further details and some results obtained with these methods. The use of tanks or bins is usually suited only to short journeys since it involves transporting a large mass of water. To reduce physical damage shrimp can be placed within individual perforated plastic cylinders capped at each end with pieces of netting, or have protective tubes or bands located on sharp rostrums and telsons. These precautions are also necessary to prevent puncturing when shrimp are transported in plastic bags.

The lightweight transport method involving chilled sawdust is particularly suited to airfreight and is regularly performed with *P. japonicus* destined for live sales in Japan (Section 3.3.1.1). By lowering the temperature the metabolic rate of the shrimp is reduced and the period of survival extended. This transport technique, however, is not suitable for all penaeids; *P. semisulcatus* for example, die rapidly out of water (Liao 1988), and *P. monodon* cannot withstand temperatures below about 12°C (Cordover 1989).

For all methods involving water attention must be paid to its quality, particularly for extended journeys. Most important are the levels of oxygen, temperature, ammonia and pH. Oxygen is essential for transport using polythene bags, and a supply of oxygen or compressed air is often needed for tanks and bins. Temperature can be controlled using chill packs and insulated boxes, and in some situations the build-up of ammonia can be delayed by starving the shrimp for 12–24 hours prior to shipping. Although various products including pH buffers and ammonia control chemicals can be added to improve water quality, their overall benefits may be negligible (Robertson *et al.* 1987). For air shipments packaging needs to conform to IATA regulations, and labels indicating 'keep cool' and 'consignee inspect on arrival' can be useful.

If broodstock are to be retained in the hatchery a period of quarantine is advisable, and the use of a dilute formalin treatment dip (200–400 ppm for 1–2 hours, then rinse) can help prevent the introduction of epifauna. Animals heavily infected with the virus *Baculovirus penaei* (BP) can sometimes be identified by observing polyhedral inclusion bodies (PIBs) in samples of faeces, but completely effective screening in this way is impossible since not all infected animals shed PIBs. Elimination of BP and other viruses may be impractical in many instances, and the only approach with any likelihood of success is to acquire specific virus-free stock and isolate them from endemic viruses (Section 12.2).

7.2.2.3 Production of broodstock in captivity

The techniques for rearing broodstock in captivity are essentially the same as those used for on-growing but are modified to promote growth beyond normal market sizes. The principal modification is a reduction in the stocking density. In semi-intensive

Table 7.2 Examples of methods and results for transporting penaeid broodstock.

Species	Duration hrs	Survival %	Size	Temp. °C	Method and density	Reference
P. chinensis	3	95	—	18	En masse in oxygenated tank, 78 shrimp m^{-3}	Zhang *et al.* 1980
	3.5	51	—	18	En masse in oxygenated tank, 400 shrimp m^{-3}	
P. setiferus	6	100*	—	18–20	20 hr fasted shrimp in 8 ℓ of water in double plastic bags topped-up with oxygen. Placed in 40 ℓ styrofoam shipping boxes with ice pack located on underside of lid. Density 59–68 g ℓ^{-1}	Robertson *et al.* 1967
P. japonicus	10	—	30–50 g	11–17	Packed in cardboard boxes (1 kg per box) between layers of chilled dry sawdust. 5 mm thick bag of frozen wet sawdust included on warm days.	Yamaha 1989; Chen 1990
P. japonicus	>14	—	—	4–10	Packed in insulated boxes between layers of chilled untreated low-resin sawdust	Richards-Rajadurai 1989
P. vannamei	24	100	13–16 cm (37 g)	—	Shrimp packed in individual PVC tubes (20 × 3.75 cm) in 30 ℓ plastic bags with minimal water (0.6–1.2 ℓ) topped-up with oxygen and placed within styrofoam boxes. Ice placed in box and separated from bag with 60 mm layer of plywood. 10 shrimp per bag	Johnson *et al.* 1984

*Suffered delayed mortality of 34–46% in the five days following shipment.

culture, *P. monodon* are normally reared at a density of 3–10 shrimp m^{-2} whereas equivalent ponds managed for broodstock production may be stocked with 0.7–1 shrimp m^{-2} (Simon 1982). To provide animals in prime condition and reduce stress, broodstock production ponds should be designed for easy harvesting and should be located near the hatchery. Depending on species, maturation may occur when the shrimp are between six and ten months old and, in theory, under ideal tropical conditions, a continuous supply of broodstock can be produced. For example, AQUACOP (1983) described a schedule for the production of 400 broodstock (25–60 g) every three months in a continuous three-stage cycle starting with 10000 hatchery post-larvae at six month intervals.

7.2.2.4 Over-wintering

It is possible to rear broodstock in sub-tropical or warm temperate regions (Section 5.2), where the cultured species are likely to be non-native or at the limit of their natural distribution. The procedure usually involves an over-wintering phase in which shrimp are retained from one season's harvests to become broodstock for the next production cycle. Shrimp are maintained in holding tanks or in ponds which are sometimes covered with transparent greenhouse structures. These structures raise water temperatures and advance the onset of maturity in the stock.

7.2.2.5 Maturation in captivity

Although most hatcheries prefer to use gravid females obtained directly from the wild (termed 'sourcing' in US literature), these females are often in short supply and some operations opt for the alternative of inducing gonad development in captive shrimp.

The facility with which broodstock will mature, copulate and spawn in captivity varies greatly between different species (Table 4.2). Even in a carefully controlled environment some important species will not mature with regularity, so, in the absence of a proven alternative, the technique of unilateral eyestalk ablation is usually applied. This simple but effective operation involves either the surgical removal of one of the two eyestalks (extirpation), each of which contains a complex of glands that functions to inhibit gonad development, or of the eyestalk contents (enucleation) (Harrison 1990). As many as nine different techniques have been described (Liao & Chen 1983).

In the most simple approach to maturation many hatcheries in south-east Asia obtain wild gravid females (usually *P. monodon*) and, after the initial spawning, induce four or five additional spawns by eyestalk ablation. Fertilization in such cases relies on sperm retained by the female throughout the inter-moult period (Section 2.2).

Indoor facilities are usually favoured for maturation systems since they permit greater control of temperature, light intensity and photoperiod, all of which influence ovarian development. Nevertheless, outdoor tanks, pens or ponds can also be used. In one such outdoor system in the Philippines (Maguire 1979), male and ablated female *P. monodon* (sex ratio 1:1) were stocked at a density of 0.8–1.6 shrimp m^{-2} in rectangular net pens measuring 250 $m^2 \times 3.5$ m deep. The shrimp received a diet of fresh mussels, and ripe females were selected weekly by lifting the net. However, the successful application of this type of maturation system depends on the availability of suitably sheltered seawater sites.

Most indoor maturation systems use circular tanks (3–5 m in diameter) made of fibreglass, plastic or cement, stocked with shrimp at 6–10 m^{-2}. Operating water depths range between 0.3 and 1 m and best results are usually obtained with a steady temperature around 28°C and a salinity close to full seawater (30–35 ppt). Facilities located near an unpolluted water source that meets these requirements usually operate with a flow-through of water, while in other circumstances much reliance may be placed on recirculation and water treatment. Recirculation can also serve to reduce the quantities of water and heat that are required in maturation facilities situated outside the tropics.

The way different maturation systems are operated varies in detail according to: type of substrate; photoperiod and light wavelength and intensity; diet; stocking rate; sex ratio; and the method of selecting and fishing ripe females. A maturation tank must provide the shrimp with sufficient area and depth of water for successful courting and mating behaviour. Simon (1982) provides details of a *P. monodon* maturation system used in the Philippines and AQUACOP (1983) describe another system employing tanks with a sand substrate that has been successfully used with *P. monodon*, *P. vannamei* and other penaeids. In all cases special attention is given to husbandry – feeding rate, cleaning routines and minimizing disturbance.

If females develop ripe ovaries but are not impregnated, it is possible to resort to artificial insemination to produce fertilized eggs. The basic technique, however, differs between penaeid species depending on whether the females possess a closed or open thelycum. With the former, for example with *P. monodon*, females must be inseminated just after they moult and before the new shell hardens. Whole spermatophores, manually extracted from the males, are transferred to the thelycum where they will be retained during the inter-moult period (Lin & Ting 1986).

Females of open-thelycum species, such as *P. vannamei* and *P. stylirostris*, require insemination prior to each spawn, and it is usual to apply only the sperm mass to the thelycum after squeezing it out from extracted spermatophores. Although artificial insemination has clearly demonstrated its viability and is routinely used in some *P. vannamei* hatcheries, most maturation units continue to rely on natural impregnation occurring within the maturation tanks since less labour is involved, mortality rates are lower, and more consistent fertilization rates are obtained.

The nutritional requirements for maturation and production of high quality eggs from shrimp broodstock are specialized, and despite considerable research effort into diet formulations, best results are only achieved with diets of natural, fresh or frozen foods incorporating items such as mussels, clams, squid, shrimp heads, adult *Artemia*, and marine polychaete worms (Harrison 1990). Commercial 'maturation' diet formulations are available but are normally used only as a supplement to fresh feeds. While most may be nutritionally adequate for egg yolk production, the claims by some manufacturers that their diets can actually induce maturation should be verified with the species to be reared.

7.2.3 Spawning and hatching

Healthy gravid female shrimp, whether taken from the wild or matured in captivity, will usually spawn viable eggs providing they have been successfully impregnated and are handled with care. Penaeid eggs, unlike those of most other farmed crustaceans, are released directly into the water rather than held beneath the abdomen for a period of incubation (Section 2.2). Hence, after spawning is completed, females can be discarded, or alternatively if a maturation system is in operation, they can be retained for re-maturation and further spawning. Larvae hatch out 12–18 hours after the eggs are spawned (Figure 2.2a).

In the most straightforward hatchery systems originally developed in Japan, gravid females release their eggs directly within the larvae rearing tank. Tanks as large as 200 mt (water capacity) may receive between 100 and 200 gravid *P. japonicus*, placed in the tanks within nets for easy removal. Alternative systems separate the spawning and hatching phases. For example, in the system described by AQUACOP (1983), females are placed into individual spawning tanks, eggs are transferred to incubators, and nauplii hatch to be separated, counted and sometimes dipped in dilute iodine-based disinfectant, before being stocked into larvae rearing tanks. This system allows the performance of individual females to be monitored and may help prevent the passage of diseases to the larval culture.

A less common approach, used in conjunction with maturation systems, leaves females to spawn undisturbed in the same tanks where they mature and copulate. Eggs are either collected in the outflowing water (Simon 1982) or left to hatch, with the nauplii being siphoned off later from near the water surface. Less handling is involved but there is a greater risk of nauplii picking up contamination from within the maturation tank.

Hatcheries which do not handle spawning females and instead acquire batches of nauplii from outside sources, are more correctly termed larvae rearing facilities. Such facilities obtain supplies of nauplii either from specialist nauplii producers, who spawn wild ovigerous females, or from other hatcheries which operate maturation facilities. In Ecuador, nauplii are produced by numerous small spawning stations many of which are based in Esmeraldas province and take advantage of seasonal supplies of wild gravid *P. vannamei*. In south-east Asia, particularly in Thailand and Taiwan, specialist nauplii producers working mostly with *P. monodon* spawn wild gravid females and also achieve repeated spawnings through eyestalk ablation.

Nauplii are sometimes traded internationally and are relatively easy to transport (Section 3.4.3). Three hundred thousand can be held in a 30 ℓ sealed plastic bag with 15 ℓ of seawater and 15 ℓ of oxygen for up to 24 hours at 18–24°C.

7.2.4 Larvae rearing

Although the biological features of penaeid larvae rearing are the same in all hatcheries, different operations can be broadly characterized depending on three factors: overall size of the operation, the size of the larval rearing tanks, and the use of western or oriental culture techniques.

Hatcheries range in size from large operations producing more than 100 million PL yr^{-1}, with a large often highly trained workforce, down to small 'backyard' family concerns producing less than 10 million PL yr^{-1}, staffed only by a handful of workers or family members. While large operations represent large investments and need to produce all year round or on a regular seasonal basis, small hatcheries tend to keep investments to a bare minimum and have very flexible production schedules (Section 10.9.1.1).

Although the distinction between western and oriental-style larvae rearing techniques is becoming increasingly blurred, the use of these terms has some historical significance and serves to distinguish between the more traditional methods originally developed in Japan, and the more intensive approaches subsequently developed in the USA and French Polynesia. Oriental methods typically rear larvae at lower densities (30–100 ℓ^{-1}) and the bulk of algal feed is produced by encouraging blooms within the larvae rearing tank. For this reason water exchange rates are kept comparatively low (5–100% day^{-1}) to avoid flushing away all the algae. The western practice is to stock 50–200 larvae ℓ^{-1} and use water exchange rates of 50–200% per day to maintain stricter control over water quality. However, the resulting higher requirement for algae leads to a far greater reliance on independently cultured algae stocks.

Tank size and design also tend to differ between the oriental and western approaches. Tanks of any size up to 200 mt can be successfully operated using oriental culture methods, but only small tanks of 0.1–15 mt are suited to the more rigorous water management required in western-style hatcheries. In addition, while oriental-style tanks typically have flat, gently-sloping bottoms, western techniques benefit from the use of cylindro-conical tanks, or tanks with V or U-shaped bottoms, which combined with strong aeration can help prevent the accumulation of organic detritus. The latest western tank designs are oval and have hyperbolic bottoms. Although the western approach typically requires greater management inputs and is sometimes labelled 'high-tech', healthy larvae can be produced by either system and regular success depends on the skill and experience of the technical staff rather than the adoption of one style of culture or another.

The use of large rearing tanks presupposes that sufficient quantities of broodstock can be obtained to fill them with nauplii. This is often the case for *P. japonicus* in Japan, *P. chinensis* in China, and in Taiwan where large numbers of *P. monodon* broodstock may be obtained from specialist importers; but in many other situations limited supplies of broodstock favour the use of smaller tanks.

Hatcheries rear shrimp through three larval sub-stages (nauplius, protozoea and mysis) to produce post-larvae. The whole rearing process may take place in a single tank or be split into two separate rearing operations: nauplius through to mysis, and mysis onwards. Such two stage operations use a smaller elevated tank (5–10 mt) from which mysis larvae are transferred by gravity to a larger (10–30 mt) tank for culture to post-larvae.

Water quality is more critical during larvae culture than at any other stage (Section 8.5). EDTA is often employed as a chelating agent to improve water quality, and antibiotics are commonly applied to limit the impact of pathogenic bacteria. However, there are several problems associated with the routine use of antibiotics (Section 11.4.4).

While a hatchery is in operation the levels and detrimental effects of pathogens, especially bacteria, increase to the point where survival rates become unacceptably low or at least highly unpredictable. In the face of this, most operators resort to the use of sanitary 'dry-out' periods in which production is suspended while all tanks, reservoirs, apparatus and water networks are disinfected and left to dry for several days or even weeks. One advantage of small hatcheries is that they can be more straightforward to disinfect than larger units, and because of their smaller capacities they can be more rapidly re-stocked when production is resumed.

During the natural planktonic existence of penaeid larvae, live phytoplankton and zooplankton are the most important components of the diet (Section 2.3). Correspondingly, in culture, best results are obtained with live feeds such as microalgae and *Artemia* nauplii. A variety of compounded particulate and microencapsulated diets have been developed to reduce dependence on live feeds, but although some are claimed to be suitable as replacements, their role in commercial units has been limited to partial replacement of live feeds, or as supplements.

Manuals and other relevant publications covering penaeid larvae culture in more detail include: McVey 1983; SEAFDEC 1985; NACA 1986; JICA 1987; Frippak 1988; Chavez 1990; Treece & Yates 1990; Treece & Fox (in press).

The stage at which post-larvae leave a hatchery for transfer to the nursery or on-growing site varies considerably between different species and countries. In Taiwan, *P. monodon* hatcheries usually produce PL_{10-15}, and in Ecuador and Panama, *P. vannamei* hatcheries typically rear post-larvae to PL_{4-15}. Some hatcheries in the Americas and Asia, however, move very young PL_{3-4} from larvae rearing tanks into larger on-site post-larvae (or pre-nursery) tanks to produce PL_{10-20}. In Chinese hatcheries, *P. chinensis* post-larvae are harvested between PL_7 and PL_{25}. Table 7.3 shows a summary of methods and results for transporting penaeid post-larvae.

7.2.5 Nursery

The inclusion of a nursery phase in shrimp culture provides farmers with hardy juveniles that have been acclimatized to the environment they are likely to encounter during on-growing. Post-larvae mortality is often most significant among hatchery reared animals produced in very artificial conditions. Nursed juveniles are usually transferred to on-growing ponds when they have reached 0.1–2 g, by which time most weak animals will have died and a realistic estimate can be made of the number stocked. Although during the on-growing phase survival rates in a pond cannot be readily determined, a good initial count of the number of juveniles stocked improves control over stocking densities and assists in the calculation of feeding levels (Section 8.3.3; 8.3.4.1).

The principal problem with nurseries arises when juveniles are transferred to on-growing units. The process usually involves draining into a netted sump and, because it involves much handling of the shrimp, substantial mortalities can occur unless great care is taken. Sensible precautions include working during cool periods of the day, and avoiding prolonged overcrowding and any marked deterioration in water quality. To simplify the whole procedure some farms incorporate nurseries adjacent to on-growing ponds, which enables transfer to be accomplished by gravity flow. Yet although this method avoids handling mortalities, it does not permit an accurate population estimate to be made. Fish pumps have been successfully employed to transfer shrimp juveniles from ponds to collecting tanks (Wyban *et al.* 1989). This pumping process resulted in minimal stress when pond levels were dropped slowly, gradually concentrating the juveniles at a rate at which they could be successfully handled by the transfer crew.

In modern semi-intensive farms, particularly in the Americas, nursery ponds are usually incorporated in the design of the whole farm. They may represent between 6% and 15% of the total culture area. They are usually made with earthen embankments, and, at sizes of 0.04–1 ha, are much smaller than the on-growing ponds. Stocking densities are typically 100–200 juveniles m^{-2}. In south-east Asia, nursery

Table 7.3 Methods and results for the shipment of penaeid post-larvae.

Duration (hours)	Post-larvae size (TL or age*)	Density (no. ℓ^{-1})	Species	Temp. (°C)	Survival (%)	Transport method	Reference/source
unspecified	unspecified	500–833	unspecified	—	—	6 ℓ water in polythene bag inside carton	Macintosh 1987
unspecified	PL_{20}	500	*P. japonicus*	—	>90	2 ton tank + O_2	Maguire 1979
unspecified	5–25 mm	260–1320	mixed***	24–25	—	Tank + O_2	Pretto 1983
unspecified	PL_{20-25}	500–1250	*P. monodon*	21–24	—	8–10 ℓ seawater in polythene bag + O_2	Primavera 1983
3	PL_{12-15}	2000	*P. vannamei*	25–27	>95	14 ℓ water in 30–35 ℓ plastic bag + O_2, ice optional	Ecuador, unpubl. data
3–6	PL_{12-15}	2000	*P. vannamei*	25–27	>95	14 ℓ water in 30–35 ℓ plastic bag + O_2, ice optional	Ecuador, unpubl. data
6	10–20 mm	70	mixed**	27.2	100	5 ℓ water in 20 ℓ polythene bags + O_2	Franklin *et al.* 1982
6	10–20 mm	100+	mixed**	27.2	>95	5 ℓ water in 20 ℓ polythene bags + O_2	Franklin *et al.* 1982
6–8	PL_2	500–1000	*P. monodon*	—	>80	20 ℓ bags + O_2	Wickins, unpubl. data
6–8	PL_{10-15}	1500	*P. monodon*	—	>80	20 ℓ bags + O_2	Wickins, unpubl. data
6–9	PL_{12-15}	1500	*P. vannamei*	25–27	>95	14 ℓ water in 30–35 ℓ plastic bag + O_2 ice optional	Ecuador, unpubl. data
<12	PL_{20-25}	375–833	*P. monodon*	24	—	6–8 ℓ plastic bags in styrofoam box + ice and sawdust	NACA 1986
12	unspecified	700	*P. japonicus*	—	—	1 ton tank with aeration	Kurian 1982
12	PL_5	833	*P. merguiensis*	28.5	—		FAO 1979
12	10–20 mm	70	mixed**	27.2	>95	20 ℓ polythene bags 5 ℓ water + O_2	Franklin *et al.* 1982
18	10–20 mm	40	mixed**	27.2	100	20 ℓ polythene bags 5 ℓ water + O_2	Franklin *et al.* 1982
18	10–20 mm	50	mixed**	27.2	>95	20 ℓ polythene bags 5 ℓ water + O_2	Franklin *et al.* 1982
18	17 mm	190	*P. vannamei*	18	>99	Double plastic bags, 12 ℓ water + O_2, styrofoam box	Smith & Ribelin 1984
24	10 mm	2500	mixed**	—	—	6 ℓ water in double polythene bags + O_2 and ice	ASEAN 1978
24	10–20 mm	40	mixed**	27.2	97–98	20 ℓ polythene bags 5 ℓ water + O_2	Franklin *et al.* 1982
24	10–20 mm	40–50	mixed**	27.2	>95	20 ℓ polythene bags 5 ℓ water + O_2	Franklin *et al.* 1982
24	17–18 mm	830	mixed**	—	—	6 ℓ water in double polythene bags + O_2 and ice	ASEAN 1978
24	20–24 mm	500	mixed**	—	—	6 ℓ water in double polythene bags + O_2 and ice	ASEAN 1978
<26	11–14 mm	500	*P. monodon*	28.5	good	18 ℓ plastic bags + O_2	Singh *et al.* 1982
27–30	11–14 mm	375	*P. monodon*	28.5	good	18 ℓ plastic bags + O_2	Singh *et al.* 1982
85	PL_{6-8}	250	*P. penicillatus*	—	60–70	10 ℓ water in plastic bag + O_2	Liao 1984
85	PL_{15}	250	*P. monodon*	—	20–30	10 ℓ water in plastic bag + O_2	Liao 1984

* days past metamorphosis, ** *P. indicus, P. monodon, P. semisulcatus*, *** *P. vannamei, P. stylirostris*.

facilities also take the form of concrete tanks, concrete walled ponds with sand bottoms, staked net pens and floating cages. These are either managed as independent operations by specialist nursery growers, or are included as an integral part of a farm or hatchery. Stocking densities are typically 50–100 post-larvae m^{-2}. Fixed or floating cages (sometimes known as 'hapas') are made from fine mesh netting (0.5 mm) attached to a wooden frame and may contain 30 000 PL_5 per cage (3.7 × 2.7 × 1.3 m deep). The shrimp are fed a paste of finely ground trash fish placed on feeding trays, but later mussel meat may be provided, suspended from an array of hooks. The juveniles are transferred for on-growing after 15–25 days (Beveridge 1987).

In extensive farms, which rely on trapping wild juveniles, nursery ponds can play an additional role since young fish predators entering alongside shrimp seed can be eradicated with rotenone or saponin (Section 8.3.5.2) before shrimp are transferred for on-growing. Some extensive operations have created nursery ponds especially for this purpose by dividing existing large ponds into smaller units. In China, hatchery-reared or wild-caught post-larvae are usually stocked directly into on-growing ponds, thus by-passing the nursery phase altogether.

Nursery management, particularly within integrated projects, requires good organization and is especially critical with the tight production schedules of intensive farms. Attention must be given to the duration of culture, the timing of output, and the quantity and size of juveniles at harvest. Table 4.4 summarizes the results of nursery rearing trials with various crustaceans, and Table 7.5 shows an example of feeding recommendations for the nursery rearing and on-growing of *P. vannamei* in Ecuador.

7.2.6 On-growing

The different techniques used for on-growing penaeids are conveniently categorized into four groups, primarily on the basis of the expected yield at harvest (Section 5.1). Table 7.4 shows typical features of the categories: extensive, semi-intensive, intensive and super-intensive. Considerable overlap can occur, however, between the different terms in published literature since some definitions attempt to include levels of feed and water management. The economic implications of growing shrimp at different levels of intensity are discussed in Chapter 10 (Sections 10.8 and 10.9.1.4) while some of the merits of the various options are discussed in Chapter 5.

7.2.6.1 Extensive

The simplest forms of penaeid shrimp culture are the traditional extensive methods of southern and south-east Asia, which involve the trapping and growing of wild shrimp juveniles in shallow earthen impoundments. Farms are built on low-lying tidal flats and marshlands along coastal and estuarine margins, and stocking relies on the influx of naturally occurring shrimp juveniles when the impoundments are filled. In general, however, the trapping of shrimp seed in this way allows little control over the quantity and species stocked and permits the entry of predatory and competitive crabs and fish. Although rotenone or saponin can be applied to kill the fish (Section 8.3.5.2), many extensive operations now make use of wild shrimp juveniles netted or trapped in nearby waters. Total stocking densities in extensive systems, however, remain low – between 0.2–5 shrimp m^{-2}.

Table 7.4 Typical features of penaeid on-growing systems.

	Extensive	Semi-intensive	Intensive	Super-intensive
General features	Low density culture often with fish and crabs, sometimes in rotation with salt or rice	Moderate density culture of mixed or single shrimp species	High density monoculture	Very high density or controlled environment monoculture
Water system	Tidal flushing via single inlet/outlet mainly on spring tides	Daily tidal and/or pumped water exchange often through separate inlet/outlet channels	Continuously controlled pumped exchange often through multiple inlets	Continuous, high rate of exchange with new or treated water
Source of juveniles	Wild juveniles trapped in pond during filling or collected from wild	Hatchery reared and wild juveniles; often separate nursery ponds	Hatchery and nursery reared juveniles	Hatchery and nursery reared juveniles
Fertilization	None or organic prior to filling	Organic or inorganic on filling and as required	Organic or inorganic as required	None
Feed	Shrimp rely solely or mostly on natural productivity	Natural productivity + supplementary feeds of low price pellets, molluscs or trash fish	Artificial formulated diets + occasional fresh fish/mollusc flesh	Artificial formulated diets
Aeration	None	Paddlewheels or propeller-aspirator pumps used when needed	Paddlewheels or propeller-aspirator pumps used regularly	Continuous aeration, sometimes compressed oxygen
Pond sizes	0.5–100 ha 0.3–1 m deep	0.2–3 ha 0.8–1.5 m deep	0.1–1.0 ha 1.2–1.5 m deep	0.03–0.1 ha 2 m deep (Japan) 0.6 m deep (raceways)
Stocking density	0.2–5.0 shrimp m^{-2}	5.0–20 shrimp m^{-2}	15–50 shrimp m^{-2}	50–250 shrimp m^{-2}
Shrimp size	10–50 g	15–40 g	15–40 g	15–25 g
Yield crop^{-1}	<1 mt ha^{-1}	0.5–5.0 mt ha^{-1}	5–15 mt ha^{-1}	10–30 mt ha^{-1}

Catches of wild shrimp juveniles vary seasonally in quantity and component species, and effective management of an extensive farm requires knowledge of the usual patterns of seed availability. In the backwaters and estuaries of India, George and Suseelan (1982) found that the post-larvae and early juvenile stages of cultivable penaeids could be found almost throughout the year. However, the greatest abundance occurred during October to May along the west and south-east coasts, and during January to April and August to December along the middle and northern regions of the east coast. *Penaeus indicus* and *Metapenaeus dobsoni* were the most abundant species and the highest concentrations of *P. monodon* juveniles were found on the middle and northern regions of the east coast. Catches are often greatest around spring tides. In Vietnam for example, the peak abundance of penaeid seed occurs from April to June and the equinoctial tide in late April is a favoured period for trapping (Quynh 1990).

Significant fisheries for wild shrimp juveniles have developed. In Ecuador a seasonal workforce of 32 000 'semilleros' (CPC 1989) provide the shrimp industry with around 5–13 billion *P. vannamei* juveniles per year. In Vietnam a typical family engaged in fishing shrimp seed may be able to collect 20 000 *P. merguiensis* juveniles in a day (Quynh 1989). The capture methods depend largely on the nature of the collection site and include:

- Boat and raft-mounted scissor nets and fine mesh trawl nets
- One man operated scissor nets (push nets) (Plate 6.1);
- Beach seines;
- Dip-nets used in conjunction with lures made from grass and twigs;
- Nets set at high tide on shallow tidal banks parallel to the shore to catch post-larvae as the tide recedes.

Unwanted larvae and juveniles caught along with the shrimp are sometimes picked out by hand, and fish can be eliminated by the use of selective poisons. Fortunately juvenile shrimp are generally more resistant than fish to the stresses of capture and transport and they tend to survive in greater numbers. Unfortunately, however, the whole process is very wasteful; Csavas (1988) estimates that for every single shrimp juvenile grown in a pond, up to 100 shrimp and fish juveniles are killed. Improvements in handling, such as keeping containers in the shade and providing oxygen and/or aeration during prolonged holding or transport, would reduce the wastage, though some buyers like to view their seed in outdoor, unshaded tanks, believing that this treatment will toughen the seed.

Healthy wild juveniles can usually be identified by observing their level of activity. Conflict between the buyers and sellers of seed arises over the quantity involved (reliable counts are difficult and laborious to obtain) and the relative abundance of preferred species. In Ecuador buyers only pay for the juveniles of *P. vannamei* and *P. stylirostris*; other shrimp, notably *P. californiensis* and *P. occidentalis*, are considered 'weed' species because they suffer high mortality in ponds and do not attain more than a few grams in weight. Identifying the different species, however, is another laborious task which can rarely be performed with precision in the field and often relies on examining the juveniles under a low power binocular microscope. Guides to the identification of post-larval shrimp have been prepared for various areas including:

Australia (Heales *et al.* 1985), Ecuador (Yoong & Reinoso 1983), The Philippines (Motoh & Buri 1981) and the Indian Ocean (Paulinose 1982).

Extensive culture systems use little or no supplementary feed, relying instead on the natural productivity of a pond to provide the shrimp with food in the form of small benthic invertebrates, algae and organic detritus. This productivity is sometimes boosted by the addition of organic fertilizers such as poultry and cow dung. Table 5.1 shows some examples of extensive cultures and the resulting yields and harvest sizes.

Extensive ponds are typically shallow (0.3–1.0 m) and often irregular in size and shape. They usually require routine maintenance of embankments and inlet/outlet sluices, and because of their low-lying positions often need repairs after storm and flood damage. Typically the pond bottoms contain undrainable pools which are difficult or impossible to dry after harvest and may have to be netted or treated with fish poisons before the pond is restocked. In some cases extensive ponds are constructed by raising the bunds of rice paddies or salt pans and sometimes, in Bangladesh and Vietnam, salt or rice is produced in seasonal rotation with shrimp (Csavas 1988).

7.2.6.2 *Polyculture*

In extensive systems, shrimp are usually harvested along with an incidental mixture of fish and crabs. Alternatively, in places such as Taiwan, the Philippines and Indonesia, shrimp ponds may be deliberately stocked with the fry of selected species such as milkfish (*Chanos chanos*) and mud crab (*Scylla serrata*) (Section 7.10.4). In Taiwan, *P. monodon* stocked at 150 000–500 000 ha^{-1} may be combined with milkfish at 200–300 ha^{-1} and mud crabs at up to 1000 ha^{-1} (Chen 1990).

7.2.6.3 *Semi-intensive*

Semi-intensive culture generates greater yields than extensive methods through the stocking of selected shrimp species at higher densities, the provision of supplementary feed, and the use of improved water exchange rates. Reliance is often placed on supplies of hatchery-reared seed, and fast growing species such as *P. monodon*, *P. chinensis* and *P. vannamei* are usually preferred. Often an intermediate phase of nursery rearing is performed (Section 7.2.5) and on-growing ponds are typically stocked at densities between five and 20 shrimp m^{-2}. A notable exception, however, is found in China where hatchery-reared *P. chinensis* (PL_{7-25}) are usually stocked directly into on-growing units, and the lower survival rates (15–38%) that result from this approach necessitate higher initial stocking densities of around 50 m^{-2}. Table 5.2a gives examples of yields from a range of different semi-intensive crustacean farming operations.

Natural productivity in ponds is often boosted by the application of organic or inorganic fertilizers (Section 8.3.5.3), but at the shrimp densities employed in semi-intensive culture this productivity alone is insufficient to provide for rapid shrimp growth, and supplementary feeding becomes essential (Section 8.3.5.4). Although compound diets are mostly used, they are sometimes combined with fresh foods such as trash fish, and in China a diet of crushed whole mussels is sometimes provided. Table 7.5 gives an example of feeding recommendations for the semi-intensive culture of *P. vannamei* in Ecuador. There an average food conversion ratio of 2.46:1 has been recorded at stocking densities around 52 000 ha^{-1} (Hirono 1986). In northern

Table 7.5 Recommended feeding levels and diet characteristics for the semi-intensive culture of *P. vannamei* in Ecuador (adapted from Lucien-Brun 1988).

Mean shrimp weight g	Daily ration* % body wt day^{-1}	Pellet size diameter × length mm	Protein content %
<0.1	50	1.0 × 1.5	40–45
0.1–0.5	35	1.0 × 1.5	40–45
0.5–1.5	20	1.0 × 1.5	40–45
1.5–4.0	15	1.5 × 2.5	30–35
4–8	9.8	1.5 × 2.5	30–35
8–12	6.4	2.5 × 8–10	30
12–16	4.7	3.2 × 13–16	25–30
16–20	3.8	3.2 × 13–16	25–30
>20	3.2	3.2 × 13–16	25–30

*To be divided into four feedings per day or a minimum of two feedings per day, one third in the morning (0600–0800 hrs) and two thirds in the evening (1800–2100 hrs).

China as much as 30 mt of crushed whole mussels may be used to produce a tonne of shrimp, and though at first this may appear excessive, for mussels of 40–45 mm this equates to a food conversion ratio of around 1.5–2:1 (dry weight mussel meat:live weight shrimp harvested).

Semi-intensive shrimp ponds are usually 0.5–1.5 m deep, rectangular in shape, and are equipped with separate inlet and outlet sluices in opposing embankments. Within one farm ponds are often built with uniform sizes because this can simplify the routine management of stocking, feeding and water exchange rates. Water exchange may rely on tides or pumps, but if ponds are built above the level of high tide, as is often the case in Ecuador, pumping becomes essential. The advantage of raised ponds is that they can be constructed without the destruction of large areas of intertidal habitat, and, because they can be drained at any state of the tide, harvesting and pond bottom preparation are simplified. In some farms an enlarged supply canal serves as a reservoir and is sometimes used to hold broodstock at low densities. Paddlewheel or other aeration devices are sometimes installed in ponds and operated at night to avoid critically low oxygen tensions in the early hours of the morning (Section 8.3.5.7).

Farms originally built for extensive culture can be converted to semi-intensive operation by deepening ponds, strengthening embankments, building independent inlet and outlet sluices, creating nursery ponds, and providing a pumped water supply (Section 8.2). This process requires capital investment, but if combined with controlled stocking and supplementary feeding it can greatly boost yields.

7.2.6.4 Cage and pen culture

Pens and cages are relatively rare for shrimp culture, although they can be used with minimal investment and as a part-time activity, if suitable sites are available. Pen culture is practised in the Philippines, Singapore and India, and in Thailand it has

Plate 7.1 Aerial view of a large Ecuadorian shrimp farm showing the pumping station (bottom right) and supply canal (centre) bounded by nursery ponds. In the background are a number of irregularly shaped on-growing ponds of 4–25 ha.

recently demonstrated high productivity and is attracting many small-scale operators. Cages are often employed as shrimp nurseries (Section 7.2.5). A significant drawback of all pens and cages, however, is their vulnerability to adverse weather conditions. Suitable sites must be relatively pollution-free, sheltered from waves and excessively strong currents, and provide a water depth of around 1–2 m on lowest tides.

Pens normally consist of plastic mesh nets supported by rectangular bamboo or wood frames, and are fastened by the corners to bamboo poles. The bases of the nets are in contact with the mud. The mesh, which should be as coarse as possible without permitting the shrimp to escape (1.4 mm for juveniles, then 5 mm after one month) allows water exchange and also provides a surface for the attachment of microalgae on which the shrimp can graze. In Thailand *P. monodon* are stocked at densities of 40–110 m^{-2}, and either fresh or pelleted feed or a combination of both are provided three or four times per day. The food is lowered into the pens on mesh feeding trays (0.5 × 0.5 m). Feeding rates for pellets are initially 10–20% body wt day^{-1} and fall steadily to 3.0% by the end of the 3–4 month culture period. The successful elimination of predators usually results in survival rates around 70%, and 100 kg of 30 g shrimp might be expected from a pen measuring 6 × 6 × 6 m (Tookwinas 1990). In the cage culture of *P. merguiensis* in Singapore, annual yields of shrimp (average live weight 12 g) equivalent to 20 000–30 000 kg ha^{-1} have been obtained (Lovatelli 1990).

Although in theory pens should allow efficient exchange of water and thereby avoid problems with oxygen depletion, the mesh can become blocked by seaweed and other fouling organisms. In such cases shrimp may require transfer to clean pens and aeration may become necessary towards the second month of the culture period. Water quality problems are often exacerbated by the large numbers of pens concentrated in suitable sites (Section 11.2.5).

Pens offer the advantage of relatively straightforward harvesting using scoop nets or cast nets, and some can simply be lifted. After the harvest, netting can be cleaned or replaced prior to re-stocking, and, if adequate supplies of nursery reared juveniles are available, three crops per year may be possible.

One particular type of pen or impoundment, often referred to as the Amakusa-style after the island in Japan where they are operated, comprises an intertidal area bounded by low intertidal concrete walls about 1 m high. The walls are topped with a mesh barrier supported by a wooden frame, to prevent the escape of the crop when the pens are flooded. Pens range in size from 0.2 to 1 ha, and when stocked with *P. japonicus* at 20–30 m^{-2} can produce 3–4 mt ha^{-1} annually (Kungvankij & Kongkeo 1988).

7.2.6.5 Intensive

Intensive shrimp culture is performed in small ponds of 0.1–1.0 ha and results in yields of between 5 and 15 mt per hectare per crop. Nursed juveniles of hatchery origin are stocked at high densities of between 15 and 50 shrimp m^{-2}, and are fed on compounded diets. With dependable supplies of juveniles, intensive farms in the tropics can produce up to three crops per year. Pond conditions are carefully controlled through water exchange, and aeration is usually provided on a continuous or regular basis.

Intensive ponds are well-made structures with concrete walls or plastic lined embankments. They are typically 1–1.5 m deep, usually rectangular in shape, and can be efficiently drained between crops to allow for the removal of accumulated organic sediment. Pond bottoms are usually natural earth or an artificial layer of 10 cm of sand or gravel. Ponds completely lined in plastic are generally deeper (up to 2 m water depth). After harvest earth pond beds may be conditioned (Section 8.3.5.1) by ploughing, sun-drying and dressing with lime (200–300 kg ha^{-1}). In Taiwan, ponds are sometimes sterilized with saponin (2.5–10 ppm), added in the form of tea seed cake in a shallow layer of water.

Natural productivity is usually encouraged by the use of fertilizers; sometimes manure (1 mt ha^{-1} annually) but more usually with daily or weekly additions of inorganic fertilizer (Section 8.3.5.3). For *P. monodon* recommended feeding rates using compound diets containing 35–39% protein are:

Daily ration (% body wt day^{-1})	Shrimp size (g)
25–15%	<2
15–7%	2–5
10–7%	6–15
5–3%	>15

These rates, however, require daily adjustment in accordance with observations of food remains on nets placed on the pond bottoms. Frequent feeding, 4–6 times per day, improves the utilization of feed, and conversion ratios of less than 2:1 are reported in Taiwan (Chen 1990).

Water is supplied to the ponds by pumping, and paddlewheels or propeller-aspirator pumps are arranged to impart a circular motion to the water. This action provides a more even distribution of dissolved oxygen, and in centrally drained ponds sweeps waste towards the centre where it can be voided. Water quality parameters are carefully monitored. If an algae bloom becomes excessively dense, denoted by a low Secchi disc reading (Section 8.3.4.2), it can be treated with an algicide such as simazine or cutrine or preferably diluted by increasing the water exchange rate (Section 8.3.5.5).

Water exchange rates recommended for intensive *P. monodon* culture in Thailand increase during the production cycle from 5% day^{-1} in the first month to 30% day^{-1} in the fourth month. By comparison, equivalent figures recommended for semi-intensive culture increase from 1% day^{-1} to 20% day^{-1} (Anon. 1990a). But for the use of aerators, the water exchange requirements in intensive systems would be considerably higher (Section 8.3.5.7).

Small plastic lined ponds (0.1–0.36 ha × 1.5–2.0 m deep) have demonstrated suitability for intensive shrimp production. Durable liners can prevent embankment erosion, simplify harvesting and facilitate pond cleaning and disinfection between crops (Plate 12.1). Trials with 0.1 ha pools in Oman, using imported *P. monodon*, have provided harvests averaging 5900 kg ha^{-1} (Fuke P., 1990 pers. comm.), and trials in Texas with *P. vannamei* have demonstrated potential for high survival rates (>90%) and productivity equalling or exceeding that of ponds with conventional soil or sand substrates (Anon. 1990b). Although more expensive (Section 10.9.1.4), lined ponds have the advantage of preventing salt intrusion into ground-water and have been successfully used in sites where sandy permeable or highly acid soils would normally be considered unsuitable for pond construction.

7.2.6.6 *Super-intensive*

Harvests in excess of 2 kg m^{-2} are possible in super-intensive systems by using very high stocking densities (50–250 shrimp m^{-2}), relying almost totally on compounded feeds, and by very careful manipulation of the culture conditions. By providing optimal conditions, assisted by high water exchange rates, high survival rates can be obtained. Occasionally controlled enclosed environments or 'battery systems' have been used.

Several distinct systems have been developed to a commercial scale. In Japan, the shortage of land suitable for building large shrimp ponds, and the existence of an exceptionally high value market for live *P. japonicus* (Section 3.3.1.1), stimulated the development of increasingly intensive culture systems. Circular concrete tanks of 1000–2000 m^2 and 2 m deep, designed by and named after Kunihiko Shigueno, are operated commercially in Kagoshima at sites with high quality seawater. Post-larvae (PL_{30}) may be stocked initially at as many as 300–400 m^{-2} and thinned out later to 140–150 m^{-2} when they reach 5 g. Sometimes, however, lower densities of 70–80 m^{-2} are employed. High-protein formulated diets (min. 55% protein) are provided in four daily feedings, initially at 12–18% body wt day^{-1} but reduced to 10% at 1 g and 3–5% at 5 g. Market size animals around 20 g can be obtained after 5–6 months, while with good quality feeds and careful adjustment of feeding levels (following routine observation of food remains), food conversion ratios almost as low as 1:1 have been recorded.

Water is pumped into the tanks from a pipe arranged across the diameter, and angled jets generate a circular flow pattern. This helps concentrate food remains, faecal waste and cast shells in the centre of the tank from which point they can be voided through a central drain. *P. japonicus* only grows well if provided with a substrate into which it can burrow, so tanks are given a 10–15 cm layer of sand on a false bottom. Water normally exits downwards through the sand, although the direction of water flow can be reversed to help clean the substrate. A high exchange rate (250–400% day^{-1}) is provided, sometimes in combination with aeration, and serves to maintain oxygen levels throughout the tank. Since water in the tank is usually quite clear, seaweed is able to grow on the bottom and this requires routine removal. To combat this problem some later tanks have been built to allow a greater depth of water (2.2–2.5 m).

In Hawaii another round pond system has been developed which, unlike the Shigueno system, employs a simple compacted earthen substrate. It has been operated with *P. vannamei*, a species which does not require sand in which to burrow and which can be reared on pellets with lower protein content than those required for super-intensive *P. japonicus* culture. Trials on both an experimental scale (0.034 ha ponds) and a commercial scale (0.2 ha ponds) have been performed (Wyban *et al.* 1989). The greatest productivity of 1.91 kg m^{-2} of shrimp averaging 15.1 g was achieved at a density of 150 shrimp m^{-2}, but a more valuable crop of 1.71 kg m^{-2} of larger (26.1 g) shrimp was achieved in a trial with a lower stocking density of 75 m^{-2}. High survival rates around 85% were normal, and feed conversion ratios averaging 2.0–2.5:1 were recorded using a 42.2% protein diet manufactured in Taiwan for the culture of *P. penicillatus* (Wyban & Sweeney 1989). A projected commercial operation stocked at 100 shrimp m^{-2} is expected to achieve three crops per year, each crop yielding 1.5 kg m^{-2} of shrimp averaging 15–17 g. However, although these ponds are cheaper to build and operate than Shigueno tanks, there is unfortunately no high price market for *P. vannamei* that is equivalent to that in Japan for live *P. japonicus*. Interestingly, the approach to water management in this Hawaiian system differs significantly from that in Shigueno tanks. Lower water exchange rates (10–100% day^{-1}, average 60% day^{-1}) are employed and blooms of diatoms are maintained since these have been observed to promote growth and survival in the crop. In addition, greater reliance is placed on paddlewheels (20 hp ha^{-1} is recommended) to oxygenate the water and maintain circulation in the tank.

Although super-intensive culture trials often yield excellent results, it should be remembered that they are usually performed in a research and development environment with well-motivated management and the back-up of laboratory facilities. Under these favourable conditions the prospects of success are greatly increased.

Another super-intensive system, developed jointly by the Universities of Arizona, USA and Sonora, Mexico, consisted of PVC lined raceways (3.4 × 6 × 0.3 or 0.6 m deep) beneath inflated plastic greenhouses ('aquacells') (Salser *et al.* 1977). High density rearing trials were performed with several penaeid species and *P. stylirostris* gave the most promising results. Water was pumped from seawater wells and sprayed on to the surface of the raceways to give turnover rates of as much as 700% per day. Only compounded diets were used and waste was routinely siphoned from the tank bottoms. Refinements of this system led to the establishment of a commercially operated *P. stylirostris* farm in Hawaii which comprised 52 raceways and an integral hatchery. Liquid oxygen was used to obtain yields greater than 50 mt ha^{-1} per crop and a weekly output totalling 3 mt. However, the facility, which represented an investment of US$25 million, has been dogged by disease problems, most notably IHHN virus, and doubts have been raised about its commercial future (New 1988).

A 'battery' operation, King James Shrimp Ltd of Illinois, USA, produced shrimp in stacked raceways within a heated 'closed' recirculating system, but went bankrupt in 1982. The system employed artificial seawater of which 10–15% was exchanged each day. Yields of 1–1.5 kg m^{-2} were obtained (Rosenberry 1982; McCoy 1986).

7.2.7 Harvesting

Shrimp can be harvested by a variety of methods including draining, seining, cast-netting, trapping and electro-fishing.

The most commonly applied technique for completely harvesting a pond is by draining. In this process the shrimp are firstly concentrated within the pond by dropping the water level and keeping screens in position in the sluices, and then the screens are removed and the shrimp are swept into a collection net. The net may be in the form of a long (3–4 m) tapering bag, a staked enclosure or an open-topped rigid cage. To handle a large amount of shrimp with minimal damage, the harvest can be spread over several consecutive nights, refilling the pond each time. Most importantly for an efficient harvest, however, a pond must drain well. Any remaining pools in channels or bottom depressions require laborious seining and hand-picking, during which shrimp deteriorate in condition and may accumulate mud beneath their carapaces.

The efficiency of a drain harvest can be improved by taking into account shrimp behaviour. Penaeids in general spend much of their time burrowed in the substrate, with just their eyes protruding, and in this position they are impossible to harvest effectively. Activity, however, does increase at night, especially during periods of new and full moon, and swimming behaviour can be encouraged by water movement; so by performing harvests during the hours of darkness and sharply lowering and raising the pond level in advance, better catches can usually be made. In addition, during a night harvest, lamps can be placed alongside the collection area to attract the shrimp. Behaviour, however, shows some variation between species; *P. monodon*, for example, is reported to be more reluctant than other penaeids to swim out of a pond during draining.

Traps rely on shrimp moving around in a pond and so are most efficient when the shrimp are active. A leader made of netting or bamboo slats is arranged at right angles to the embankment and guides shrimp into bamboo or net catching chambers. Frequent collections may be necessary to avoid overcrowding and keep the shrimp alive. An advantage of traps is that slat spacing or mesh size can be set to harvest the largest shrimp.

Electro-fishing makes use of an electric current to stimulate shrimp to jump. Electrodes are located either on the base of a drag-net, which is drawn across the pond bottom, or at the ends of hand-held poles. In the latter system the cathode forms the ring of a scoop net and the operator, equipped with an accumulator, wades through the pond collecting the shrimp as they jump. Electro-fishing is conveniently performed in small ponds or tanks and is often chosen for intensive and super-

Plate 7.2 A disgruntled and mud-covered Thai farm worker displaying a meagre catch of exhausted and dying shrimp taken from an undrainable pond.

intensive operations. It enables shrimp to be collected in excellent condition and has proved to be popular with professional harvesting teams in Taiwan.

Since the value of the crop is reduced by the presence of newly moulted soft shelled shrimp (because of their susceptibility to physical damage during collection and processing) harvesting should be timed to avoid the peak occurrence of moulting, which may coincide with full moons. The period two to three days after new moon has been recommended for harvesting, although in practice cast-net samples are usually taken to check the condition of shrimp before any decision is taken to commence a full-scale harvest. Sometimes flushing a pond with new water when harvesting (to wash shrimp to outlet) has been found to precipitate moulting.

Harvesting may be performed in a single complete operation or as a series of partial harvests spread over a period of several weeks or months. The harvesting strategy adopted by any particular farm will be determined by processing capabilities and the market requirement for either bulk or small quantities (Section 3.2.2.1).

7.2.8 Processing

Large amounts of shrimp are sold as peeled or unpeeled tails after washing, grading and de-heading. Some are sold whole, either chilled, individually quick frozen (IQF) or bulk frozen in blocks with water. Small amounts are sold live. *P. japonicus*, for example, destined for specialist Japanese markets (Section 3.3.1.1), is packed in boxes with sawdust after washing, grading and chilling. The many market forms of shrimp are summarised in Chapter 3 (Figure 3.2), and a range of value-added and speciality shrimp products is listed in Section 3.2.3.1.

7.2.9 Hatchery supported fisheries, ranching

Shrimp release programmes have been undertaken in several countries to enhance fishery stocks, though in Taiwan releases are made more with a view to improving broodstock availability.

In Japan *P. japonicus* are liberated either as post-larvae (PL_{20}) or as nursery-reared juveniles (3 cm TL). Nursery rearing is performed primarily to reduce the predation losses associated with the direct release of small hatchery-reared post-larvae. Three types of nursery are employed: fenced enclosures, man-made lagoons and artificial tidelands (Kurata 1981).

Fenced enclosures are supported by sticks, floats and anchors but a major problem remains their vulnerability to damage from wind, waves and tidal currents. Prior to stocking, fish predators within the enclosure are either netted or poisoned. After two to three weeks of rearing, juveniles are released simply by removing the fence.

Lagoons are created by constructing temporary embankments along the shoreline, and so can only be sited where suitable embankment-building material is available. Stocked juveniles are released when they reach 3 cm in length, by rupturing the embankment. In both fenced enclosures and lagoons, the juveniles are very vulnerable to adverse weather, especially heavy rains and typhoons. Red tides (see Glossary) can also be problematic and can contribute to the general unpredictability of survival rates (range 0–70%).

Artificial tidelands consist of a series of low walls, which retain water to a depth of about 5 cm when the tide recedes. They are protected by a breakwater and a seaward fence to exclude fish, and are specially designed to provide conditions which are favourable to shrimp juveniles but not to their predators. For example, a comparatively hard substrate is provided which is unsuitable for the burrowing habit of gobies. Juveniles are stocked at a maximum density of 100 m^{-2} and they rely on natural productivity for their sustenance. Although the substrate sometimes becomes completely exposed between tides, the shrimp juveniles can survive for a few hours in just a few millimetres of water. A pumped water supply may be provided during low tide. Survival rates are variable and particular problems may arise with the excessive growth of seaweed.

In Japan an average of 293 million *P. japonicus* juveniles were released each year over the period 1981–1985 (Liao 1988). Estimates of overall recapture rates (in the shrimp fishery) for juveniles released on to artificial tidelands in Ohmi Bay, Yamaguchi, have been placed at 6.1–8.4% (Kurata 1981).

Figures for shrimp releases in China indicate that significant efforts are being made in *P. chinensis* stock enhancement in and around the Yellow Sea and Gulf of Bohai. In 1986 an estimated total of 4 billion hatchery-reared shrimp were liberated (Shang 1989). In the semi-enclosed Jiaozhou Bay, the release of over 350 million juveniles from 1984–86 appears to have had a dramatic effect on the depleted local fishery. Stocks increased by factors of between 4.7 and 7.3 during the three year release period and declined immediately afterwards. Estimated average survival was a very impressive 32% (Liu 1990).

Over the period 1972–79, 120 million post-larvae, mostly *P. semisulcatus*, were released into Kuwaiti waters, and although recapture rates were unknown, Farmer (1981) estimated that the operation would break even at a recapture rate of 2% (Section 10.9.1.6). In Taiwan, sea ranching is considered a promising method of increasing supplies of wild gravid females, and recapture rates as high as 15% have been estimated (Chiang & Liao 1985).

One ranching operation, set up in the 1970s as an enclosed, stocked fishery by Marifarms Inc. in Florida, USA, incorporated a netted embayment of 1000 ha and two lagoons of 120 ha each. A total of 19 km of netting was required, which was subject to frequent damage from storms and tides. Within the embayment, hatchery reared seed was stocked into a two stage nursery which consisted of staked net enclosures. Juveniles were released and fed with trash fish and pellets, but no attempt was made to eliminate predators and competitors. Harvesting was performed with a trawler and in the first year yielded a total of 95 mt of shrimp (after processing) (Hanson & Goodwin 1977). However, the operation eventually proved to be uneconomic, in part due to legislative constraints which required access by leisure craft, and to an undertaking to restock the fishery from the hatchery.

7.2.10 References

Akiyama D.M. (Ed.) (1989) *Proceedings of the south-east Asia shrimp farm management workshop*. 26 Jul.–11 Aug. 1989, Philippines, Indonesia, Thailand. American Soybean Association, Singapore.

Anon. (1990a) Intensive sea system works with *monodon*. *Fish Farmer*, **13** (5) 7.

Anon. (1990b) Plastic liner trial shows encouraging results: Initial tests showed liner-use potential. *Newsline*, **3** (3) 4–5. Oceanic Institute Honolulu, Hawaii.

AQUACOP (1983) Constitution of broodstock, maturation, spawning, and hatching systems for penaeid shrimps in the Centre Océanologique du Pacifique. In *Handbook of Mariculture, Vol. 1, Crustacean aquaculture* (Ed. by J.P. McVey), pp. 105–27. CRC Press, Boca Raton, Florida.

ASEAN (1978) *Manual on pond culture of penaeid shrimp*. ASEAN National Coordinating Agency of the Philippines, Ministry of Foreign Affairs, Manila.

Beveridge M. (1987) *Cage aquaculture*. Fishing News Books, Blackwell Scientific Publications, Oxford.

Chavez C. (Ed.) (1990) *The aquaculture of shrimp, prawn, and crayfish in the world: basics and technologies*. Haworth Press, Food Products Press Inc., Binghamton, New York.

Chen L.-C. (1990) *Aquaculture in Taiwan*. Fishing News Books, Blackwell Scientific Publications, Oxford.

Chiang P. & Liao I.C. (1985) The practice of grass prawn (*Penaeus monodon*) culture in Taiwan from 1968 to 1984. *J. World Maricult. Soc.* **16**, 297–315.

Cordover R. (1989) Prawn farming the Australian way. *Austasia Aquaculture Magazine*, **3** (7) 8–10.

CPC (1989) *Libro blanco del camarón*. Cámara de Productores de Camarón, Guayaquil, Ecuador.

Csavas I. (1988) Shrimp farming developments in Asia. In *Shrimp '88, Conference proceedings*, 26–28 Jan. 1988, pp. 63–92. Bangkok, Thailand. Infofish, Kuala Lumpur, Malaysia.

FAO (1979) *Brackishwater shrimp and milkfish culture, applied research and training project, Indonesia*. FAO, Rome.

Farmer A.S.D. (1981) Prospects for penaeid shrimp culture in arid lands. In *Advances in food producing systems for arid and semi-arid lands*, pp. 859–97. Academic Press Inc.

Franklin T., Sondara Raj R. & Prabhavathy G. (1982) Transport of prawn seed. In *Proc. Symp. Coastal Aquaculture 1982*, Pt. 1, pp. 395–6. Mar. Biol. Assn. of India, Cochin.

Frippak (1988) *Frippak hatchery manual*. Frippak Feeds, Batley, W. Yorks.

George M.J. & Suseelan C. (1982) Distribution of species of prawns in the backwaters and estuaries of India with reference to coastal aquaculture. In *Proc. Symp. Coastal Aquaculture 1982*, Pt. 1, pp. 273–284. Mar. Biol. Assn. of India, Cochin.

Hanson J.A. & Goodwin H L. (Eds.) (1977) *Shrimp and prawn farming in the western hemisphere*. Dowden, Hutchinson and Ross Inc., Stroudsburg, Pennsylvania.

Harrison K. (1990) The role of nutrition in maturation, reproduction and embryonic development of decapod crustaceans: a review. *J. Shellfish Res.*, **9** (1) 1–28.

Heales D.S., Polzin H.G. & Staples D.J. (1985) Identification of the commercially important *Penaeus* species in Australia. In *Second Aust. Nat. Prawn Sem.* (Ed. by P.C. Rothlisberg, B.J. Hill & D.J. Staplos), pp. 41–6. NPS2, Cleveland, Australia.

Hirono Y. (1986) Shrimp pond management. In *Acuacultura del Ecuador* (in Spanish and English). Cámara de Productores de Camarón, Guayaquil, Ecuador.

JICA (1987) *A manual for the mass propagation of the Japanese prawn (kuruma ebi) post larvae*. Japan International Cooperation Agency, Aichi Prefecture Fish Culture Centre.

Johnson S.K., Cichra C.E. & Chamberlain G.W. (1984) *Wet transport of mature marine shrimp in sealed containers*. FDDL-S14, Fish Disease Diagnostic Lab., Texas Agricultural Extension Service, Texas A&M University, College Sta., Texas.

Kungvankij P. & Chua T.E. (1986) *Shrimp culture: pond design, operation and management*. NACA training manual series No. 2, NACA, Bangkok.

Kungvankij P. & Kongkeo H. (1988) Culture system selection. In *Shrimp '88, Conference proceedings*, 26–28 Jan. 1988, pp. 123–36. Bangkok, Thailand. Infofish, Kuala Lumpur, Malaysia.

Kurata H. (1981) Shrimp fry releasing techniques in Japan with special reference to the artificial tideland. In *Proceedings of the International shrimp releasing marking and recruitment workshop*, 25–29 Nov. 1978, (Ed. by A.S.D. Farmer), (2) 117–47. Salmiya, Kuwait, Kuwait Bulletin of Marine Science.

Kurian C.V. (1982) Some constraints in prawn culture. In *Proc. Symp. Coastal Aquaculture 1982*, Pt. 1, pp. 98–102. Marine Biological Association of India, Cochin.

Liao I-C. (1984) A brief review on the larval rearing techniques of penaeid prawns. Presented at *First Int. Conf. on the Culture of Penaeid Prawns/Shrimps*, 4–7 Dec. 1984, Iloilo City, Philippines.

Liao I-C. (1988) History, present status and prospects of prawn culture in Taiwan. In *Shrimp '88, Conference proceedings*, 26–28 Jan. 1988, pp. 195–213. Bangkok, Thailand. Infofish, Kuala Lumpur, Malaysia.

Liao I.-C. & Chen Y.-P. (1983) Maturation and spawning of penaeid prawns in Tungkang Marine Laboratory, Taiwan. In *Handbook of Mariculture, Vol. 1, Crustacean aquaculture* (Ed. by J.P. McVey), pp. 155–60. CRC Press, Boca Raton, Florida.

Lin M.-N. & Ting Y.-Y. (1986) Spermatophore transplantation and artificial fertilization in grass shrimp. *Bull. Jap. Soc. Sci. Fish.*, **52** (4) 585–9.

Liu J.Y. (1990) Resource enhancement of Chinese shrimp, *Penaeus orientalis. Bull. Marine Sci.*, **47** (1) 124–33.

Lovatelli A. (1990) *Regional seafarming resources atlas.* FAO/UNDP Regional seafarming development and demonstration project, RAS/86/024 Jan. 1990.

Lucien-Brun H. (1988) *Guía para la producción de camarón en el Ecuador* (in Spanish). Expalsa, Guayaquil, Ecuador.

Macintosh D.J. (1987) *Aquaculture production and products handling in ASEAN.* ASEAN Food Handling Bureau, Kuala Lumpur, Malaysia.

Maguire G.B. (1979) *A report on the prawn farming industries of Japan, the Philippines and Thailand.* Brackish Water Fish Culture Research Station, C/-P.O., Salamander Bay, N.S.W. 2301.

Main K.L. & Fulks W. (Eds.) (1990) *The culture of cold-tolerant shrimp: Proceedings of an Asian-US workshop on shrimp culture.* Oceanic Institute, Honolulu, Hawaii.

McCoy H.D. II. (1986) Intensive culture systems past, present and future: parts I, II and III. *Aquaculture Magazine*, **12** (6) 32–5; **13** (1) 36–40; **13** (2) 24–9.

McVey J.P. (1983) (Ed.). *Handbook of Mariculture, Vol. 1, Crustacean aquaculture.* CRC Press, Boca Raton, Florida.

Motoh H. & Buri P. (1981) Identification of postlarvae of the genus *Penaeus* appearing in shore waters. *Researches on Crustacea*, **11**, 3–11.

NACA (1986) *A prototype warm water hatchery.* Technology series, selected publication No. 4, NACA, Bangkok, Thailand.

New M.B. (1988) Shrimp farming developments in other areas. In *Shrimp '88, Conference proceedings*, 26–28 Jan. 1988, pp. 102–22. Bangkok, Thailand. Infofish, Kuala Lumpur, Malaysia.

Paulinose V.T. (1982) Key to the identification of larvae and postlarvae of the penaeid prawns (Decapoda: Penaeidea) of the Indian Ocean. *Mahasagar Bull. of Nat. Inst. Oceanog.*, **15**, 223–9.

Pretto R. (1983) Penaeus shrimp pond grow-out in Panama. In *Handbook of Mariculture, Vol. 1, Crustacean aquaculture* (Ed. by J.P. McVey), pp. 169–78. CRC Press, Boca Raton, Florida.

Primavera J.H. (1983) Prawn hatcheries in the Philippines. *Asian Aquaculture*, **5**, (3) 5–8.

Quynh V.D. (1989) Viet Nam, coastal aquaculture in the southern provinces. *World Aquaculture*, **20** (2) 22–8.

Quynh V.D. (1990) Shrimp larviculture in Viet Nam. *Larviculture and Artemia Newsletter*, (16) 22–6. State (University of Ghent, Belgium.

Richards-Rajadurai P.N. (1989) Live storage and distribution of fish and shellfish. *Infofish International*, (3) 23–7.

Robertson L., Bray W.A. & Lawrence A.L. (1987) Shipping of penaeid broodstock: Water quality limitations and control during 24 hr shipments. *J. World Aquacult. Soc.*, **18** (2) 45–56.

Rosenberry R. (1982) Shrimp Systems International Ltd. *Aquaculture Digest*, **7** (12) 1–5.

Salser B., Mahler L., Lightner D., Ure J., Danald D., Brand C., Stamp N., Moore D. & Colvin B. (1977) *Controlled Environment Aquaculture of Penaeids*. Environmental Research Laboratory, University of Arizona, Tucson International Airport, Tucson, Arizona.

SEAFDEC (1985) *A Guide to Prawn Hatchery Design and Operation*. Extension manual No. 9, South-East Asia Fisheries Development Centre, Tigbauan, Iloilo, Philippines.

SEAFDEC (1988) *Biology and culture of* Penaeus monodon. SEAFDEC, Tigbauan, Iloilo, Philippines.

Shang Y.C. (1989) Marine shrimp farming in China. *Infofish International* (2) 16–17.

Simon C.M. (1982) Large-scale, commercial application of penaeid shrimp maturation technology. *J. World Maricult. Soc.*, **13**, 301–12.

Singh H., Chowdhury A.R. & Pakrasi B.B. (1982) Experiments on the transport of post-larvae of tiger prawn *Penaeus monodon* Fabricius. In *Proc. Symp. Coastal Aquaculture 1982*, Pt. 1, pp. 232–5. Marine Biological Assoc. of India, Cochin.

Smith T.I.J. & Ribelin B. (1984) Stocking density effects on survival of shipped post-larval shrimp. *Progv. Fish Cult.*, **46** (1) 47–50.

Tacon A.G.J. (1990) *Standard methods for the nutrition and feeding of farmed fish and shrimp*, **1**, 1–117; **2**, 1–129, **3**, 1–208. Argent Laboratories Press, Redmond, USA.

Tookwinas S. (1990) Pen culture techniques of marine shrimp in Thailand. *Infofish International* (2) 38–40.

Treece G.D. & Fox J.M. (in press). *Design, operation and training manual for an intensive culture shrimp hatchery*. Marine Advisory Service, Texas A&M Sea Grant College Program, Galveston, Texas.

Treece G.D. & Yates M.E. (1990) *Laboratory manual for the culture of penaeid shrimp larvae*. Marine Advisory Service, Sea Grant College Program, Texas A&M University, Texas.

Tseng W.-Y. (1987) *Shrimp mariculture – a practical manual*. Chien Cheng, 38 Chien Kuo 3rd Road, Kaohsiung, Rep. of China.

Villalon J.R. & Treece G.D. (in press). *Practical manual for semi-intensive commercial production of marine shrimp*. Sea Grant College Program, Texas A&M University, Texas.

Wyban J.A. & Sweeney J.N. (1989) Intensive shrimp grow-out trials in a round pond. *Aquaculture*, **76** (3/4) 215–25.

Wyban J.A., Sweeney J.N., Kanna R.A., Kalagayan G., Godin D., Hernandez H. & Hagino G. (1989) Intensive shrimp culture management in round ponds. In *Proceedings of the south-east Asia shrimp farm management workshop*, 26 Jul.– 11 Aug. 1989, (Ed. by D.M. Akiyama), pp. 42–7. Philippines, Indonesia, Thailand. American Soybean Association, Singapore.

Yamaha (1989) Prawn culture. *Yamaha Fishery Journal* (30). Yamaha Motor Co., Shizuoka-ken, Japan.

Yoong F. & Reinoso B. (1983) *Manual practico para la identificación de post-larvas y juveniles de cuatro especies de camarones marinos* (in Spanish). Boletín Científico y Técnico VI (II), Instituto Nacional de Pesca, Guayaquil, Ecuador.

Zhang W., Dengong C., Rujie L., Naiyu Z., Hengzu G. & Xuanyuan L. (1980) Notes on factors affecting the successful artificial rearing of the Chinese prawn (*Penaeus orientalis* Kishinouye) from eggs to postlarvae. (In Chinese, translated by Chau-ling Chan, Lancaster Polytechnic.) *Transac. Oceanology and Lim. Sinica* (1) 39–44.

7.3 *Macrobrachium*

7.3.1 Species of interest

Giant river prawn (*Macrobrachium rosenbergii*); monsoon river prawn (*M. malcolmsonii*); African river prawn (*M. vollenhovenii*); cinnamon river prawn (*M. acanthurus*).

Although several species of freshwater prawn have demonstrated potential as candidates for aquaculture, none has challenged *Macrobrachium rosenbergii*'s complete domination of commercial freshwater prawn culture. The techniques for rearing the other species of *Macrobrachium* differ only slightly from those described below for *M. rosenbergii*. The manuals by McVey (1983) and New and Singholka (1985) provide details of the operating techniques for both prawn hatcheries and farms. Reviews of the status and techniques of freshwater prawn culture include Malecha (1983) and New (1988; 1990), while a comprehensive collection of papers concerning developments in many aspects of prawn farming (principally covering the period up to 1980) has been compiled by New (1982).

7.3.2 Broodstock, incubation and hatching

Where *M. rosenbergii* occurs naturally, local fisheries can provide a source of broodstock females. However, because this species will readily mature, copulate and spawn in on-growing ponds, they are more commonly and conveniently selected from farm production ponds, or in some cases from broodstock holding ponds. If temperatures are adequate (over 23°C), ponds can provide a year-round supply of egg-carrying or 'berried' females. These females, instantly recognizable by the clutch of eggs held in the brood chamber on the underside of the abdomen, are selected for the hatchery on the basis of large size, healthy appearance and egg colour. The eggs are bright orange when spawned, gradually darken to a brownish orange and eventually turn dark grey two or three days prior to hatching. Usually only females with eggs at the late stage of development (dark grey) are chosen. When the eggs have hatched, the females can be sold for food or returned to broodstock or on-growing ponds.

The fecundity of female *M. rosenbergii* is influenced by size. Fully mature large females produce well over 100 000 eggs per brood whereas smaller animals, taken from production ponds, can be expected to produce between 10 000 and 30 000 eggs. Maturation of *M. rosenbergii* in captivity often occurs at a small size (<25 g) which minimises the effort and facilities required for broodstock management. The use of small female broodstock, however, does not take advantage of the potential fecundity of the species and if repeated over many generations may have the unwanted effect of selecting for early maturity at the cost of growth (Section 12.10).

Careful handling of berried females is essential to minimise egg loss. Shipping in aerated tanks or bins is convenient for short journeys of one hour or less. For longer periods, transport in plastic bags with water and oxygen gives good results. To protect

Table 7.6 Methods and results for shipping post-larval, juvenile and adult *Macrobrachium rosenbergii*.

Stage and size	Duration (hours)	Density (no. ℓ^{-1} or g ℓ^{-1})	Temp. °C	Survival %	Transport method	Reference/source
PL	short	750	—	—	40 ℓ container	Macintosh 1987
PL	<1	750	—	—	Aerated tank	New 1990
PL	<12	300	—	—	Plastic bags with water + O_2	New 1990
PL	<24	833	—	—	40 ℓ bags in cartons, 6 ℓ water	SICA 1988
PL < 20 mg	<24	200	18–23	90	24 hr starved PLs*	Smith & Wannamaker 1983
PL	24–36	100	—	—	Plastic bags with water + O_2	New 1990
PL < 20 mg	24–48	150	18–23	90	24 hr starved PLs*	Smith & Wannamaker 1983
Juvenile 6 g	24	18 g	19–20	>90	24 hr starved juveniles, 40 per box*	Smith & Wannamaker 1983
Juvenile 6 g	48	9–11 g	19–20	>90	24 hr starved juveniles, 20–25 per box*	Smith & Wannamaker 1983
Adult 17 g	24	12–15 g	19–20	100	24 hr starved adults, 10–12 per box*	Smith & Wannamaker 1983
Adult 17 g	48	12–15 g	19–20	75	24 hr starved adults, 10–12 per box*	Smith & Wannamaker 1983

* styrofoam box (38 × 38 × 20 cm deep) containing double plastic bags with 13.6 ℓ water plus oxygen

bags from punctures, blunting sharp rostrums and telsons with scissors has been recommended. However, immobilising prawns by wrapping them individually with mesh or webbing is not advisable since it can significantly reduce survival rates (Smith & Wannamaker 1983). The overall effects of stress can be reduced by lowering the temperature to between 18° and 23°C using crushed ice applied externally to the bags. Details of shipping methods and some results for shipping small adults (17 g) are included in Table 7.6. After arrival at the hatchery, the use of a disinfectant dip (0.2–0.5 ppm copper or 15–20 ppm formalin for 30 min.) is advisable to reduce the likelihood of introducing fouling organisms.

Berried females are kept in water at 12–15 ppt so that newly emerged larvae do not require subsequent acclimatization to the brackish water used for rearing. The females may be held in larvae rearing tanks or in independent hatching containers. In the former, common in Thailand, a stocking rate of three prawns (measuring 10–12 cm TL) per cubic metre of water has been recommended to achieve a density of 30–50 larvae per litre (New & Singholka 1985). In Taiwan, hatching tanks equipped with numerous shelters may be stocked with 20–40 g berried females at a rate of about 13 kg m^{-2} (Chen 1990). Newly hatched larvae are subsequently siphoned into separate rearing tanks.

7.3.3 Larvae rearing

Techniques for large-scale rearing of *Macrobrachium* larvae were successfully developed by Takuji Fujimura in Hawaii in the mid 1960s. His method is known as the green-water technique because a population of algae, usually a green coloured *Chlorella* spp., is maintained in the culture water. It differs from the so-called clear-water technique, subsequently developed in mainland USA and French Polynesia, which, as the name suggests, uses water without added algae that is frequently changed or recycled. The algae in green-water cultures do not represent a source of nutrition for the prawn larvae and are only ingested incidentally with other items of prey or food. They do however, provide feed for *Artemia*, rotifers or other living prey, and help condition the culture water through the removal of toxic substances, particularly ammonia. The disadvantages of the green-water technique are that the algae often grow erratically and can cause the pH to fluctuate widely. In order to maintain 'green-water' conditions, water renewal must not flush away more algae than can be replaced by reproduction within the tank or by topping-up from separate algae stocks.

The clear-water system avoids these drawbacks, and water quality can be maintained by greater exchange rates and controlled addition of live and compounded feeds. If a water quality or disease problem does occur in any particular tank, water exchange can be increased in response. These advantages have led to the usual preference for clear-water techniques and nowadays green-water hatcheries are relatively rare.

Larvae rearing tanks can range in size from a few hundred litres to 30 000 litres, but most are at the lower end of this range. Rectangular concrete tanks with gently sloping flat bottoms are most common, though alternatives include fibreglass cylindro-conical vessels, modified, circular, concrete drainage pipe sections, and plastic-lined tanks made of wood or bamboo. Operating water depths are usually between 0.4 and 1 m. Shading is provided to protect larvae from the harmful effect of direct sunlight known as 'sun-cancer'.

Rearing densities usually lie between 30 and 100 larvae ℓ^{-1}, but may be initially pushed as high as 200 ℓ^{-1} in systems in which the larvae are later to be divided between two or more culture vessels. In different hatcheries the average salinity employed may range between 8 and 20 ppt., but 12 ±2 ppt., is generally recommended (New & Singholka 1985). Optimal temperatures and pH ranges are 26–31°C and pH 7.0–8.5, respectively.

Larvae receive a combination of live and prepared feeds. Newly hatched *Artemia* are provided (to maintain a concentration between two and five nauplii ml^{-1}) along with egg custard and minced flesh of fish such as tuna. Mussel and fish flesh are sometimes incorporated within egg custards and soybean curd is also used. In all cases, prepared feeds are sieved through a mesh prior to feeding to ensure that appropriately small particles are created which can be easily maintained in suspension in the rearing tank. As a rough guide, for each production cycle a total of 1.2–1.6 kg of prepared feed may be needed per cubic metre of culture (stocked at 30–50 larvae ℓ^{-1}) (New & Singholka 1985). Daily feeding levels are adjusted to satisfy the demand of the larvae yet avoid water pollution through over-feeding. Commercially produced diets for crustacean larvae are available but their use has not become standard in *Macrobrachium* hatcheries (New 1990).

Aeration is used to maintain dissolved oxygen concentrations, and if sufficiently strong helps keep particles of food and detritus suspended in the water column. This enables much particulate waste to be flushed away with outgoing water during a water exchange and slows down the accumulation of organic material within the larvae rearing tank. Nevertheless, in many systems it is necessary to siphon out settled waste at least once every two days. Daily water exchange rates, which normally range between 10 and 100%, are increased as the prawn larvae grow. Recirculating water systems incorporating rotating biodiscs or other biological filters have been developed to operate where supplies of good quality fresh or seawater are limited or where heat needs to be conserved (AQUACOP 1983; Cange *et al.* 1987)

Larvae survival rates usually average between 30 and 60% and result in yields of between 10 and 35 post-larvae ℓ^{-1}. Production of 90 post-larvae ℓ^{-1} and survival rates above 90% have been achieved in small intensively managed cylindro-conical rearing vessels (of 0.8 and 2 m^3 capacity) (AQUACOP 1982). However, for most hatcheries, high survival rates are not essential to maintain a viable production level because supplies of berried females are usually abundant. The use of antibiotics to control outbreaks of bacterial disease is reported to be widespread (New 1990) (Section 11.4.4).

Before metamorphosis into post-larvae, *M. rosenbergii* larvae pass through 11 successive moult stages over a period of, usually, 18 to 23 days. Sometimes, however, adverse nutrition or water quality can prolong larval life and lengthen the time the population is metamorphosing. This increases the opportunity for cannibalism and can greatly reduce survival. Post-larvae have the appearance of miniature adult prawns and switch from a planktonic to a largely bottom-dwelling existence. They also acquire the euryhaline capability characteristic of the species, and can be safely exposed to a drop in salinity in preparation for the freshwater conditions they will encounter during nursery rearing and on-growing.

7.3.4 Nursery

The nursery rearing of prawn post-larvae is usually carried out in concrete tanks, or in ponds which may have concrete or brick walls or an entirely earthen construction. To

Plate 7.3 Handpicking the last last few juvenile freshwater prawns from a nursery pond in Mauritius. Netting material, previously suspended within the pond to provide additional substrate, has been removed prior to harvesting and can be seen piled on the far embankment.

accommodate the aggressive and territorial behaviour of juveniles and reduce losses through cannibalism, additional substrate material may be provided in tanks in the form of draped netting or in ponds by rows of netting arranged lengthwise (Plate 7.3).

In Thailand, post-larvae are stocked in concrete tanks at densities between 1000 and 5000 m^{-2} and held for periods of seven to 28 days (New & Singholka 1985). Earthen nursery ponds in the same country may be stocked at lower densities of 20–25 post-larvae m^{-2} and harvested after longer periods of 75–90 days (New 1988). In West Malaysia post-larvae are stocked at 100 m^{-2} for between 28 and 42 days (Lee 1982). Higher densities of 800 post-larvae m^{-2} have been reported in Taiwan for rearing cycles lasting 30–50 days in ponds of around 400 m^2 (Chen 1990).

Prawn post-larvae readily accept the same type of prepared feeds that are used for on-growing. For feeding young post-larvae in concrete tanks, the use of floating diets, such as those used for catfish, can help in the visual assessment of food consumption rates and thereby assist in adjusting feeding levels to demand (New & Singholka 1985).

In Mediterranean and warm temperate climates the use of enclosed nurseries for the production of juveniles in early spring enables maximum benefit to be obtained from a limited on-growing season. Systems typically rely on greenhouses to raise temperatures, and water is usually biologically filtered and recycled. In enclosed system trials in south Carolina, USA, cylindrical fibreglass and rectangular concrete or aluminium tanks (3.8–9.5 m^3 water capacity) were used, equipped with strip layer habitats of plastic netting or fibreglass screen. The use of these habitats permitted densities of over 6000 post-larvae m^{-2} (of tank bottom) although the most consistent survival rates (60–90%) were generally obtained in the range 1200–5400 post-larvae m^{-2}. At mean temperatures of 25.2–28.6°C, post-larvae stocked at a size of 0.01 g reached 0.2–0.44 g after 77–92 days. Feed was supplied three times per day *ad libitum*, and consisted of a compounded diet, chopped fish, squid, spinach and egg. Water plants (*Egeria* sp. and *Lemna* sp.) provided supplemental food and substrate, and assisted in water purification (Smith *et al.* 1983). Both greenhouses and geothermal heat have been employed to raise water temperatures in Israeli prawn nurseries (Cohen & Barnes 1982).

7.3.5 On-growing

7.3.5.1 Extensive

The operation of extensive prawn ponds that rely entirely on natural productivity is not widely reported. Such ponds however are known to be capable of yielding around 200–300 kg ha^{-1} annually; more when fertilizers are applied.

7.3.5.2 Polyculture

Although commercial polyculture involving freshwater prawns is quite rare, *Macrobrachium rosenbergii* has repeatedly shown potential in mixed species cultures both as the main crop or as a valuable secondary crop in fish ponds. It has been noted that it integrates well with tilapia, carps (common, silver, grass) or mullet in Israel (Cohen 1984) and has led to efficient utilization of pond resources when cultured with fish species, such as milkfish (*Chanos chanos*) and grey mullet (*Mugil cephalus*) in Taiwan (Liao & Chao 1982).

In Israel typical stocking figures have been given as 0.5–1.5 prawns m^{-2} combined with about 1.2 fish m^{-2} (carp and Tilapia) with resulting yields of 6700–10 100 kg of fish ha^{-1} per crop and 220–780 kg ha^{-1} per crop of large prawns (averaging between 45 and 90 g). Food conversion ratios for the whole system were between 0.87:1 and 1.7:1, and conveniently, prawns and fish could be harvested and processed with the same equipment (Cohen 1984).

Experiments have demonstrated the feasibility of prawn polyculture with catfish, (providing prawn juveniles are stocked large enough to avoid predation) (Cohen 1984; Lamon & Avault 1987), and with bait fish, golden shiners (*Notemigonus crysoleucas*) (Perry & Tarver 1987). Low yields (13–16 kg ha^{-1}) of *M. lanchesteri* have been obtained with rice in paddies in the Philippines (Guerrero *et al.* 1982). In extensive polyculture trials, relying only on natural productivity enhanced by swine manure, Malecha *et al.* (1981) stocked four carp species at a combined density of 0.55 fish m^{-2} with freshwater prawns at 7.9 m^{-2}. Resulting prawn yields averaged 322 kg ha^{-1}, although mean size was only 12.7 g.

7.3.5.3 Semi-intensive

Nearly all prawn farming operations can be categorized as semi-intensive. Typically, annual yields of between 1 and 5 mt ha^{-1} are achieved by stocking prawns at densities in the range 5–20 m^{-2} and applying supplemental feeds.

The on-growing ponds are usually of earthen construction and range in size from a few hundred square metres up to 2 or 3 ha. Larger units are difficult to harvest efficiently (Section 7.3.6). Various materials such as bamboo, bricks and pipes have been used in production ponds to provide shelters for prawns and reduce aggressive interaction. The benefit gained, however, is usually outweighed by the practical disadvantages of increased labour and interference with harvesting. Most operators therefore favour uncluttered pond beds or, in a few cases, adopt the approach used in nurseries, of using suspended netting which can be easily removed before harvesting takes place.

Water is obtained from rivers, wells, irrigation channels or reservoirs and pumped or gravity-fed to ponds individually or sequentially. Some ponds, for example in Hawaii, operate with flow-through systems, but in Thailand and elsewhere water addition rarely represents more than just topping-up to replace evaporative or seepage losses. On-growing ponds in Honduras have been operated with eight to 12 water exchanges per year, which only represented a daily exchange rate of 2.2–3.3% of pond volume, approximately half of which compensated for evaporation (Wulff 1982a).

In experimental on-growing trials *Macrobrachium* has been shown to grow and survive well in water up to a mean salinity of 16 ppt. (Popper and Davidson 1982; Smith *et al.* 1982), and some farmers take advantage of its euryhaline nature to utilize tidal water with a fluctuating but low salinity. Nevertheless, *M. rosenbergii* is predominantly cultured in freshwater, and a reliable supply is normally a prerequisite for prawn farms. To achieve optimum growth and survival, water temperatures in the range 26–31°C are recommended. In temperate zones, short growing seasons can only support single annual crops unless some source of thermal effluent or geothermal water can be exploited. Such heat sources have demonstrated potential to support prawn culture (New 1988), though density limitations due to aggression preclude intensive cultures. Optimum water hardness levels appear to lie between 40 and 100 mg $CaCO_3$ ℓ^{-1} (New & Singholka 1985). Some very hard well waters (registering 305–638 mg $CaCO_3$ ℓ^{-1}) have been found not only to depress growth but also to lead to prawns becoming encrusted with bryozoans and protozoa (Cripps & Nakamura 1979). Elsewhere, Bartlett & Enkerlin (1983) found that growth was not adversely affected by hardness levels between 940–1060 mg ℓ^{-1} $CaCO_3$, but the water they used had relatively low alkalinity. It therefore seems that the effects of hardness on prawn growth can be greatly influenced by the associated level of alkalinity in the water (Brown *et al.* 1991).

Optimal stocking densities are partly determined by water exchange rate. Ten prawns m^{-2} is considered suitable for ponds in Hawaii with a more or less constant flow-through of water (Fujimura 1982) while 3–5 m^{-2} is more appropriate for many ponds in Thailand which contain more static and hence warmer water. Stocking strategies must also be closely tied to the harvesting regime (Sections 7.3.6 and 7.3.7).

Effective predator control usually leads to survival rates of 50% or more during on-growing. In addition to basic precautions (Section 8.3.5.2), in south-east Asia small, 60 cm high netting fences may need to be fixed around ponds to keep out invading catfish and snakehead fish. In addition jars sunk into pond banks can serve as crab traps. Some *Macrobrachium* ponds are operated on a continuous basis and are rarely drained, so the eradication of fish predators cannot be performed as part of pond preparation (Section 8.3.5.1) and must be done during seine harvests. The stocking of nursery reared juveniles helps to reduce early losses to predation.

The growth of vegetation is usually promoted on embankments to reduce erosion, even though this can reduce the efficiency of seine harvesting. Plant growth within the ponds has the same drawback, but can be prevented by filling ponds to a depth of around 1 m and by maintaining a bloom of algae sufficiently dense to shade the bottom. In semi-intensive *Macrobrachium* ponds, fertilizer is rarely needed to maintain algae growth since nutrient release through the breakdown of supplemental feeds is usually more than adequate for this purpose.

Aeration equipment is sometimes used on prawn farms in Taiwan to permit increased stocking densities and productivity, and it can be valuable as a temporary response to critically low oxygen concentrations when the alternative of rapid water renewal is unavailable (New & Singholka 1985). Aspects of water quality management in crustacean ponds are considered in Section 8.3.

A range of different fresh foods, prepared diets and commercially produced pelleted feeds are used for prawn on-growing. The latter include pellets produced for chickens, pigs and shrimp as well as prawns, though non-aquatic animal feeds usually display poor stability in water. Various recipes exist for the preparation of feeds at the farm, which take advantage of locally available foodstuffs such as rice bran, cooked broken rice, chopped trash fish, fishmeal, cornmeal, soybean meal or cake, meat, bonemeal, alfalfa and brewery waste. New and Singholka (1985) give various recipes for producing moist pellets containing between 15 and 35% protein, which incorporate binding agents such as high-gluten wheat or guar gum to provide water stability (protein requirements for *M. rosenbergii* are lower than for some penaeid shrimp). After grinding, mixing and extrusion, these diets are sun-dried to a moisture content of less than 10% in order to extend storage life. The direct application of fresh foods such as chopped trash fish, chicken carcasses, animal bones with flesh remains, soft shelled snails, and mussels is reported along with cautionary notes about the dangers of severe pollution associated with overfeeding these items.

In many artisanal operations, feeding levels are determined solely in response to water quality changes and observations of the quantity of uneaten food remaining in shallow pond margins (ideally they should also be based on estimates of the crop biomass (Sections 8.3.4.1 and 8.3.5.4)). If sufficient care is taken to adjust feeding rates to suit demand, the worst effects of pollution are avoided, but if dissolved oxygen concentrations (usually measured at their lowest just before dawn) become critically low, feeding sometimes needs to be suspended altogether until conditions improve. Food conversion ratios between 2 and 4:1 for dry diets and 7 and 9:1 for wet feeds such as trash fish, are reported (Wulff 1982b; New & Singholka 1985).

7.3.5.4 Intensive

Although *M. rosenbergii* is not among species suited to intensive or super-intensive culture, largely because of its aggressive territorial behaviour, some attempts have been made to assess its performance in high density on-growing trials.

In Great Britain, very small (0.62 m^2) indoor tanks stocked with 162 juveniles m^{-2} yielded 486 g m^{-2} (equivalent to 4860 kg ha^{-1}) after 112 days of culture. Prawns averaged only 7 g at harvest and the survival rate was 45% (Wickins & Beard 1978). Although these results were comparable with those obtained under the same conditions with some species of penaeid shrimp, they were notably inferior to the productivity of 1908 g m^{-2} and the survival rate of 91% achieved with *Penaeus monodon*.

An outdoor concrete tank of 173 m^2 in South Carolina, USA, stocked with nursed juveniles at 83 m^{-2} (mean size 1.0 g) yielded a crop equivalent to 4700 kg ha^{-1} after 110 days, but the mean harvest weight of prawns again was low at 8.5 g. In a parallel trial, larger animals averaging 16.5 g were produced using a lower stocking density of 32 individuals m^{-2} and a culture duration of 138 days. However, total yield under this regime was reduced to 3828 kg ha^{-1}. In both trials, survival rates were high at 66.5 and 73.2% and the tanks were equipped with a compressed air supply and artificial substrate in the form of draped netting. Using a compounded diet containing 25% protein, feed conversion ratios of 2.3:1 and 1.4:1 were obtained in the higher and lower density trials respectively (Sandifer *et al.* 1982).

New (1988) has noted that pens are sometimes used for commercial prawn farming in Thailand, but no indication of yields was given.

7.3.6 Harvesting

Harvesting techniques rely on draining or seining or a combination of both. Seining employs a net (about 2 m high when stretched vertically) extended across the pond and drawn along the banks usually by a team of workers. Wulff (1982a) described a method using two such nets positioned across the centre of the pond and drawn independently towards opposite ends. This process shortens the duration of each haul and reduces stress and physical damage to the catch. Narrow pond design (max. 30–50 m) facilitates the seining procedure and, as a result, ponds to be harvested in this way rarely exceed 1 ha in size.

Losordo *et al.* (1986) describe a mechanised seine harvest system incorporating a boom mounted on a tractor. In trials in heavily silted ponds a tractor operator and three workers were able to collect an average of 63.5% of the marketable prawns (sizes not given).

During drain harvesting by gravity, prawns are carried towards the pond exit gate where they collect in a sump, net bag or net enclosure. While the prawns are being concentrated they can be manually scooped-up or transferred by pump to containers on the pond bank, but great care must be taken to avoid damage. Large ponds (> 2–3 ha) do not usually drain sufficiently well for efficient drain harvests. Prawns remaining in the pond require laborious netting or hand-picking which must be performed quickly if the animals are to be collected in marketable condition. Unfortunately *Macrobrachium* have a tendency to swim upstream against any flow of water intended to sweep them to the collection point.

7.3.7 Stocking and harvesting regimes

The management of stocking and harvesting in freshwater prawn ponds is greatly complicated by the heterogeneous growth rates characteristic of cultured populations. Within a pond stocked with a single batch of post-larvae some animals have been observed to grow fifteen times faster than others (Barnes 1982). In one typical example, after six months of on-growing, prawns reached an average size of 47.8 g, yet ranged from 10 g to 110 g (Menasveta & Piyatiratitivokul 1982).

The great bulk of the size heterogeneity is found among the male prawns and has been shown to arise from group interactions within a population, rather than from differences in genetic growth potential. Early maturing males develop large blue claws (BC), take dominant positions in a hierarchy and at the same time undergo a reduction in growth rate. Meanwhile other males, characterized by smaller orange claws (OC), grow more rapidly and eventually transform at a single moult to become even larger, newly dominant BC males. In addition, a significant proportion of small males (SM) experience stunted growth and remain at the bottom of the dominance hierarchy, and although they possess the potential to become OC and eventually BC males, they remain as runts unless the density of the larger and more aggressive BC and OC males is reduced. The practical result of this is that the growth potential of much of the prawn population in a pond can only be released by repeatedly harvesting the largest animals.

In straightforward batch culture, a single stocking operation is followed by a single total harvest at the end of the on-growing period and then the pond is drained and prepared for the next cycle. This approach, however, produces a wide size range among the prawns and complicates the processes of size grading and marketing. As a result the alternative strategies of multiple harvesting and continuous culture are often employed.

The multiple harvest approach involves a single stocking followed about five months later by a series of partial seine harvests at monthly or fortnightly intervals. By the use of a seine with an appropriate mesh size (3.8–5 cm), the larger prawns (>25–40 g) are selected while the smaller ones remain to continue growing at a reduced density. After about eight months the pond is drained and all remaining prawns are harvested. This approach is efficient and popular in Thailand where it results in the production of manageable quantities of prawns at regular intervals and maximizes the productivity of a growing period often limited by the onset of the dry season. In situations where the growing season is not limited, after harvesting is completed the pond can be immediately prepared for treatment, refilling and restocking.

Continuous culture, which involves multiple stocking as well as multiple harvesting, is often chosen to take advantage of a year-round water supply and growing season. In theory it has the potential to maximize annual output from a prawn farm. Proponents of the approach have recorded annual yields of up to 4000 kg ha^{-1} and Taiwanese operations are reported to produce even higher yields when good pond management is applied (Liao & Chao 1982). Hawaiian ponds managed for 'continuous' culture are stocked with 16–22 prawns m^{-2} and can be expected to produce around 2500 kg ha^{-1} annually if they have been in operation for more than three years (Shang & Mark 1982).

A theoretical continuous culture schedule for a 1 ha prawn pond is outlined in New and Singholka (1985), in which a stocking density of around 15 prawns m^{-2} is maintained for a total cycle lasting over 36 months. Monthly harvesting produces yields of 1700 kg for the first year and 3375 kg yr^{-1} thereafter until the ponds are drained and revitalized after five years of continuous use (Section 8.3.5.1). Interestingly though, Wulff (1982b), who operated ponds in Honduras under both simple batch culture and continuous culture, noted no significant difference in annual yields between the two systems. He produced 3000 kg ha^{-1} of prawns averaging 30 g under either regime.

The average harvest size of prawns is basically a function of the duration and density of the rearing process. For example, to produce prawns averaging 70 g in an eight month on-growing season in Thailand, a stocking density of 5 m^{-2} is recommended (New & Singholka 1985). In temperate climates with only 6–7 months of growing season, D'Abramo *et al.* (1989) recommended stocking densities of four prawns m^{-2} or fewer in order to harvest animals with a mean weight of 25 g or more and at the same time minimize stocking and feeding costs. They also recorded that the proportion of stunted males was much reduced at low densities. At a stocking density of 4.0 prawns m^{-2} the percentage of individuals in the small male category was 4.5%, but at a stocking density of 11.9 m^{-2} it was 21.8%.

To reduce size heterogeneity at harvest and increase overall yields, the possibility of culturing monosex populations and size-graded sub-populations has been investigated (Ra'anan & Cohen 1983; Cohen *et al.* 1988). Despite some encouraging results, the practicalities of dividing juvenile prawns on the basis of size or sex may not be economically attractive, and attention is now being paid to producing all female or largely female broods by mating females with sex reversed females or 'neomales' (Section 12.3).

7.3.8 Processing

Freshwater prawns are usually sold to local markets either live or whole on ice. Some farmed product is exported from Thailand in the form of frozen tails, although the processing yield of headless prawns, at around 40%, is inferior to that of marine shrimp (57–68%). Large claw (BC) males yield 5–8% less meat than OC males. More information on the processing and marketing of prawns is provided in Section 3.3.2.

7.3.9 Hatchery supported fisheries, ranching

Many wild stocks of *M. rosenbergii* and other species of freshwater prawn have been depleted through pollution, overfishing and the interruption of migration routes with irrigation schemes. Hence, in some cases, the possibility exists of augmenting fisheries through the release of hatchery reared juveniles. New and Singholka (1985) note some success with this in Thailand in providing additional food and income for local fishermen. In one stocking programme lasting three years, three million juveniles were released in a lake measuring 410 km^2 × 15 m deep. Four thousand families fished the lake and recapture rates were estimated at 2% (NACA 1986).

7.3.10 References

AQUACOP (1982) Mass production of juveniles of freshwater prawn *Macrobrachium rosenbergii* in French Polynesia: Predevelopment phase results. In *Proc. Symp. Coastal Aquaculture 1982*, Pt. 1, pp. 71–5. Mar. Biol. Association of India, Cochin.

AQUACOP (1983) Intensive larval rearing in clear water of *Macrobrachium rosenbergii* (de Man, Anuenue stock) at the Centre Oceanologique du Pacifique. In *Handbook of Mariculture, Vol. 1, Crustacean aquaculture* (Ed. by J.P. McVey), pp. 179–88. CRC Press, Boca Raton, Florida.

Barnes A. (1982) Discussion session – research. In Giant prawn farming (Ed. by M.B. New). *Dev. Aquacult. Fish. Sci.* **10**, 449–68.

Bartlett P. & Enkerlin E. (1983) Some results of rearing giant prawn, *Macrobrachium rosenbergii* in asbestos asphalt ponds in hard water and on a low protein diet. *Aquaculture*, **30** (1–4) 353–6.

Brown J.H., Wickins J.F. & Maclean M.H. (in press). The effect of water hardness on growth and carapace mineralization in juvenile freshwater prawns, *Macrobrachium rosenbergii* de Man. *Aquaculture*, **95** (3/4) 329–45.

Cange S.W., Pavel D.L., Lamon L.S. & Avault J.W. Jr. (1987) Development of larval rearing systems for the Malaysian prawn *Macrobrachium rosenbergii* in southern Louisiana. *NOAA Tech. Rep. 47*, pp. 43–9.

Chen L.-C. (1990) *Aquaculture in Taiwan*. Fishing News Books, Blackwell Scientific Publications, Oxford.

Cohen D. (1984) Prawn production in catfish ponds: proposed strategy and test trials. *Aquaculture Magazine*, **10** (2) 14–20.

Cohen D. & Barnes A. (1982) The *Macrobrachium* programme of the Hebrew University, Jerusalem. In Giant prawn farming (Ed. by M.B. New). *Dev. Aquacult. Fish. Sci.*, **10**, 381–5.

Cohen D., Sagi A., Ra'anan Z. & Zohar G. (1988) The production of *Macrobrachium rosenbergii* in monosex populations. III. Yield characteristics under intensive monoculture conditions in earthen ponds. *Bamidgeh*, **40** (2) 57–63.

Cripps M.C. & Nakamura R.M. (1979) Inhibition of growth of *Macrobrachium rosenbergii* by calcium carbonate water hardness. *Proc. World Maricult. Soc.*, **10**, 575–80.

D'Abramo L.R., Heinen J.M. Randall Robinette H. & Collins J.S. (1989) Production of the freshwater prawn *Macrobrachium rosenbergii* stocked at different densities in temperate zone ponds. *J. World Aquacult. Soc.*, **20** (2) 81–9.

Fujimura T. (1982) Discussion session – practical farming. In Giant prawn farming (Ed. by M.B. New). *Dev. Aquacult. Fish. Sci.*, **10**, 469–98.

Guerrero L.A., Circa A.V. & Guerrero R.D. III (1982) A preliminary study on the culture of *Macrobrachium lanchesteri* (de Man) in paddy fields with and without rice. In Giant prawn farming (Ed. by M.B. New). *Dev. Aquacult. Fish. Sci.*, **10**. 203–6.

Lamon M.S. & Avault Jr., J.W. (1987) Polyculture stocking strategies for channel catfish, *Ictalurus punctatus*, and the prawn, *Macrobrachium rosenbergii*, using one catfish density and three prawn densities with two prawn sizes. Abstract in *J. World Aquacult. Soc.*, **18** (1) 23A.

Lee C.L. (1982) Discussion session – practical farming. In Giant prawn farming (Ed. by M.B. New). *Dev. Aquacult. Fish. Sci.*, **10**, 469–98.

Liao I.-C. & Chao N.-H. (1982) Progress of *Macrobrachium* farming and its extension in Taiwan. In Giant prawn farming (Ed. by M.B. New). *Dev. Aquacult. Fish. Sci.*, **10**, 357–9.

Losordo T.M., Wang J.-K., Mark J.B. & Lam C.Y. (1986) A mechanised seine harvest system for freshwater prawns. *Aquacultural Engineering*, **5** (1) 1–16.

Macintosh D.J. (1987) *Aquaculture production and products handling in ASEAN*. ASEAN Food Handling Bureau, Kuala Lumpur, Malaysia.

Malecha S.R. (1983) Commercial pond production of the freshwater prawn, *Macrobrachium rosenbergii*, in Hawaii. In *Handbook of Mariculture, Vol. 1, Crustacean aquaculture* (Ed. by J.P. McVey), pp. 231–60. CRC Press, Boca Raton, Florida.

Malecha S.R., Buck D.H., Baur R.J. & Onizuka D.R. (1981) Polyculture of the freshwater prawn *Macrobrachium rosenbergii*, Chinese and common carps in ponds enriched with swine manure. 1. Initial trials. *Aquaculture*, **25** (2/3) 101–16.

McVey J.P. (Ed.) (1983) *Handbook of mariculture, Vol. 1, crustacean aquaculture.* CRC Press, Boca Raton, Florida.

Menasveta P. & Piyatiratitivokul S. (1982) Effects of different culture systems on growth, survival and production of the giant freshwater prawn (*Macrobrachium rosenbergii* de Man). In Giant prawn farming (Ed. by M.B. New). *Dev. Aquacult. Fish. Sci.*, **10**, 175–89.

NACA (1986) *Giant freshwater prawn breeding and farming in Thailand: an introduction.* FAO Network of Aquaculture Centres in Asia, NACA TV Video production.

New M.B. (Ed.) (1982) Giant prawn farming. *Dev. Aquacult. Fish. Sci.*, **10**, 1–532.

New M.B. (1988) Freshwater prawns: status of global aquaculture, 1987. *NACA Technical Manual 6*, World Food Day 1988. Publication of the Network of Aquaculture Centres in Asia, Bankok, Thailand.

New M.B. (1990) Freshwater prawn culture: a review. *Aquaculture*, **88** (2) 99–143.

New M.B. & Singholka S. (1985) Freshwater prawn farming. A manual for the culture of *Macrobrachium rosenbergii. FAO Fish. Tech. Pap. 225, Rev. 1.*

Perry W.G. & Tarver J. (1987) Polyculture of *Macrobrachium rosenbergii* and *Notemigonus crysoleucas. J. World Aquacult. Soc.*, **18** (1) 1–5.

Popper D.M. & Davidson R. (1982) An experiment in rearing freshwater prawns in brackishwater. Abstract in Giant prawn farming (Ed. by M.B. New). *Dev. Aquacult. Fish. Sci.*, **10**, 173.

Ra'anan Z. & Cohen D. (1983) Production of the freshwater prawn *Macrobrachium rosenbergii* in Israel: II. Selective stocking of size subpopulations. *Aquaculture*, **31** (2/3/4) 369–79.

Sandifer P.A., Smith T.I.J., Stokes A.D. & Jenkins W.E. (1982) Semi-intensive grow-out of prawns (*Macrobrachium rosenbergii*): Preliminary results and prospects. In Giant prawn farming (Ed. by M.B. New). *Dev. Aquacult. Fish. Sci.*, **10**, 161–72.

Shang Y.C. & Mark C.R. (1982) The current state-of-the-art of freshwater prawn farming in Hawaii. In Giant prawn farming (Ed. by M.B. New). *Dev. Aquacult. Fish. Sci.*, **10**, 351–6.

SICA (1988) *General information about our hatchery.* Trade brochure. Sica Guadeloupéene D'Aquaculture, Les Plaines, Pointe Noire, Guadeloupe, FWI.

Smith T.I.J. & Wannamaker A.J. (1983) Shipping studies with juvenile and adult Malaysian prawns *Macrobrachium rosenbergii* (de Man). *Aquacultural Engineering*, **2** (4) 287–300.

Smith T.I.J., Sandifer P.A. & Jenkins W.E. (1982) Growth and survival of prawns, *Macrobrachium rosenbergii*, pond reared at different salinities. In Giant prawn farming (Ed. by M.B. New). *Dev. Aquacult. Fish. Sci.*, **10**, 191–202.

Smith T.I.J., Jenkins E.W. & Sandifer P.A. (1983) Enclosed prawn nursery systems and effects of stocking juvenile *Macrobrachium rosenbergii* in ponds. *J. World Maricult. Soc.*, **14**, 111–25.

Wickins J.F. & Beard T.W. (1978) *Prawn culture research.* Lab. Leafl. (42), MAFF Direct. Fish Res., Lowestoft.

Wulff R.E. (1982a) Practical farming discussion session. In Giant prawn farming (Ed. by M.B. New). *Dev. Aquacult. Fish. Sci.*, **10**, 469–98.

Wulff R.E. (1982b) The experience of a freshwater prawn farm in Honduras, Central

America. In Giant prawn farming (Ed. by M.B. New). *Dev. Aquacult. Fish. Sci.*, **10**, 445–8.

7.4 Other caridean shrimps and prawns

7.4.1 Species of interest

Spot prawn, (*Pandalus platyceros*); common prawn, (*Palaemon serratus*); freshwater prawns and shrimps, *Atya* spp.; ornamental carideans.

7.4.2 Broodstock, larvae culture and nursery

The species that has attracted the most attention in this group is *Pandalus platyceros* (Rosenberry 1987) but it is a temperate water species and cannot compete with the penaeids or *M. rosenbergii* in terms of growth rate or with penaeids in terms of meat yield (Kelly *et al.* 1977; Wickins 1982). One particular difficulty with *P. platyceros* and indeed many of the other large carideans such as *Sclerocrangon boreas*, all of which have low fecundity, would be the need to maintain large broodstock, incubation and nursery facilities in support of a farming enterprise.

Techniques for broodstock maintenance, larvae culture and nursery are similar to those described for *Macrobrachium*, albeit conducted at lower temperatures, and are described further by Forster & Wickins (1972), Wickins (1972) and Wickins & Beard (1978).

7.4.3 On-growing

Attempts to grow *P. platyceros* in raceways in power station effluents were made in Britain in the early 1970s (Wickins 1982). The best results suggested that only one crop per year of 6 to 8 g prawns could be obtained and that culture was unlikely to be profitable. It has been suggested that the species has promise for polyculture, perhaps with abalone (Kelly *et al.* 1977) or in salmon cages (Rensel & Prentice 1979; Oesterling & Provenzano 1985), but a major problem with the latter was the need to stock prawns large enough to be retained by the mesh used for the salmon.

7.4.4 Other prospects

Smaller species that have been considered for cultivation because of their value as gourmet food items, bait for sport anglers, laboratory bioassay animals or as ornamental shrimp in display aquaria include *Palaemon serratus*, (Wickins 1982), *Atya lanipes* and *A. scabra* (Cruz-Soltero & Dallas 1990), and *Palaemonetes kadiakensis*, *P. pugio*, *P. paludosus*, *Hippolysmata wurdemanni* and *Stenopus hispidus* (Oesterling & Provenzano 1985). Although generally small in size, atyids differ from other cultivable shrimps and prawns in that the post-larval stages are able to filter particulate material from the water column. The largest atyid shrimp are *A. gabonensis* of west Africa and *A. innocus* from the West Indies, which may attain 90 to 124 mm total length. Ornamental prawns are easy to culture and very valuable. The potential market may be substantial but there may be a need for a marketing campaign, and an increase in supply may depress prices. Possible reasons why no commercial culture is performed include the high capital cost of a hatchery and the high labour costs of larval rearing and food culture when compared to the ease of transporting specimens through the established channels of the ornamental fish trade.

7.4.5 References

Cruz-Soltero S. & Dallas E.A. (1990) Status report on research with *Atya lanipes* and *A. scabra* in Puerto Rico. *Abstract from 42nd Ann. Mtg. Gulf and Caribbean Fish. Inst. Larviculture and Artemia Newsletter*, (16) 87–8. State University of Ghent, Belgium.

Forster J.R.M. & Wickins J.F. (1972) *Prawn culture in the UK: its status and potential.* Lab. Leafl. (27), MAFF Direct. Fish. Res., Lowestoft.

Kelly R.O., Haseltine A.W. & Ebert E.E. (1977) Mariculture potential of *Pandalus platyceros* Brandt. *Aquaculture*, **10** (1) 1–16.

Oesterling M.J. & Provenzano A.J. (1985) Other crustacean species. In *Crustacean and mollusk aquaculture in the United States* (Ed. by J.V. Huner & E. Evan Brown), pp. 203–34. AVI Inc., Westport, Connecticut.

Rensel J.E. & Prentice E.F. (1979) Growth of juvenile spot prawn, *Pandalus platyceros*, in the laboratory and in net pens using different diets. *US Fish Wildl. Serv. Fish. Bull.*, **76** (4) 886–90.

Rosenberry R. (1987) Western Aquaculture Enterprises. *Aquaculture Digest*, **12** (11) 1–2.

Wickins J.F. (1972) Experiments on the culture of the spot prawn *Pandalus platyceros* Brandt and the giant freshwater prawn *Macrobrachium rosenbergii* (de Man). *Fish. Invest., London, Ser.* 2 **27** (5).

Wickins J.F. (1982) Opportunities for farming crustaceans in western temperate regions. In *Recent advances in aquaculture* (Ed. by J.F. Muir & R.J. Roberts), pp. 87–177. Croom Helm, London.

Wickins J.F. & Beard T.W. (1978) *Ministry of Agriculture, Fisheries and Food prawn culture research.* Lab. Leafl. (42), MAFF Direct. Fish. Res., Lowestoft.

7.5 Crayfish: USA

7.5.1 Species of interest

Red swamp crayfish (*Procambarus clarkii*); white river crayfish (*P. acutus*); bait crayfish, *Orconectes* spp. The signal crayfish (*Pacifastacus leniusculus*) is also cultured to a small extent but is considered more fully under Europe (Section 7.6). Of the two crayfish raised for the table, *P. clarkii* makes up 85% of production and in many respects the 'farms' more closely resemble well-managed, private fisheries than aquaculture operations. Both wild and farmed crayfish may also be used as bait by sport anglers.

This section includes information from trade articles and technical papers. It may be extended through reference to Huner & Avault (1981); LaCaze (1981); Avault (1983); Avault & Huner (1985); Dellenbarger *et al.* (1988); Momot (1988); Brunson (1989); Huner (1989a,b); Roberts & Dellenbarger (1989).

7.5.2 Broodstock

Red swamp crayfish broodstock are obtained from natural fisheries or from managed ponds. Most farmers only stock once as self-maintaining populations provide the basis for most production. The adults are transported in sacks or plastic trays by road, and stocking of broodstock into new ponds occurs from April to June at around 20–65 kg ha^{-1}, animals being evenly distributed all along levees at a sex ratio of about 1:1. The animals burrow well below water level and after two weeks (longer if no old burrows are present) the ponds are drained leaving the crayfish submerged in their burrows.

7.5.3 Hatchery and nursery

Procambarus clarkii breeds year round at the high temperatures in the deep south (Louisiana) but less frequently further north. Hatchery production or stocking with known numbers of juveniles is neither practised nor required on a commercial scale. It is likely that photoperiod control could extend the breeding season in *Orconectes* spp.

7.5.4 On-growing

7.5.4.1 Natural/extensive

The crayfish in the southern United States are produced primarily in Louisiana where extensive on-growing may be grouped into four categories:

(1) Marsh and swamp ponds (up to 100 ha in area) which exist by the coast are filled and drained by pumping. Typically they are on soils with a high peat content and contain slightly brackish water. The ponds are filled in autumn and drained by the following June. Oxygen depletion is a major problem and circulation lanes are mown through the vegetation prior to flooding, to assist water movement and harvesting. The ponds are flooded at intervals to gradually increasing depths to discourage decomposition of plant material. Flooding is however delayed in hot weather.

(2) Wooded ponds are similar to the above but conditions are poorer as the water circulation is reduced by the trees and shrubs. Annual yields from marsh and wooded ponds are about 200–600 kg ha^{-1}.

(3) Rice fields where the banks are raised from about 0.1 to 0.3–0.5 m. These may be operated by double cropping rice and crayfish. Rice is planted in March/April but care must be taken if fungicides and herbicides are applied to protect the rice since many are toxic to crayfish; it is likely that all insecticides are toxic to crayfish. The rice is harvested after 100–120 days and water that is on the rice field to reduce weed growth is drained and the field re-flooded in mid-September to mid-October. The crayfish emerge and release young into the rice stubble that should be left standing. Trapping begins in December and is continued until April when the ponds are drained and rice planted again. Alternatively, trapping may continue until June when soybeans are planted instead of rice. Annual yields range from 1–2 mt ha^{-1} with this method.

(4) Open ponds of 8–20 ha are typically built in low lying areas of marginal agricultural quality and may look as if they are full of grass. However, they often

have boat or trapping lanes which aid water circulation. The soil is typically heavy clay and ponds with smooth bottoms can be readily constructed. It may be necessary to place anti-seep collars of metal, fibreglass or plastic round drainpipes to reduce the risk of leaks from some ponds. Feeding is effective but adds cost and is not much practised. Instead, volunteer or intentionally cultivated cover crops (grasses, rice) are grown in the dry summer season while crayfish are quiescent or spawning in their burrows, to provide the basis for detritus in the food chain. When the vegetation decomposes in autumn the ponds are flushed with water. As a forage rice is better than many natural grasses, particularly varieties that produce a lot of foliage. This provides nutrition through decomposition but may be depleted by March when compounded feeds (cattle range pellets) or hay (rice, alfalfa or wheat straw) may be added. The majority of Louisiana farmers do not plant forage. Yields may be around 500–1500 kg ha^{-1}. About 65–70% of the crayfish production area in Louisiana falls in this category.

A typical production schedule is:

May	Stock 50–60 kg adults per ha.
May–June	Slowly drain pond over 1–2 weeks.
June–August	Plant rice or other vegetation for forage.
October	Re-flood the pond.
November–May/June	Continuous harvesting.
May	Drain pond and repeat cycle (de la Bretonne & Romaire 1989).

Predator control is not usually a problem; screens are sometimes used against fish, birds and mammals, while rotenone may be used against fish larvae entering with the water. In contrast to the European situation, no major disease or parasite problems have been reported.

Polyculture with fish is not widely practised, but in south Carolina crayfish and waterfowl are sometimes combined (Eversole & Pomeroy 1989). The shooting season is autumn which delays trapping and results in lower annual yields of crayfish.

7.5.4.2 *Intensive*

No more than two or three commercial US firms are believed to grow *P. clarkii* in controlled-environment conditions (Avault & Huner 1985) either for stocking or to adult size. Prices are too low to justify the costs of such systems although intensive indoor systems for soft-shell crayfish production seem profitable (Section 7.5.7).

7.5.5 Harvesting

Most harvesting is done by traps, with two types commonly used. The first is a 'stand-up' funnel placed at the pond bottom, with an open top above the surface. It is about 90 cm tall × 45 cm dia. & made of 20 mm poultry netting coated in plastic. The 'pillow' type of trap lays on the pond bottom and has one or more funnel entrances. Often 25 traps per ha are inspected daily for 100 days but the economics can be better if 75 traps per ha are inspected three times per day for 150 days (Pfister 1982

M.Sc. thesis cited in Avault & Huner 1985). Trapping accounts for 35–40% of total operating costs (Section 10.9.2) but in some years not enough of the population is caught and forage becomes depleted. This loss of food results in stunting or reduced growth. Although significant improvements in harvesting boats and machinery have been made since 1980, trap designs could still be improved. Unfortunately the prospects for using water flows to induce *P. clarkii* to concentrate in areas where they can be easily netted do not seem promising (Romaire & Lawson 1990). A useful review of harvesting techniques is to be found in Romaire (1989).

The fishery and culture industries together use some 15 000–30 000 tonnes of bait annually. Baits typically include gizzard shad, striped mullet and pollack but the development of artificial (pelleted) baits has allowed significant reduction in costs of food preparation and storage (Section 3.4.1). Harvesting often occurs in late November but traps are set periodically until late May to check the condition of the stock (for example to see if they are emaciated or ovigerous). Drop nets are also useful when checking the stock.

7.5.6 Transportation

Clean crayfish placed in 16–20 kg mesh sacks or plastic trays and held in a high humidity, cool container, may be transported alive for journeys lasting several days. Good air circulation around the sacks and minimal vibration will enhance survival rates.

7.5.7 Soft shelled crayfish

Soft shelled crayfish are prized as bait (*Procambarus* and *Orconectes*) and also as food (*Procambarus*). To keep the latter in perspective, 50 mt were marketed by 100 producers in 1988 in comparison with 60 000 mt of hard shelled crayfish. Huner (1988a) predicted that the potential total soft-shell food market in the USA was about 1500 mt. Indeed growth has been rapid in the past five years and there are now about 300 producers.

In essence, crayfish are held captive in shallow trays until they moult. They are then removed, processed and sold. Competent identification and careful handling of sexually immature, inter-moult crayfish of 70 mm total length or over is a necessary prerequisite of soft-shell crayfish production (Huner 1988b; Culley & Duobinis-Gray 1989; Homziak, 1989). An electro-trawl has been developed to catch soft- and paper-shell crayfish from ponds, since these do not feed and therefore enter traps less readily. Those caught up to 48 h after moulting are suitable for some markets.

Alternatively pre-moult and immature inter-moult animals are placed in 'shedding trays' at densities of 250 to 500, 10–20 g crayfish m^{-2} (about 5 kg m^{-2}). Typical tray systems described by Huner (1988a) and Culley & Duobinis-Gray (1989) comprised a series of trays (0.91 × 2.74 × 0.15 m deep; water depth 5 cm) which received water at 0.8–2.1 ℓ water kg^{-1} min^{-1}. Control over water quality may be improved through the incorporation of a recirculation system (Malone & Burden 1988). High protein feeds may be given at 1–3% of body weight per day in two to three feeds per day.

Two labour-saving intensive soft-shell systems have been proposed and were expected to be in operation in the 1988–9 season. (See US patents of Bodker 1984; Malone & Culley 1988).

The soft crayfish are identified by a change of colour and are removed from the trays after they have moulted. Three to four per cent of the population moult each day for the four to seven month season. It is also possible to hold them overnight in deionised or chilled water which prevents the shell from hardening and reduces the need for all-night collection.

7.5.8 Processing

Most crayfish are sold alive for boiling in seasoned water, although 30% of the Louisiana crop is processed, i.e. beheaded, peeled then packed fresh or frozen. The crop is placed in cooled sacks at 4–6°C for shipment. They may also be held in purging vats for 24–48 h to evacuate the gut (Moody 1989). In the case of soft shelled crayfish, the New Orleans and Washington markets take whole, newly moulted crayfish with only the gastroliths removed, which represents about 92% edible product. If the internal organs are also removed, about 72–82% remains. The soft shelled crayfish are sold frozen or in vacuum packs. Processing wastes are potentially valuable additions in crustacean diets but at present the resource seems under-utilized (Section 3.3.7).

7.5.9 *Orconectes* spp.

These are mainly small species cultured for bait (incidentally raised with fish in some areas). The adults are stocked into mud-bottomed ponds in autumn at 1500 to 2500 per ha or as berried females at 750–1200 per ha. The young hatch in spring and reach bait size by July when seining begins. The ponds may be fertilized with agricultural fertilizer (NPK ratio of 6-12-6 or 0-12-0 at 220 kg ha^{-1}) at three week intervals or as required. Some supplemental feeding is practised, for example with cracked corn and potatoes. Yields of 56 000–156 000 animals ha^{-1} have been recorded.

7.5.10 References

Avault J.W. Jr. (1983) Crayfish species plan for the United States: Aquaculture. In *Freshwater Crayfish 5*. (Ed. C.R. Goldman), pp. 528–33. AVI Inc. Westport, Connecticut.

Avault J.W. Jr. & Huner J.V. (1985) Crawfish culture in the United States. In *Crustacean and Mollusk Aquaculture in the United States* (Ed. by J.V. Huner & E. Evan Brown), pp. 1–61. AVI Inc. Westport, Connecticut.

Bodker J.E. Jr. (1984) *Method and apparatus for raising softshell crawfish*. U.S. Patent No. 4,475,480, Washington, D.C.

Brunson M.W. (1989) Forage and feeding systems for commercial crawfish culture. *J. Shellfish Res.*, **8** (1) 277–80.

Culley D.D. & Duobinis-Gray L. (1989) Soft-shell crawfish production technology. *J. Shellfish Res.*, **8** (1) 287–91.

de la Bretonne L.W. Jr. & Romaire R.P. (1989) Commercial crawfish cultivation practices: a review. *J. Shellfish Res.*, **8** (1) 267–75.

Dellenbarger L.E., Schrupp A.R. & Avault J.W. Jr. (1988) Louisiana's crayfish industry: an economic perspective. In *Freshwater Crayfish 7* (Ed. by P. Goedlin de Tiefenau), pp. 231–7. Musée Zoologique Cantonal, Lausanne, Switzerland.

Eversole A.G. & Pomeroy R.S. (1989) Crawfish culture in South Carolina: an emerging aquaculture industry. *J. Shellfish Res.*, **8** (1) 309–13.

Homziak J. (1989) Producing soft crawfish: is it for you? *Aquaculture*, **15** (1) 26–32.

Huner J.V. (1988a) Soft shell crawfish industry. In *Proc. 1st. Australian Shellfish Aquacult. Conf.*, Perth, 1988, (Ed. by L.H. Evans & D. O'Sullivan), pp. 28–42. Curtin University of Technology.

Huner J.V. (1988b) *Procambarus* in North America and elsewhere. In *Freshwater Crayfish, biology, management and exploitation* (Ed. by D.M. Holdich & R.S. Lowery), pp. 239–61. Croom Helm, London.

Huner J.V. (1989a) Overview of international and domestic freshwater crawfish production. *J. Shellfish Res.*, **8** (1) 259–65.

Huner J.V. (1989b) Culture of White River (*Procambarus acutus acutus*) and Red Swamp (*Procambarus clarkii*) crawfishes: an update. *Proc. Ann Meeting Western Australian Marron Growers Assoc.*, Oct. 1988. (mimeo).

Huner J.V. & Avault J.W. Jr. (1981) Producing crawfish for fish bait (revised). *Sea Grant Publication No. LSU-T1-76001.* Centre for wetland resourses, Louisiana State University, Baton Rouge, Louisiana.

LaCaze C.G. (1981) Crawfish Farming (revised). *Fish. Bull No. 7.* Louisiana Wildlife and Fisheries Commission (now Louisiana Dept. of Wildlife and Fisheries), Baton Rouge, Louisiana.

Malone R.F. & Burden D. (1988) *Design manual for intensive soft-shell crawfish production.* Louisiana Sea Grant program, Louisiana State University, Baton Rouge, Louisiana, USA, not seen, cited in Huner (1988a).

Malone R.F. & Culley D.D. (1988) *Method and apparatus for farming soft-shell aquatic crustaceans.* U.S. Patent No. 4,726,321, Washington, D.C.

Momot W.T. (1988) *Orconectes* in North America and elsewhere. In *Freshwater Crayfish, biology, management and exploitation* (Ed. by D.M. Holdich & R.S. Lowery), pp. 262–82. Croom Helm, London.

Moody M.W. (1989) Processing of freshwater crawfish: a review. *J. Shellfish Res.*, **8** (1) 293–301.

Roberts K.J. & Dellenbarger L. (1989) Louisiana crawfish product markets and marketing. *J. Shellfish Res,.* **8** (1) 303–7.

Romaire R.P. (1989) Overview of harvest technology used in commercial crawfish aquaculture. *J. Shellfish Res.*, **8** (1) 281–6.

Romaire R. & Lawson T. (1990) Evaluation of water circulation to improve crayfish (*Procambarus* spp.) harvest efficiency. *Abstract from World Aquaculture 90*, p. 92, June 10–14, 1990, Halifax, Nova Scotia, Canada.

7.6 Crayfish: Europe

7.6.1 Species of interest

Signal crayfish (*Pacifastacus leniusculus*); Turkish crayfish (*Astacus leptodactylus*); noble crayfish (*Astacus astacus*); white-footed crayfish (*Austropotamobius pallipes*).

The red swamp crayfish (*Procambarus clarkii*) is also cultured in southern Europe but the methods are similar to those used in the USA (Lorena 1983a, b; 1986) (Section 7.5). Otherwise the methods currently used for rearing crayfish are similar throughout Europe (Hofmann 1980; Arrignon 1981; Wickins 1982; Laurent 1984; Groves 1985; Mancini 1986; Holdich & Lowery 1988; Alderman & Wickins 1990). Most young are produced for restocking but some are reared for bait or, especially in Britain, for on-growing for the table.

7.6.2 Broodstock

Broodstock are obtained either from cultured stocks or from wild stock brought into the hatchery; often the eggs are removed to be incubated separately (see below). They are fed a variety of natural foods including small crustaceans and water plants. Detritus forms a large part of the diet in nature.

7.6.3 Mating and spawning

Dimensions of outdoor broodstock units seem to be based on individual preferences or existing facilities, for example they may be tanks (2–10 m long, 0.5–1 m wide, 0.5–1 m deep) or ponds (100–500 m^2 and about 1 m deep). They are stocked with berried females or mature wild-caught broodstock taken in autumn. Density is typically 4 m^{-2} with a sex ratio of 1 male:3 females; hides are considered essential (Koksal 1988). After mating, males are removed from tank systems and sold, but in ponds the sexes are not so readily separated. Alternative systems, used for example in France, include cages (floor to roof height 30–40 cm for adults and 10 cm for juveniles). These are floated in etangs (see Glossary) and are also used for holding juveniles prior to restocking.

7.6.4 Incubation and hatching

Hatcheries may be simple or complex depending on whether or not eggs are incubated artificially and temperature is controlled. They may house broodstock, mating and incubation tanks which are often rectangular (10 m × 2 m × 0.4 m deep) but can also be circular (0.8–1 m dia. and 0.8 m deep). They are often fitted with hides and may also be used to hold young prior to stocking.

Ponds or tanks containing ovigerous females are drained in April or early May and the females transferred to individual mesh cages or held communally in boxes (40 × 20 × 10 cm) fitted with a 1 cm mesh floor. Alternatively, shallow nursery ponds or tanks (approximately 3 × 0.5 × 0.6 m deep) fitted with perforated floors and containing one or more shelters per female are employed. The perforations allow the juveniles to escape from the females as soon as they become independent (Section 2.2).

Hatching occurs in spring or early summer when temperatures rise to 14°C and over. Towards the end of the incubation period hatching may be encouraged if the temperature is raised to 18–24°C. In some hatcheries the water exchange rate is set at once per 24 hrs (Alderman & Wickins 1990), in others at 15 ℓ min^{-1} for tanks containing up to 6–9 females m^{-2} (Koksal 1988).

7.6.4.1 *Artificial incubation*

The water supplied to a hatchery must be of high quality if eggs are to be removed from the females and incubated artificially. Temperature control is advantageous and recirculation systems employing biological filters are sometimes installed to conserve heated water. Eggs that have reached the eyed stage are carefully removed from the female by siphon and stocked into conical bottomed vessels (approximately 6–10 ℓ capacity) in which a gentle upwelling current of water keeps the eggs in suspension. The flow is about 1 ℓ min^{-1} but is regulated at hatching so that the young are maintained 1–2 cm from the bottom. Pieces of sponge may be added to which the young attach. As soon as hatching begins it is advisable to separate the juveniles from the unhatched eggs and put them into another incubator at a density of 6000–10 000 per vessel, with a flow 0.7 ℓ min^{-1} so that they can cling on to one another in a ball 1–2 cm from the bottom (Arrignon 1981). When they moult to stage two the flow is reduced to 0.4–0.5 ℓ min^{-1}. At this stage the density must be reduced to below 6000 per vessel and feeding can begin. Control of infestations may be achieved by immersing the juveniles in a solution of malachite green (10 ppm) for 15–20 minutes.

7.6.5 Nursery

Keller (1988) recommends lifting broodstock females every 10 days and shaking those with attached juveniles in a bucket of water to facilitate separation of the young. By this means he found that up to 10 000 juveniles per hour could be collected. Normally juveniles detach or leave the mother after two to 21 days, depending on species, and escape predation by going through the mesh floor. Nursery facilities are frequently shallow troughs or long rectangular tanks fitted with hides, especially around the margins. The juveniles are fed a range of foods including detritus, *Cladophera* (a green filamentous alga), live and frozen zooplankton, especially crustaceans.

Widely different stocking densities have been reported for rearing juveniles to a size at which they may be on-grown (generally 2–3 cm TL). Figures range from 50 m^{-2} (Westman 1973) to 2500 m^{-2} (Arrignon 1981) although Keller (1988) claims a density of 400 m^{-2} was the most profitable (70–80% survival) for *A. astacus* juveniles (3.5 months old, 2–3 cm long). During this time they may be thinned to 100 m^{-2} and fed for 3–4 months until they are big enough to be stocked in ponds. He also suggested that culture of juveniles for restocking could usefully be done in salmonid farm facilities during their off-season. If, for example, salmonid rearing occurred from December to June, crayfish would be reared from July to October.

7.6.6 On-growing

Three levels of culture may conveniently be defined: natural or extensive, semi-intensive and intensive or controlled-environment culture. Some examples of extensive and semi-intensive systems are shown in Figure 7.1.

7.6.6.1 *Natural/extensive*

In the European context, this category clearly embraces the concepts of both restocking (hatchery supported fisheries) and ranching as defined in Sections 5.6.1 and 2, as

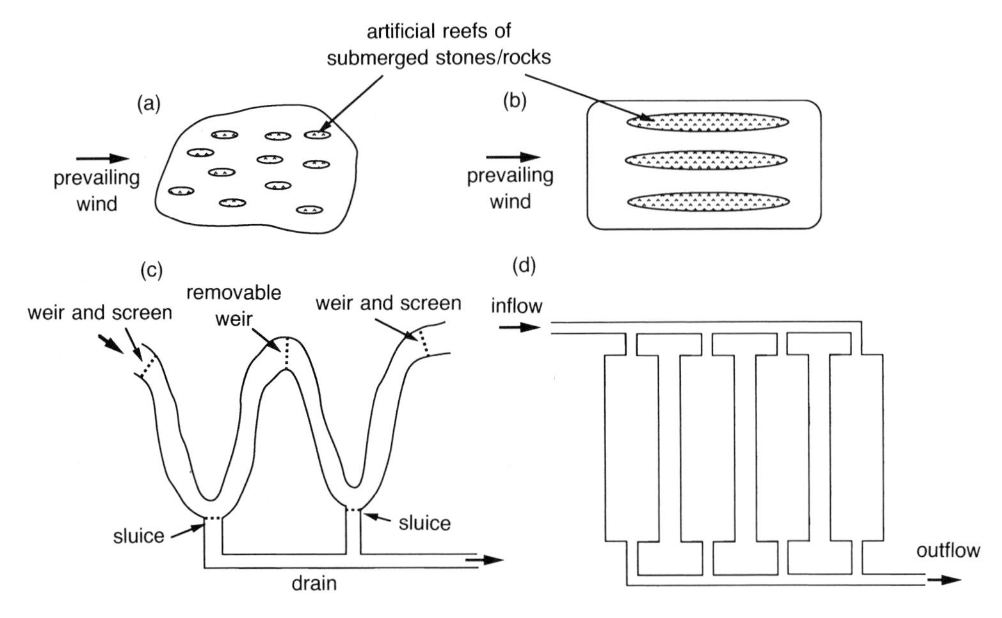

Fig. 7.1 Examples of crayfish farm designs: (a) natural lake; (b) earthen pond; (c) simple stream/channel; (d) semi-intensive canal or 'raceway' culture system (see also Plate 7.4).

well as the more traditionally defined extensive culture methods. Fisheries in rivers and streams, as well as those in lakes and etangs (often covering 200 ha or more), may be stocked with hatchery-reared young. While very little management other than conventional fisheries regulation may be practised, slightly more management influence may be applied to crayfish grown in smaller lakes (10 ha × 0.3–5 m deep) where vegetation or hides such as stone outcrops or islands may be provided to create refuges (Figure 7.1a and b). Lakes may be stocked with 200 to 1000 crayfish ha^{-1} for three to five years using groups of 50 crayfish placed at intervals along the bank. This type of 'stock & forget' culture in natural or semi-natural enclosed waters, such as irrigation ponds or gravel pits in Britain, can produce annual yields of 60–500 kg ha^{-1}. In some shallow waters (e.g. ponds and lakes 1.5–5 m deep) fertilization with agricultural fertilizers can increase natural production while the increased turbidity associated with it can provide protection from birds and sunlight, albeit sometimes at the cost of increased oxygen consumption at night.

7.6.6.2 Semi-intensive

In this category crayfish are introduced into prepared ponds, at densities up to 6–10 individuals per habitable square metre of underwater surface. Management is minimal, although fertilization (up to 50 kg N ha^{-1}) may be practised. Fertilization and water exchange need to be carefully controlled to prevent excessive growth of filamentous algae. Feeds, which may include potatoes, apples or other vegetable matter, are sometimes given but care is taken not to cause fouling and subsequent deoxygenation. Mesh enclosure of parts of lakes or etangs has been tried, but, in some areas of France for example, cages seem to be preferred as they allow better control of stock. Existing ponds and lakes can also be modified to increase the underwater 'bank' area

Plate 7.4 A small freshwater crayfish farm in southern England. The canals are stocked with imported signal crayfish. In the background, members of the local fox hunting fraternity are just visible.

by constructing reefs of rock and stones. Careful planning is necessary to ensure good water circulation can be maintained (Figure 7.1a and b). Rectangular earth ponds (10–15 × 3 × 1 m deep), situated 2 m apart, are also employed but in these, vegetation and water exchange are strictly controlled. Stretches of suitable streams may also be used for production. Flow control is accomplished by sluices, screens and by-pass or drainage canals (Figure 7.1c).

Many new commercial ventures use canals or raceways. Some utilize former watercress beds which are 2–3.5 m wide and 10–50 m long but must be deepened from 0.75 m deep to 1.5 m. In Britain new semi-intensive farms have been built around specially constructed canals, which should give greater yields but only at greater input costs (labour and feed). The canals are typically 1.5–2 m deep with sloping sides to avoid collapse (Figure 7.1d). The width is 3–10 m and canal lengths range from 50 to 150 m with banks of 2–3 m width in between. Barriers are required to prevent escapes of crayfish and entry of eels into the farm. Shelter is all important in most culture systems, both to reduce cannibalism and exposure to light, and to permit increased stocking density. Habitats are provided in the form of rough stone lining, hardcore piles at 10 m intervals or short lengths of weighted plastic pipes. Water flow on most British farms seems likely to give 50% exchange in 54–150 hrs.

Stocking may be with juveniles (spring) or one summer old juveniles (late summer/ autumn) at three crayfish per metre of canal bank each year for at least three and up to five years. The diet of adults in the wild consists of up to 70% vegetable matter, so most farmers encourage good peripheral plant growth. Young crayfish feed on the macro-invertebrates (worms, crustaceans and molluscs in particular) that abound in calcium-rich waters. The effect of supplemental and compounded feeds on farmed stock has yet to be determined.

Production targets are around six crayfish per running metre of canal bank but this could not be achieved without supplemental feeds; natural feed might support three crayfish per running metre and it must be remembered that unless canals are wider than 9 m, only one bank is counted.

7.6.6.3 *Intensive*

No known truly intensive culture systems are in operation (Taugbøl 1989; Taugbøl *et al.* 1989). Generally speaking the term intensive is currently applied to nursery systems for the production of large numbers of juveniles for restocking. Artificial (e.g. concrete) ponds are typical and can be circular (0.8–1 m dia. × 0.8 m deep), fitted with hides and have a circular water distribution pattern. Initial stocking can be as high as 750–1000 m^{-2}, but after the first winter the density is halved to 300–500 m^{-2}. After the following winter the density is reduced to 50 m^{-2}, but this prolonged nursery phase is only practised when large crayfish are required for stocking.

7.6.7 Feeding

In extensive cultures, vegetation and naturally occurring worms, crustaceans and molluscs are consumed, but very little information is available on successful compounded feeds for European crayfish (Hessen 1989). Arrignon (1981) refers to diets made in 1972 which were bound with alginate and based on fish or shrimp meal and cereals. Protein levels were at around 18–44%, lipids 1–5%, and minerals between 7–10%. Low feeding levels were reported: 1–4% body weight d^{-1} for juveniles, 0.3–1% for adults which reduced to less than 1% after mating.

7.6.8 Harvesting

Crayfish may be caught in fyke nets or traps. Of the latter, baited funnel traps are commonly used and set at 5 m intervals in canals or at 25–50 ha^{-1} in ponds. Traps are set before dusk and fished at dawn, starting two to three years after the initial stocking.

7.6.9 Transportation

Fished crayfish are frequently transported in expanded polystyrene boxes, in layers separated by damp foam, cloths or moist algae and mosses to maintain high humidity. A few weeks after hatching, juveniles (1–2 cm TL) are shipped by air in 20 cm long perforated plastic tubes with a spiral insert to which they cling. The tubes are placed in a cool, moist container. In another method, 500 juveniles are transported in a 12 ℓ container of $\frac{1}{3}$ water $\frac{2}{3}$ oxygen and can survive for up to 40 hrs provided the temperature remains below 10°C. One summer old crayfish are carried as described above but at a lower density (300 per container). Adult broodstock are carried in 7 kg boxes with 60–80 adults in two layers in a box (40 × 40 × 15 cm) and will remain in good condition for up to 20 hrs if kept cool. Ice or cooling packs can be added to the box if required. During the journey they must never be turned on their backs or exposed to chlorinated tap water (Brown 1982). At the water's edge crayfish must be allowed to walk backwards into the pond after they have been exposed to air for any length of time.

7.6.10 Processing

Prior to shipment or cooking, crayfish are best held in clean running water for 24 to 48 hrs to purge the gut. They are then sold live, frozen or, less commonly, cooked and frozen. Edible meat yield is around 10–26%.

7.6.11 References

Alderman D.J. & Wickins J.F. (1990) *Crayfish culture*. Lab. Leafl (62). MAFF Direct. Fish. Res., Lowestoft.

Arrignon J. (1981) *L'écrivisse et son elevage*. Gauthier-Villars, Paris.

Brown M. (1982) The transportation of crayfish. Sparsholt College of Agriculture, Hampshire, UK. *Crayfish Bulletin*, **1** (1) 11–12 (mimeo).

Groves R.E. (1985) *The crayfish: its nature and nurture*. Fishing News Books, Blackwell Scientific Publications, Oxford.

Hessen D.O. (1989) Crayfish food and nutrition. In *Crayfish Culture in Europe* (Ed. by J. Shurdal, K. Westman & P.I. Bergen), pp. 164–74. Norwegian Directorate for Nature Management, Trondheim, Norway.

Hofmann J. (1980) *Die flusskrebse: biologie, haltung und wirtschaftlich bedeutung*. Paul Parey, Berlin.

Holdich D.M. & Lowery R.S. (Eds.) 1988. *Freshwater crayfish, biology, management and exploitation*. Croom Helm, London.

Keller M. (1988) Finding a profitable population density in rearing summerlings of European crayfish *Astacus astacus* L. In *Freshwater crayfish* 7 (Ed. P. Goeldin de Tiefenau), pp. 259–66. Musée Zoologique Cantonal, Lausanne, Switzerland.

Koksal G. (1988) *Astacus leptodactylus* in Europe. In *Freshwater crayfish* 7 (Ed. by P. Goeldlin de Tiefenau), pp. 365–400. Musée Zoologique Cantonal, Lausanne, Switzerland.

Laurent P.J. (1984) Crayfish farming in France. Sparsholt College of Agriculture, Hampshire, UK. *Crayfish Bull.*, **1** (4) 15–16 (mimeo).

Lorena A.S.H. (1983a) Some observations on crawfish farming in Spain. In *Freshwater crayfish 5* (Ed. by C.R. Goldman), pp. 549–51. AVI Publishing Co. Westport, Conn.

Lorena A.S.H. (1983b) Socioeconomic aspects of the crawfish industry in Spain. In *Freshwater crayfish 5* (Ed. by C.R. Goldman), pp. 549–51. AVI Publishing Co., Westport, Conn.

Lorena A.S.H. (1986) The status of the *Procambarus clarkii* population in Spain. In *Freshwater crayfish 6* (Ed. by P. Brinck), pp. 131–3. Int. Assoc. Astacology, Lund, Sweden.

Mancini A. (1986) *Astacicoltura allevamento e pesca dei gambari d'aqua dolce*. Edizione Calderini, Bologna.

Taugbøl T. (1989) Crayfish culture in Norway. In *Crayfish Culture in Europe* (Ed. by J. Skurdal, K. Westman & P.I. Bergen), pp. 101–9. Norwegian Directorate for Nature Management, Trondheim, Norway.

Taugbøl T., Gydemo R., Haug J., Huner J.V. & Jarvenpåa T. (1989) Bioengineering and cultivation environment. In *Crayfish Culture in Europe* (Ed. by J. Skurdal, K. Westman & P.I. Bergen), pp. 10–17. Norwegian Directorate for Nature Management. Trondheim, Norway.
Westman K. (1973) Cultivation of the American crayfish *Pacifastacus leniusculus*. In *Freshwater crayfish 1* (Ed. by S. Abrahamsson), pp. 211–20. Studentlitteratur, Lund.
Wickins J.F. (1982) Opportunities for farming crustaceans in western temperate regions. In *Recent advances in aquaculture* (Ed. by J.F. Muir & R.J. Roberts), pp. 87–177. Croom Helm, London.

7.7 Crayfish: Australia

7.7.1 Species of interest

Red claw (*Cherax quadricarinatus*); marron (*C. tenuimanus*); yabbie (*C. destructor*). Considerable interest is currently being shown in the commercial culture of red claw, both in Australia and elsewhere. The semi-intensive culture methods employed in earthen ponds are similar to those used for shrimp and prawns. Several intensive culture trials have also been tried with battery systems based on those used for clawed lobsters. This section includes information from published and unpublished sources. Further details may be found in Morrissy (1984, 1987); Kowarsky *et al.* (1985); Aiken (1988); Alon *et al.* (1988, 1989); Anon. (1988b); Hutchings (1988); Huner (1988c, 1989); Villarreal (1988); Jones (1989a,b; 1990); O'Sullivan (1990) and various articles in *Austasia Aquaculture Magazine*.

7.7.2 Broodstock

Broodstock may be obtained both from the wild and from dealers. However, close seasons and other legislation prevent the taking of wild spawning or egg-carrying female marron in some areas and seasons. Populations in permanent waters can become self-sustaining and provide a seasonal supply of stock for extensive operations. Larger projects depend on wild broodstock as well as hatchery reared juveniles. However, Mills (1983) noted that year-round breeding may be possible in yabbie through photoperiod control, while Sammy (1988) reported that red claw breeds readily throughout the year. Holker (1989) described a marron broodstock facility of circular tanks fitted with individual housing and stocked at up to 12 crayfish m^{-2}, and recommended that broodstock marron over four years old (over 80 mm CL) be discarded to reduce the risk of poor egg quality or hatch rates.

7.7.3 Mating and spawning

Males and newly-moulted females held separately in hatchery tanks may be put together for breeding unless fighting occurs, in which case the male should be replaced. Mating normally takes place within a few hours or days, after which either the male is removed and the female left undisturbed to spawn and incubate her eggs, or the females are transferred to a similar tank with a mesh false bottom and stocked at 10–12 females m^{-2}. In trials with red claw over 80% of captive broodstock spawned, and sex ratios of from 1:1 to 1:4 males to females were satisfactory (Jones 1990).

7.7.4 Incubation and hatching

Egg-bearing females may be held in tanks or submerged cages in ponds, preferably with some form of shelter until the eggs hatch. After spawning, the young are collected or escape through mesh to imitation weed hides. In the case of the yabbie, incubation lasts 40 days and the young stay with the mother for about eight weeks. These small juveniles concentrate in bunches of twigs, rope fibres or straws placed in the ponds, or are harvested directly from the tanks for on-growing. It is best to keep crayfish of a common size together as cannibalism increases with disparity of sizes, especially in crowded populations. Morrissy (1976) calculated that 1000 marron females of 50 mm CL would release sufficient juveniles to stock a 1 ha pond at around 5–10 m^{-2} after allowing for up to 75% mortality.

7.7.5 Nursery

It seems likely that many growers and perhaps hatcheries use nursery ponds or tanks to acclimatize, or increase the size of, juveniles prior to stocking. During this phase they are fed crayfish pellets, compost red worms, sorghum and lucern. Densities may reach 200 m^{-2} (Holker 1989). Survival of red claw juveniles stocked in laboratory tanks at 980–1842 m^{-2} ranged from 4–84% (mean 46.3%) over 25 to 50 days (Jones 1990).

7.7.6 On-growing

Techniques for on-growing crayfish in Australia are conveniently described under four categories: extensive (including farm dam) culture; backyard 'hobby' culture; commercial semi-intensive culture; and experimental intensive 'battery' or controlled-environment culture.

7.7.6.1 Extensive

This method is more akin to the 'stock and forget' method used in parts of Europe. Either natural or man-made water reservoirs constructed for watering agricultural livestock, are utilized. These may be excavated ponds (farm dams) filled with run-off water in winter, or gully dams built in hilly regions. Gulley dams are considered best as they usually have a stream flow through them, although they can still be polluted by agricultural run-off. On the other hand farm dams may be susceptible to overloading with detritus and organic pollution. It is recommended that ponds have hard clay bottoms and are designed to be completely drainable for cleaning.

7.7.6.2 Backyard culture

This method appears widespread in Australia but is not likely to attract substantial investment. It has merit in that it allows entrepreneurs to experience some of the demands of an aquaculture operation without serious loss of capital. Ponds may be plastic swimming pools or specially constructed tanks (2–4 m dia., 35–90 cm deep) provided with shelters and shade cloth to control temperature and protect against bird predation. Aeration may be continuous or applied for about 10 minutes 2–3 times per day. A shallow area is sometimes provided in case oxygen levels fall to critical levels. Stocking density is 10 m^{-2}. Feed is commonly chicken or trout pellets at up to 30 g m^{-2} wk^{-1} but compost red worms are best. The maximum yield is likely to be 300 g m^{-2} with survivals of 2–3 crayfish m^{-2}.

7.7.6.3 Semi-intensive

The transition from extensive to semi-intensive cultures is not a clear-cut process but basically involves the following changes:

Extensive	*Semi-intensive*
deep ponds	shallow ponds
mixed age, small biomass	single age, large biomass
slow, continuous production	complete harvest every 18–24 months
no aeration	aeration

Ponds suitable for commercial operation are 50–250 × 20 × 1–1.5 m deep, with excavated top soil lining the banks down to the water line and planted with Kikuyu grass for stability (Holker 1989). Pond sides are pitched at angles of not less than 45° (1:1 horizontal:vertical) on long sides and over 60° (1:1.7) on the short sides. An anti-escape barrier of planking or similar material may be necessary, as may shade or bird netting. The drains must be able to empty the pond in five hours for harvesting, but the supply should be capable of filling a pond in 24 hrs. Ponds are cleaned annually and treated with agricultural limestone at 100 kg ha^{-1} to control algae. After five years the ponds are dried, allowed to crack and a further 200–300 kg ha^{-1} of limestone are added to neutralise bottom deposits.

Fertilization may be employed to increase natural productivity. Paddlewheel aeration is used and feeding is done at sunset along pond margins. Cheaper feeds are more often used in on-growing ponds than in ponds holding early juveniles or potential broodstocks. Proprietary crayfish and chicken pellets, lucern and sorghum hay are given as food. A water exchange every two to three days is recommended but may be increased to maintain quality as required. Adults are stocked at 2 m^{-2} if required, otherwise five to 12 juveniles of uniform size are initially stocked per square metre. Young crayfish can survive up to 80 hrs out of water under favourable conditions, but after acclimatization to pond temperatures they should be released on to the bank so that they can walk in. Annual yields of about 2000–4000 kg ha^{-1} are a reasonable expectation after 12 months (40–70 g animals) to two years (100–200 g animals) depending on species and temperature.

Published information (Anon. 1988a) about commercial practices in Australia for the red claw, marron and yabbie indicates that farms typically operate from four to 40 ponds of 0.4 to 7.9 ha each and 1–2 m deep. Most use pumped bore and well water and eight out of ten use aerators, but exchange rates vary widely from 10–15% per day to once per year. Six out of ten farms provide shelters and most need predator control screens. Staff number between one and seven full-time, and up to three part-

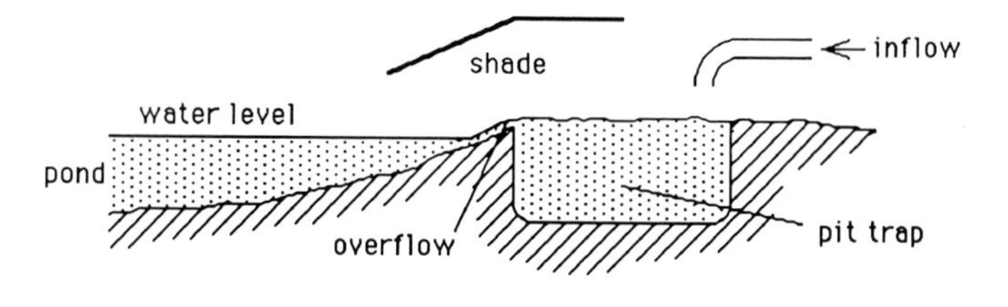

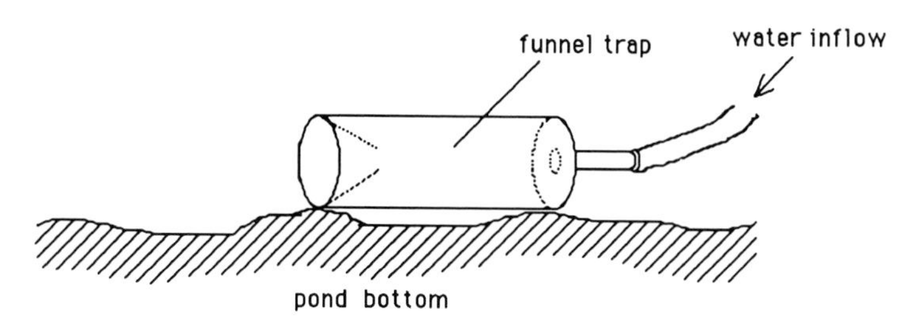

Fig. 7.2 Pit and funnel traps for catching Australian crayfish (*Cherax* spp.) that move towards a flow of water.

time. Harvesting is mostly by pump or by draining rather than by trapping. Most have specific nursery ponds, although the red claw breeds in on-growing ponds.

7.7.6.4 *Intensive*

There seems little doubt that battery farming of individually confined *Cherax* species is technically possible. Economic feasibility, however, is unlikely until the twin problems of developing cost effective diets and low-labour rearing systems have been solved. Several technical investigations have been made (Morrissy 1984; Kowarsky *et al.* 1985; Cogan 1987; Anon. 1988c; O'Sullivan 1990) but as yet do not seem to have reached the degree of reliability achieved with techniques for *Homarus* (Section 7.8.9).

7.7.7 Harvesting

Most harvesting is accomplished by draining the pond or using baited traps. Red claw crayfish, however, respond strongly to a current of water coming from outside the pond and can be caught in flow-traps (Figure 7.2), which may be portable or built into the pond floor (Jones 1990).

7.7.8 Transportation

Attempts have been made to culture marron and red claw outside Australia using semi-intensive methods similar to those described above. In one instance juveniles and adults survived shipment to Dominica, when packed conventionally in polythene bags containing 1 ℓ of water and oxygen. On arrival a quarantine dip in nitrofurazone solution (an antibiotic) and then salt water was given to remove external infestations. Shipping mortality was 50–100% and it was felt that storage at 15°–20°C was necessary at transfer points during the flight (Alon *et al.* 1988; 1989; Anon. 1988b). Further details of acclimatization and disinfection procedures may be found in Rubino *et al.* (1990).

7.7.9 Processing

Australian crayfish are purged in clean running freshwater and sold live or cooked and frozen. Some are exported as frozen headless product. It is worth noting that it is conventional in Australia to include the shell when determining meat yield. Thus the figures of around 40% often found in the literature really equate to about 25% meat yield if the shell is excluded (Morrissy *et al.* 1990).

7.7.10 References

Aiken D. (1988) Marron farming. A real industry or just great promotion? *World Aquaculture*, **19** (4) 14–17.

Alon N.C., Rubino M.C. & Wilson C.A. (1988) Australian marron lobster (*Cherax tenuimanus*) aquaculture feasibility in the Caribbean. *Program and Abstracts, East meets West*, p. 14. 19th Annual Conference and Exposition, World Aquaculture Soc., Hawaii.

Alon N.C., Rubino M.C., Wilson C.A. & Armstrong J.M. (1989) Pond culture of the Australian marron lobster, *Cherax tenuimanus*, in the Eastern Caribbean: survival and growth. *Aquaculture '89 Abstracts*, p. 56, from World Aquaculture Society Meeting 1989, Los Angeles.

Anon. (1988a) Local crayfish 'saves' Qld crayfish industry. *Austasia Aquaculture Magazine*, **2** (9) 8–11.

Anon. (1988b) Lessons from shipping marron overseas. *Austasia Aquaculture Magazine*, **2** (9) 16–17.

Anon. (1988c) Battery culture research in W.A. *Austasia Aquaculture Magazine*, **2** (9) 16.

Cogan P. (1987) Marron battery study proves encouraging. *FINS*, **20** (3) 5–6. Fishing Industry News Service, Perth, W. Australia.

Holker D.S. (1989) Marron – *Cherax tenuimanus*. Unpublished report.

Huner J.V. (1988) Status of crayfish transplantations. In *Freshwater Crayfish* 7 (Ed. by Goeldlin de Tiefenau), pp. 29–34. Musée Zoologique Cantonal, Lausanne Switzerland.

Huner J.V. (1989) Overview of international and domestic freshwater crawfish production. *J. Shellfish Res.*, **8** (1) 259–65.

Hutchings R. (1988) A review of the Australian freshwater crayfish fauna with reference to aquaculture. In *Freshwater Crayfish* 7 (Ed. by P. Goeldlin de Tiefenau), pp. 13–18. Musée Zoologique Cantonal, Lausanne Switzerland.

Jones C.M. (1989a) Aquaculture potential of *Cherax quadricarinatus*. *Queensland Dept. of Primary Industries Leaflet*, 6 July 1989 (mimeo).

Jones C.M. (1989b) *Aquacultural potential of* Cherax quadricarinatus*: current reaearch developments*. Paper presented at the 1989 Freshwater Aquaculture Association, University of Queensland, Brisbane, 8 April 1989.

Jones C.M. (1990) *The biology and aquaculture potential of the tropical freshwater crayfish* Cherax quadricarinatus. Queensland Department of Primary Industries, Information Series QI90028.

Kowarsky J., Gazey P. & Rippingale R. (1985) Intensive culture potential of freshwater crayfish – a research update (March 1985). *Marron Growers Bulletin*, **7** (1) 8–15.

Mills B. (1983) Breeding and reproduction in Yabbies. *SAFIC* **7** (2) 28-30. South Australian Fishing Industry Council.

Morrissy N.M. (1976) Aquaculture of marron, *Cherax tenuimanus* (Smith). 1. Site selection and the potential of marron for aquaculture. *Fish. Res. Bull. West. Aust.*, **17** (1) 1–27.

Morrissy N. (1984) Assessment of artificial feeds for battery culture of a freshwater crayfish, marron (*Cherax tenuimanus*) (Decapoda: Parastacidae). *Dept. Fish. Wildl. West. Aust. Rept. No. 63*, 1–43.

Morrissy N.M. (1987) Marron pond management-aeration. *FINS* **20** (3) 7–9. Fishing Industry News Service, Perth, W. Australia.

Morrissy N.M., Evan L.E. & Huner J.V. (1990) Australian freshwater crayfish: aquaculture species. *World Aquaculture*, **21** (2) 113–22.

O'Sullivan D. (1990) Intensive freshwater crayfish system tested. *Austasia Aquaculture Magazine*, **5** (4) 3–5.

Rubino M., Alon N., Rouse W.D. & Armstrong J. (1990) Marron aquaculture research in the United States and the Caribbean. *Aquaculture Magazine*, **16** (3) 27–44.

Sammy N. (1988) Breeding biology of *Cherax quadricarinatus* in the Northern Territory. In *Proc. 1st. Australian Shellfish Aquacult. Conf.* (Ed. by L.H. Evans & D. O'Sullivan), pp. 79–88. Perth, 1988, Curtin University of Technology.

Villarreal H. (1988) Culture of the Australian freshwater crayfish *Cherax tenuimanus* (marron) in Eastern Australia. In *Freshwater Crayfish* 7 (Ed. by P. Goeldlin de Tiefenau), pp. 401–8. Musée Zoologique Cantonal, Lausanne, Switzerland.

7.8 Clawed lobsters

7.8.1 Species of interest

American lobster (*Homarus americanus*); European lobster (*H. gammarus*). Although clawed lobsters are technically straightforward to culture, the need to individually confine them during the on-growing phase is a serious constraint to economic viability. Current interest centres on the prospects for ranching hatchery-reared juveniles on artificial reefs.

7.8.2 Broodstock

Substantial trade in live lobsters from the east coast of North America and western Europe often provides opportunities for the purchase of egg-bearing (berried) females for use as broodstock. In some areas, however, the landing of berried females is prohibited and special dispensation would be required to utilize this source of material for a commercial hatchery.

The ease with which ovigerous females can be selected from commercial sources, and the risk of introducing disease to a broodstock facility, are very much dependent on the live handling practices of the fishermen and lobster merchants. Common practices include keeping lobsters at sea in moored keep boxes or in intertidal ponds (both of which make access and sorting difficult), in land-based shallow tanks (lobster pounds) and tray systems which may or may not utilize recirculated natural or artificial seawater (Beard & McGregor in press). Claws are either banded (Europe) or held closed by a small wooden peg forced into the articulation joint (parts of north America) to prevent fighting and claw loss during storage. Pegging however increases the risk of Gaffkaemia, the most important disease of lobsters in captivity (Section 8.9). Overcrowding and large fluctuations in water quality e.g. low salinity, high ammonia and turbidity, are factors which can adversely affect the quality of larvae that eventually hatch, without noticeably affecting the marketability of the adult lobsters (Aiken & Waddy 1986). Occasionally egg masses are heavily infested with worms and epizootic organisms (Aiken *et al.* 1985) which can easily spread to larvae culture systems.

Transportation of lobsters from the fishery to the storage merchants may be done either in water (vivier transport systems, Section 8.4.6) or 'dry' when lobsters are packed between layers of damp seaweed, foam or sacking. The effects of such treatments on egg viability are likely to be detrimental. After the eggs hatch the females may be re-sold for consumption.

7.8.3 Maturation and mating

The feasibility of rearing broodstock from wild, immature and hatchery-bred lobsters to meet the demands of a putative lobster farm has been investigated in North America (Hedgecock 1983; Waddy & Aiken 1984a,b; Aiken & Waddy 1985a,b). Lobsters grown rapidly to maturity at 20°C often did not perform well as broodstock. Egg production was poor (only about 5% of *H. americanus* spawned) and attachment was weak. Males produced fewer sperm and spermatophores. Most workers concluded that at present better results are obtained from wild caught egg-bearing broodstock. However, the ability to condition wild pre-ovigerous females with a high degree of reliability has been developed (see below) and minimises the social and legal opposition to taking berried females from the fishery (Aiken & Waddy 1985c).

Control over mating of captive broodstock utilizes either natural copulation between selected animals (which sometimes involves mating with inter-moult females) or artificial insemination. The latter offers little control, however, since egg extrusion and fertilization follow mating, but often not for several months. Additionally, eggs spawned after artificial insemination often do not attach well to the female.

7.8.4 Spawning

The environment in which captive females spawn has a marked influence on the success with which the eggs are attached to the pleopods for incubation. For example, if the females do not properly position themselves on their backs, or if they do not remain so long enough for the attachment glue to set, the newly extruded eggs fail to attach securely (Talbot *et al.* 1984). Waddy (1988) reported a lengthy conditioning technique for improving egg production and attachment in *H. americanus*. Animals were held for five months (December to April) at 0–5°C. Spawning and attachment increased with the time spent at these winter temperatures. Unfortunately three to four years of normal summer-winter temperature cycles were required before most females spawned predictably, and additional time was required (3–4 spawning cycles) before proper attachment occurred reliably. No such in-depth study has been made on *H. gammarus*, but in captivity this species frequently displays an annual spawning cycle rather than the biennial rhythm characteristic of *H. americanus* (Bertran & Lorec 1986).

7.8.5 Incubation

Incubation may take four to 18 months according to temperature, and losses can occur at any time, perhaps as a result of abnormal attachment stalk formation (Talbot & Harper 1984), adverse water quality, infestation (Harper & Talbot 1984) or because of abnormal egg aeration and grooming behaviour by the female. Exposure of eggs that are close to hatching, to salinity below about 24 ppt is also likely to be detrimental (Charmantier & Aiken 1987).

The development of the eggs is monitored by changes in colour and later in the size of the eye of the developing larva (Perkins 1972; Richards & Wickins 1979). By this means, the time of hatching may be predicted to within a few weeks from a small sample of eggs.

As with many other crustacean species, artificial incubation of lobster eggs removed from the female prior to hatching is possible but disease risks and consequent losses are often unacceptably high.

7.8.6 Hatching

The hatching period typically lasts 3–5 days and most, but by no means all eggs hatch overnight. Differences in viability of larvae hatched in different seasons have been reported (Anger *et al.* 1985; Eagles *et al.* 1986) and it is likely that larvae hatching first from a brood are more robust than those hatching later (Eagles *et al.* 1986; O'Donoghue 1989). In practice this indicates that only those larvae hatching during the first two to three days should be used in cultures.

The number of larvae that hatch from a female varies with female size. For example female *H. gammarus* weighing 450 to 1500 g may release from 800 to 13 000 larvae each (Beard & Wickins in press). The free-swimming larvae are easily collected by allowing them to pass in a current of water from the female's incubation chambers to a separate container fitted with a 1.5 mm bar mesh screen. There the larvae may be washed with clean seawater and counted before being transferred to the larvae rearing vessels.

The hatching of eggs from wild caught broodstock is convenient on a small scale but a major commercial venture would probably require at least monthly supplies of larvae for year-round production. Studies with *H. gammarus* demonstrated early progress towards this by hatching batches of larvae at regular three month intervals over a four year period (Richards & Wickins 1979; Beard *et al.* 1985). This was achieved through careful selection of broodstock females (from the north and south of Britain), with eggs at an appropriate stage of development. The females were fed at 2–3% of their body weight per day, with a 1:1 mixture of fresh mussel (*Mytilus*) gonad and fresh-frozen shrimp (*Crangon*). Finer manipulation of hatching time was achieved through temperature control during incubation, but some batches 'forced' at higher temperatures or held captive for several months seemed to yield less viable larvae.

Similar results were obtained in a larger study with wild caught *H. americanus* (Waddy & Aiken 1984a). An alternative approach was tried by Hedgecock (1983) who, by photoperiod manipulation, attempted to control time of spawning rather than time of hatching. One advantage of this method was that incubation could be done at the culture temperature (provided this was constant), eliminating the need for additional, controlled low-temperature facilities. Disadvantages were a lack of flexibility, poor spawning rate (60%) and excessive egg loss. He calculated that 79 broodstock animals would be needed for the production of 80 000 lbs of lobsters per month and this figure was used by Coffelt & Wikman-Coffelt (1985) in preparing a model for a battery farm unit capable of producing one million one pound lobsters per year. In a similar exercise based on manipulation of both spawning and incubation, Aiken & Waddy (1985a) suggested that 200–800 pre-ovigerous females would be required to produce one million marketable lobsters per year.

7.8.7 Larvae culture

The most commonly used rearing container is the Hughes 40 ℓ capacity 'kreisel' (Hughes *et al.* 1974). It is designed to maintain an homogeneous distribution of larvae and their food by means of a spiral, upwelling flow pattern. Sizes range up to 80 litres capacity (Beard & Wickins in press), rarely larger, and several may be linked together in recirculation systems (Figure 7.3). Other simpler systems have been used with success in the laboratory. The larval phase in *H. gammarus* lasts 14 to 18 days at

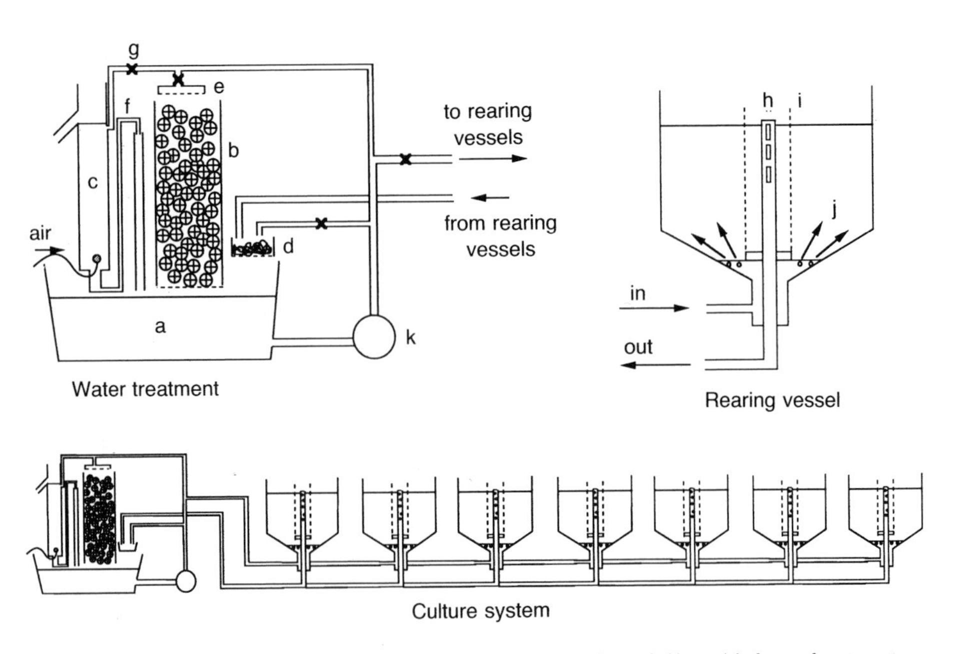

Fig. 7.3 Lobster larvae culture system: (a) reservoir; (b) biological filter; (c) foam fractionation column; (d) washable strainer; (e) sprinkler; (f) constant head pipe; (g) valve; (h) overflow; (i) screen; (j) upwelling water current; (k) pump.

about 18°C and survival rates are typically 20–40%; with *H. americanus*, however, survival can be as high as 60–70% at densities of about 40 larvae ℓ^{-1}. Metamorphosis may be spread over a 10 day period in cultured populations of either species.

The diet may be live or frozen adult *Artemia* (around 48 ml per 1000 larvae daily, Eagles *et al.* 1986), frozen mysid shrimp or other zooplankton, chopped molluscs or prepared feeds. However, choices are constrained in practice since fouling, feeding regime and survival rates are intimately linked. Accumulation of suspended and dissolved organic material encourages the growth of epibiotic infestations on larval exoskeletons and can seriously interfere with moulting. This is the main problem associated with the use of prepared and non-living foods. Live adult *Artemia* (maintained in cultures at a concentration of about four *Artemia* per larva) is probably the best food available, although its culture is expensive, particularly when it is fed with live algae or complex compounded feeds.

A convenient alternative has proved to be frozen mysid shrimp supplemented with live, newly-hatched *Artemia* nauplii for six hours per day, on three days a week (Beard & Wickins, in press). During the time the nauplii are in the kreisel, flow is reduced and a fine mesh screen is temporarily placed over the outlet screen to retain the nauplii.

If poor quality food is fed at the start of the culture period, recovery seldom occurs even if a good diet is subsequently fed. Lighting also influences survival, dim natural light being better than darkness. Heavy mortality (up to 20%) commonly occurs among larvae and post-larvae between stages four and six. Studies with *H. gammarus* have shown that better survival rates among early juveniles can be obtained if only the earliest larvae to reach stage four are kept for further culture (Galindo 1985). One of the most laborious tasks is to separate the newly developed stage four lobsters from the remaining stage three larvae. If this is not done several times each day during the period of metamorphosis, extensive cannibalism occurs.

7.8.8 Nursery

In captivity lobsters are cannibalistic throughout their larval and post-larval life. Losses are unavoidable during the larval phase, but after settlement they become economically unacceptable. Some authors have calculated that the losses when juveniles are reared communally may be tolerable for a short period (perhaps one to six months post-metamorphosis), in relation to the savings made in labour and feed expenditure. However, communal rearing is unlikely to be commercially acceptable because the intimidation and fighting which occurs, even when shelters are provided, increases size heterogeneity and the proportion of damaged or crippled individuals in the population. Stunting may follow communal rearing, with differences in size of up to three times between the biggest and smallest individuals after just three months (Van Olst *et al.* 1975; Aiken & Waddy 1988). This indicates there would be a need to grade the lobsters at least once during a period of communal rearing so that small individuals could resume rapid growth. Grading, however, has proved difficult and laborious because of the need to first separate all the lobsters from their shelters. Incidentally, if hides are provided experience indicates that they must have two exits so that lobsters cannot get trapped by intruders.

The drive to solve the problems caused by the lobster's intolerance of crowding has led to some bizarre experiments. Indeed, some of these indicated that periodic removal, mutilation or immobilisation of claws may be worthwhile. However, the labour costs and risks of disease following such operations seem unlikely to appeal to investors. The types of communal rearing tanks employed in laboratory and pilot studies range from large 59 m^2 outdoor tanks containing a 10 cm deep layer of oyster shells (Henocque 1983c) to small aquaria containing various type of hides configured in two or three dimensions (Van Olst *et al.* 1975; 1980). The best survival obtained in Henocque's large tank system was 67% from an initial stocking density of 28 lobsters m^{-2}. After one year at ambient seawater temperatures the lobsters had, however, only grown to a mean weight of 5 g. The diet used was a shrimp pellet but there was also much natural production of macroalgae and benthos in the tank.

There is more variety in the types of individual holding systems that have been built and tested. Several of these are also suitable, when rescaled, for the further on-growing of larger lobsters (Section 7.8.9). The most important features of individual compartments are that they must each receive a supply of well oxygenated water, be self-cleaning and amenable to automatic feeding. Approaches have included blocks of mesh cages suspended in deep tanks of moving seawater containing a suspension of live *Artemia* which can pass through the mesh, and shallow trays containing mesh floored compartments suspended over rotating sprinkler bars or subjected to a tidal rise and fall of water (Van Olst *et al.* 1980; Beard *et al.* 1985).

Diets are at present very expensive and frozen or live adult *Artemia*, if used, would need to be fed at a rate of approximately 10 g d^{-1} per 50 g of lobsters.

Survival in individual confinement systems is good (>80%) after about stage six. Transfer to larger containers becomes necessary when the size of the compartment begins to restrict growth (Table 7.7) (Richards & Wickins 1979).

7.8.9 On-growing

Most studies of on-growing made in research and pilot units involved holding each lobster separately. In fact the whole future of battery culture (both for lobsters and for Australian freshwater crayfish – Section 7.7.6.4) depends primarily on the advent of cost-effective system designs. Four interrelated factors are critical:

(1) Calculation of the minimum sizes of container which do not inhibit growth in selected sizes of lobster, and the minimum number of different container sizes (and hence transfers) that need to be employed for the 2–3 year growth period;

Table 7.7 Suggested container floor areas for individually confined clawed lobsters.

Lobster size mm CL	Age months	Area of container cm^2
5–10	1	25
11–25	4	115
26–40	12	310
41–60	24	620
61–85	30	1058

(2) Configuration of, and materials used for, the containers must be cost-effective;
(3) Satisfactory removal of waste (and dead lobsters) and adequate water exchange in each container;
(4) Accurate and rapid distribution of food to each container.

The basic requirements for the sizes and numbers of transfers are known, as are the general tolerance limits of lobsters to water quality parameters (Section 8.5). In addition, a variety of ingenious container/system configurations and materials have been developed and tested in a number of countries over the past 15 years. Most fulfil the requirements of space per lobster and water exchange, but not all cope satisfactorily with the waste removal, feeding and cost criteria. Current thinking is that the containers will be designed in such a way as to save space, and in a limited range of sizes to reduce manufacturing costs. The design must also ensure that lobsters can be moved, or the containers expanded easily (though as infrequently as possible) during the culture period. Some of the experimental and pilot systems have been described by Mickelsen *et al.* 1978; Van Olst *et al.* 1980; Wickins 1982; Beard *et al.* 1985; D'Abramo & Conklin 1985; Ingram 1985; McCoy 1986; Waddy 1988). Examples include:

(1) Simple troughs divided into compartments by wooden or plastic slats and mesh screens through which water continuously flows;
(2) Rectangular tanks or troughs containing blocks of containers fitted with perforated floors through which water is distributed by specialised flushing or tidal systems;
(3) Various experimental systems tested in North America and Norway for example:
    (a) Deep tanks containing horizontal stacks of tubes or corrugated sheets of plastic, or modular cages fitted to a supporting framework;
    (b) Deep tanks containing vertical or horizontal stacks of perforated trays. These are serviced by sequential removal of the stacks or by continuous slow rotation of trays. According to Waddy (1988), the latter represents the best design currently available for holding large numbers of large juveniles over 25 mm CL. However, a US$5 million plant incorporating this technology was partly constructed on Prince Edward Island, Canada but went into receivership before commercial viability could be established (Campbell 1989);
    (c) Shallow, round tanks containing a revolving group of floating mesh-bottomed containers. Water is jetted upwards or downwards into the containers as they revolve. (A notable example of this exists at Tiedemann's Norwegian lobster plant which has a capacity of 120 000 one year old lobsters (Anon. 1987; Grimsen *et al.* 1987));
    (d) A novel, cylindrical stacked system built and tested by Sanders Associates, Inc. at Kittery, Maine, Moss Landing, California, and in Hawaii at a cost of US$3 million. The company indicated that the initial capital investment would be US$10 million (1986 prices) (McCoy 1986).

Of all these systems, those using horizontally stacked shallow trays of containers are perhaps the most easily constructed and serviced. They may also be the least expensive for laboratory scale studies.

The supply of food to individually confined lobsters involves separate techniques for small (20 mm CL, four months post-metamorphosis) and larger lobsters. If it can be made economically attractive to grow and feed live adult *Artemia* to juveniles up to four months of age, it should be possible to house the small lobsters in mesh or perforated containers stacked horizontally or vertically in a deep-water tank in such a way that live *Artemia* would swim freely through all the chambers. From an engineer's viewpoint the design of such a system would be straightforward, but care would be needed to ensure even food distribution and to prevent fighting and subsequent limb loss through the meshes. The main consideration would be the cost of culturing the food and the labour of transferring the lobsters to different containers after four months. One or two further changes would also be required.

Much of the current research on lobster culture in North America and Europe revolves around the development of compounded feeds for on-growing lobsters to market size. The development of a purified diet that supports good growth and high survival in *H. americanus* represents significant progress and will serve as a foundation for the formulation of commercial diets (Conklin *et al.* 1983). At present diets are expensive and some formulations are thought to be responsible for producing severe moulting difficulties, a problem being studied on both sides of the Atlantic. Normal moulting can be restored by the addition of frozen or fresh natural foods once or twice a week (Ali & Wickins in prep.). Recently designed prototype feeders have facilitated the precise and rapid provision of both wet and dry diets to trays of individually held lobsters (Wickins *et al.* 1987), while in Norway the Tiedemann group uses a computerised food distribution system (Grimsen *et al.* 1987; Schjetne 1987).

It has been calculated (Conklin *et al.* 1983) that a unit producing one million marketable lobsters annually would require 24 mt of food per day. Storage and preparation of this food would be a major operation. As far as is known, no one has reared lobsters from metamorphosis to commercial size in two to three years solely, or even largely, on a compounded diet. Claims that growth rates achieved with artificial feeds over a limited period of time can be extrapolated to indicate that market sized lobsters could be obtained in this time, must therefore be regarded as speculative.

Not all farmed lobsters would reach a marketable size at the same time. The estimated percentage of European lobsters reaching 80 mm CL in 0.75–3.5 yrs is shown in Table 4.5.

Most culturists agree that the homarid lobsters are remarkably resistant to disease. Problems chiefly arise when wild-caught lobsters are crowded together (even mutilated to prevent fighting) in storage pounds. Apart from this, few serious losses have been reported among individually held juveniles or indeed larvae, that could not be explained in terms of poor water, diet or husbandry. As with many other species, the true significance of disease will probably not be known until commercial culture becomes a reality. It may be noted here that artificial sea water can be successfully used to store lobsters (Beard & McGregor in press) but it is unlikely that maximum growth rates would occur in lobsters cultured in the simple salt solutions used in most live storage systems.

7.8.10 Harvesting and processing

The harvesting of cultured lobsters from indoor battery units is likely to be highly mechanised or automated. Grading according to size and appearance would be followed by packing live for shipment to market either in vivier trucks (Section 8.4.6) or out of water in boxes for air freight. Other processing (shelling, cooking, freezing in brine) would only be necessary for sub-size or substandard lobsters (Shortall 1990).

7.8.11 Hatchery supported fisheries, ranching

In the absence of demonstrably profitable culture operations, interest in the prospects for ranching and developing hatchery-supported lobster fisheries has revived in North America, Europe and Japan (Section 5.6).

7.8.11.1 Juvenile production

In essence the techniques involve the culture of juveniles from wild caught broodstock to either the first benthic stage (North America) or, in European experiments, to between three and 12 months from hatching (Henocque 1983b; Grimsen *et al.* 1987; Lorec 1987; Beard & Wickins in press). The juveniles are then taken to sea and released on to selected lobster grounds. Culture methods are almost identical to those described above, although in one British experiment the juveniles were given special

Plate 7.5 The mass culture of clawed lobster larvae in modified Hughes 'kreisels' at the Fisheries Laboratory, Conwy, UK.

treatment to ensure the best chance of survival on the sea bed. Specifically they were fed supplemented natural diets to enhance their natural pigmentation and improve moulting success (Ali & Wickins in prep.). In addition live bivalve spat (2–4 mm) were provided to stimulate crusher claw development, thus increasing the range of food items accessible to each lobster on release (Wickins 1986). Laboratory tests indicated that acclimatization to sea temperatures prior to release would increase the speed with which they constructed defensible burrows (Wickins in prep.).

7.8.11.2 Tagging

The ability to distinguish hatchery-reared lobsters from their wild counterparts is essential if the success of any pilot release is to be evaluated. Three methods have been used to distinguish released animals:

(1) Release outside the normal geographic range of the species. For example, *H. americanus* released in Japanese waters (Henocque 1983a). Any lobster subsequently caught is easily recognised and must have come from the release programme;
(2) Culture of hybrids (*H. americanus* × *H. gammarus*) or unusually coloured lobsters (Syslo, 1986). Recaptured hybrids would be moderately easy to identify, colourmorphs more so;

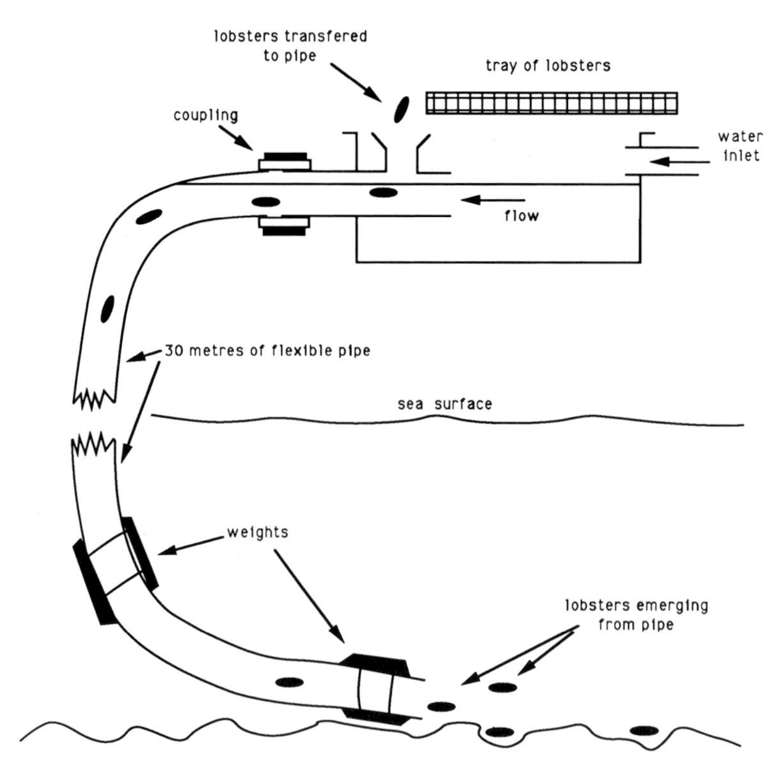

Fig. 7.4 The arrangement used to release juvenile lobsters directly on to suitable sea bed habitat through a 30 metre flexible pipe. (Redrawn from Beard & Wickins (in press) and adapted from the original idea of W. Cook and P. Oxford, North West and North Wales Sea Fisheries Committee.)

(3) Tagging with an internal, coded microwire tag (Wickins *et al.* 1986). With this method, however, it is impossible to recognise a tagged animal without specialized detection equipment, and a rigorous sampling approach is essential (Bannister & Wickins 1989).

Releases under the first two techniques may be made using newly metamorphosed stage four or five juveniles, but under the third, consistent placement of micro-tags in such small individuals may be hard to achieve. Once success has been established it would not normally be necessary to tag further juveniles released in that area.

7.8.11.3 Transport and release

Juveniles are transported individually in perforated tubes placed in tanks or plastic bags of seawater, or in compartmented trays contained in tanks of recirculating seawater. Juveniles released in Norwegian waters after transportation in damp seaweed or newspaper did not behave normally for some time after reaching the sea bed and suffered undue predation (Van der Meeren, 1991).

To survive, lobsters must be released directly on to suitable sea bed substrate. This is most reliably accomplished by divers or, where extensive areas of suitable substrate exist, through a pipe reaching from a surface vessel to the sea bed (Figure 7.4). In France (Y. Henocque, 1987 pers. comm.), lobsters were sealed in individual tubes fitted with a soluble paper cap which dissolved soon after reaching the sea bed.

7.8.12 References

Aiken D.E. & Waddy S.L. (1985a) The uncertain influence of spring photoperiod on spawning in the American lobster (*Homarus americanus*). *Can. J. Fish. Aquat. Sci.*, **42** (1) 194–7.

Aiken D.E. & Waddy S.L. (1985b) Periodic control of vitellogenesis in the American lobster (*Homarus americanus*): comment on a recent report. *Can. J. Fish. Aquat. Sci.*, **42** (1) 198–9.

Aiken D.E. & Waddy S.L. (1985c) Production of seed stock for lobster culture. *Aquaculture*, **44** (2) 103–14.

Aiken D.E. & Waddy S.L. (1986) Environmental influence on recruitment of the american lobster *Homarus americanus*: a review. *Can. J. Fish. Aquat. Sci.*, **43** (11) 2258–70.

Aiken D.E. & Waddy S.L. (1988) Strategies for maximizing growth of communally-reared juvenile American lobsters. *World Aquaculture*, **19** (3) 61–3.

Aiken D.E., Waddy S.L. & Uhazy L.S. (1985) Aspects of the biology of *Pseudocarcinonemertes homari* and its association with the American lobster, *Homarus americanus*. *Can. J. Fish. Aquat. Sci.*, **42** (2) 351–6.

Ali Y. & Wickins J.F. (in prep.). The use of fresh food supplements to ameliorate moulting difficulties in lobsters (Homarus gammarus (L.)) destined for release to the sea.

Anger K., Storch V., Anger V. & Capuzzo J.M. (1985) Effects of starvation in moult cycle and hepatopancreas of stage 1 lobster (*Homarus americanus*) larvae. *Helgoländer Wiss. Meeresunters*, **39**, 107–16.

Anon. (1987) Tobacco firm starts smolt production and launches in lobsters. *Fish Farming International*, **14** (9) 18–20.

Bannister R.C.A. & Wickins J.F. (1989) A new perspective on lobster stock enhancement. *Proc. 20th Ann. Shellfish Conf.*, 16–17 May 1989, pp. 38–47. Shellfish Association of Great Britain.

Beard T.W. & McGregor, D. (in press.). *The care and live storage of lobsters.* Lab. Leafl. MAFF Direct. Fish. Res., Lowestoft.

Beard T.W. & Wickins J.F. (in press). Techniques for the production of juvenile lobsters (*Homarus gammarus* (L.)). *Fish. Res. Tech. Rep.*, MAFF Direct. Fish. Res., Lowestoft.

Beard T.W., Richards P.R. & Wickins J.F. (1985) The techniques and practicability of year-round production of lobsters, *Homarus gammarus* (L.) in laboratory recirculation systems. *Fish. Res. Tech. Rep.* (79), MAFF Direct. Fish. Res., Lowestoft.

Bertran R. & Lorec J. (1986) *Observations sur le cycle de reproduction du Homard European (*Homarus gammarus*) en captivité.* International Council for Exploration of the Sea, Shellfish Committee, ICES CM 1986/K:12 (mimeo).

Campbell M. (1989) Prince Edward Island. Lobster culture technology on hold. *Atlantic fish farming*, 20 Aug. 1989.

Charmantier G. & Aiken D.E. (1987) Osmotic regulation in late embryos and prelarvae of the American lobster *Homarus americanus* H. Milne-Edwards, 1837 (Crustacea, Decapoda) *J. exp. mar. Biol. Ecol.* **109**, (10) 101–8.

Coffelt R.J. & Wikman-Coffelt J. (1985) Lobsters: one million one pounders per year. *Aquacultural Engineering*, **4** (1) 51–8.

Conklin D.E., D'Abramo L.R. & Norman-Boudreau K. (1983) Lobster nutrition. In *Handbook of Mariculture, Vol. 1, Crustacean aquaculture* (Ed. by J.P. McVey), pp. 413–23. CRC Press, Boca Raton, Florida.

D'Abramo L.R. & Conklin D.E. (1985) Lobster Aquaculture. In *Crustacean and Mollusk Aquaculture in the United States* (Ed. by J.V. Huner & E.E. Brown), pp. 159–201. AVI Inc. Westport, USA.

Eagles M.D., Aiken D.E. & Waddy S.L. (1986) Influence of light and food on larval american lobsters, *Homarus americanus*. *Can. J. Fish. Aquat. Sci.*, **43**, 2303–10.

Galindo C.G. (1985) *Direct and indirect effects of natural diets on cultured lobster (*Homarus gammarus *(L)) larvae and their subsequent performance as juveniles.* M.Sc. Thesis, University of Stirling.

Grimsen S., Jaques R.N., Erenst V. & Balchen J.G. (1987) Aspects of automation in a lobster farming plant. *Modeling, Identification and Control*, **8** (1) 61–8.

Harper R.E. & Talbot P. (1984) Analysis of the epibiotic bacteria of lobster (*Homarus*) eggs and their influence on the loss of eggs from the pleopods. *Aquaculture*, **36** (1/2) 9–26.

Hedgecock D. (1983) Maturation and spawning of the American lobster, *Homarus americanus*. In *Handbook of mariculture, Vol. 1, Crustacean aquaculture* (Ed. by J.P. McVey), pp. 261–86. CRC Press, Boca Raton, Florida.

Henocque Y. (1983a) Adaptability and propagation of lobster seedlings transplant experiments in Japan. *Symp. fr-japon. Aquacult.*, **1**, 123–8. Montpellier 16 Dec. 1983.

Henocque Y. (1983b) Lobster aquaculture and restocking in France. In *Proc 1st Int. Conf. on Warm Water Aquaculture – Crustacea*, 9–11 Feb. 1983 (Ed. by G.L. Rogers, R. Day & A. Lim), pp. 235–7. Brigham Young University, Hawaii.

Henocque Y. (1983c) *Techniques d'élevage du homard et experience d'implantation au Japon.* Unpublished report to La Maison Franco-Japonaise.

Hughes J.T., Shleser R.A. & Tchobanoglous G. (1974) A rearing tank for lobster larvae and other aquatic species. *Progv. Fish Cult.*, **36** (3) 129–32.

Ingram M. (1985) *Clearwater guide to intensive culture of lobsters and other species.* Clearwater Publishing, Isle of Man.

Lorec J. (1987) L'aquaculture du homard (*Homarus americanus* et *Homarus gammarus*) en France. *Aqua Revue*, (10) 26–30.

McCoy H.D. II. (1986) Intensive culture systems past, present and future. Pt 1. *Aquaculture Magazine*, **12** (6) 32–5.

Mickelsen R.W., Infanger R.C. & Heckmann R.A. (1978) Culturing the American lobster (*Homarus americanus*) using a vertically stacked cage system. *Proc. 9th Ann. Mtg. World Maricult. Soc.* 3–6 Jan. 1978, pp. 723–30. Atlanta, Georgia. Louisiana State University Press.

O'Donoghue F.P. (1989) *Variations in the composition of eggs and newly hatched larvae of the European lobster (*Homarus gammarus *(L)).* M.Sc. Thesis, University of Stirling.

Perkins H.C. (1972) Developmental rates at various temperatures of embryos of the Northern lobster (*Homarus americanus* Milne-Edwards). *Fishery Bull. Natl. Oceanic and Atmos. Adm. (U.S.)* **70**, 95–99.

Richards P.R. & Wickins J.F. (1979) *Ministry of Agriculture, Fisheries and Food lobster culture research.* Lab. Leafl. (47) MAFF Direct. Fish. Res., Lowestoft.

Schjetne P. (1987) Allevamento dell' astice – un sogno o un' industtria vitale. *Riv. Ital. Piscic. cttiopatol*, **22** (4) 121–4 (in Italian).

Shortall D. (1990) Canadian lobster – resource management and market development. *Proc. 21st. Ann. Shellfish Conf.*, 15–16 May, pp. 21–34. Shellfish Assoc. of Great Britain.

Syslo M. (1986) Getting the 'bugs' out. *Massachusetts Wildlife*, **36** (3) 4–10.

Talbot P. & Harper R.E. (1984) Abnormal egg stalk morphology is correlated with clutch attrition in laboratory-maintained lobsters (*Homarus*). *Biol. Bull.*, **166**, 349–56.

Talbot P., Thaler C. & Wilson P. (1984) Spawning, egg attachment and egg retention in captive lobsters (*Homarus americanus*). *Aquaculture*, **37** (3) 239–44.

Van der Meeren, G.I. (1991) Out-of-water transportation effects on behaviour in newly-released juvenile atlantic lobsters (*Homarus gammarus*). *Aquacultural Engineering*, **10** (1) 55–64.

Van Olst J.C., Carlberg J.M. & Ford R.F. (1975) Effects of substrate and other factors on the growth, survival and cannibalism of juvenile *Homarus americanus* in mass rearing systems. *Proc. 6th Ann. Wkshop.* 27–31 Jan., pp. 261–74. World Maricult. Soc., Seattle, Washington.

Van Olst J.C., Carlberg J.M. & Hughes J.T. (1980) Aquaculture. In *The biology and management of lobsters, Vol, 2, Ecology and management* (Ed. by J.S. Cobb & B.F. Phillips), pp. 333–84. Academic Press, London.

Waddy S.L. (1988) Farming the Homarid lobsters: state of the art. *World Aquaculture*, **19** (4) 63–71.

Waddy S.L. & Aiken D.E. (1984a) Broodstock management for year-round production of larvae for culture of the American lobster. *Can. Tech. Rep. Fish. Aquat. Sci.*, (1272).

Waddy S.L. & Aiken D.E. (1984b) Seed stock for lobster culture: the role of temperature in synchronizing the moult and reproductive cycle of cultured American

lobsters. *J. World Maricult. Soc.*, **15**, 132–7.
Wickins J.F. (1982) Opportunities for farming crustaceans in western temperate regions. In *Recent advances in aquaculture* (Ed. by J.F. Muir & R.J. Roberts), pp. 87–177. Croom Helm, London.
Wickins J.F. (1986) Stimulation of crusher claw development in cultured lobsters, *Homarus gammarus* (L.). *Aquaculture and Fisheries Management*, **17**, 267–73.
Wickins J.F. (in prep.). *The effect of temperature on initial burrowing in newly released cultured lobsters (*Homarus gammarus *(L.)).*
Wickins J.F., Beard T.W. & Jones E. (1986) Microtagging cultured lobsters for stock enhancement trials. *Aquaculture and Fisheries Management*, **17**, 259–65.
Wickins J.F., Jones E., Beard T.W. & Edwards D.B. (1987) Food distribution equipment for individually held juvenile lobsters, *Homarus* sp. *Aquacultural Engineering*, **6** (3) 277–88.

7.9 Spiny lobsters

7.9.1 Species of interest

Western rock lobster (*Panulirus cygnus*); California spiny lobster (*P. interuptus*); mud spiny lobster (*P. polyphagus*); Japanese spiny lobster (*P. japonicus*); scalloped spiny lobster (*P. homarus*); Caribbean spiny lobster (*P. argus*); red and green rock lobsters (*Jasus edwardsii*, *J. verreauxi*). At present the prospects for the commercial culture of spiny and rock lobsters are constrained by two factors: the technical difficulties of rearing the phyllosome larvae (Figure 2.2c) and the unpredictable availability of wild pre-juveniles (pueruli) or juveniles for fattening. Prospects for attracting spiny lobsters and their juveniles to selected seabed areas by providing artificial habitats are being investigated in several countries interested in developing ranching programmes.

7.9.2 Broodstock, incubation and hatching

Little difficulty is expected in obtaining and transporting sufficient wild-caught broodstock to supply commercial hatcheries. Mature spiny lobsters can be successfully maintained on diets of bivalve mollusc and crustacean flesh; indeed the general husbandry techniques and conditions that are applied to *Homarus* broodstock are also likely to be suitable for spiny lobsters. Spawning, incubation and hatching have occurred in a number of captive species, although no detailed studies have been reported on which expectations of hatchery performance could be based.

7.9.3 Larvae culture

The main problems with rearing the larvae arise from the unusual larval morphology, prolonged larval life and a fundamental lack of information on larval diet and feeding habits. Japanese researchers were the first to rear a palinurid lobster from egg to puerulus (Kittaka 1988; Kittaka *et al.* 1988; Kittaka & Kimura 1989; Yamakawa *et al.* 1989), but the effort involved during the lengthy larval phase (often over 300 days) suggests that it would be very difficult to maintain hygienic culture conditions for so long in a commercial hatchery. An ample exchange of good quality water was found to be essential throughout the period.

The species cultured were *Panulirus japonicus*, a warm-temperate water species, *Jasus lalandii*, a cool temperate species and various *Jasus* hybrids. Few have reached or lived beyond the puerulus stage. The larvae are cultured in 40 to 100 ℓ capacity cylindro-conical vessels in upwelling seawater, not unlike the vessels used in the culture of *Homarus* larvae. At high culture densities the long appendages of the phyllosomes readily become entangled with debris suspended in the water (cast shells, filamentous bacteria and algae) and particularly high standards of husbandry are essential. Many diets have been tested (Lellis 1989), including the marine algae *Phaeodactylum* and *Chlorella* at concentrations of 1–20 million cells per ml, rotifers, newly hatched and partially grown *Artemia*, and mussel flesh. Larvae often cling to the sides of the vessel but we do not know if they graze when in this position, or if their survival could be improved through the provision of additional flat surfaces.

7.9.4. Nursery

The pueruli of some species, for example *P. argus*, congregate in large numbers in suitable habitat (mangrove roots) and can be caught in artificial floating 'bushes' or 'habitat traps' that resemble the clumps of algae that the pueruli seek for refuge (Ingle & Witham 1968; Serfling & Ford 1975a,b; Marx & Herrenkind 1985). These may then be transferred to protected areas for on-growing (Section 12.8). Experience to date has shown, however, that concentration and reliability of occurrence of the young stages are inadequate for commercial exploitation in many regions. Nevertheless feeding does not seem to be a problem, since in laboratory cultures the post-larvae accept live *Artemia* during the first and second moults and thereafter will take artificial diets.

7.9.5 On-growing

In Taiwan wild caught juveniles ranging in size from 0.5 to 250 g are grown to 300 to 800 g in small 200 m^2 ponds. On average, animals stocked at 25 g mean weight reach 330 g in 16 months. Growth rate is heterogeneous and animals are graded every two to three months. About five to six ponds are needed to accommodate this practice. Water exchange is about 10% per day and survival about 80%. The largest Taiwanese farm is reported to have 13 ha of ponds and the capacity to produce 150 000 marketable lobsters per year (Chen 1990). Juvenile spiny lobsters readily accept and grow well on natural foods such as abalone, mussels, crab and squid (Ting 1973; Serfling & Ford 1975b; Chittleborough 1976; Phillips *et al.* 1983). Several other methods for on-growing have been proposed elsewhere (e.g. Ingle & Witham 1968) but few have been evaluated comprehensively (Oesterling & Provenzano 1985).

Since 1983 wild-caught juvenile *Panulirus polyphagus* have been increasingly cultured (fattened) in floating fish farm cages in Singapore (Table 5.2c). The cages are typically wooden framed, 2–5 m square and about 2–3 m deep. Regular cleaning of the synthetic mesh netting is necessary to remove fouling organisms. The lobsters are fed daily with chopped trash fish and mussels. Output is around 24 mt per year (Lovatelli 1990). There is a lack of published information on the effects of density on growth rate and survival in captive animals. One Australian study (see Van Olst *et al.* 1980) determined that the cost of culture plus the expense of capture of the pueruli made the price of the farmed animal twice that of the fished product. A similar but more recent study on *Jasus edwardsii* was reported from New Zealand (Anon. 1987; 1989) but results are not yet known.

7.9.6 Transportation

Spiny lobsters may be harvested by seining and are transported live using the same techniques and precautions that are used for clawed lobsters. Animals being transported live to consumer outlets and processing plants are purged for three days in clean, flowing seawater chilled for 4 hrs to 12°C before being packed between layers of crushed ice (Anon. 1989).

7.9.7 Processing

Most processed spiny lobsters are beheaded, washed, graded and frozen shell-on. Small spiny lobsters are shelled and canned.

7.9.8 Habitat modification

Fishermen in several Caribbean countries construct fields of artificial shelters (called 'casas Cubanas' or 'casitas'), within which wandering palinurids take shelter and can be conveniently captured (Miller 1983; Fee 1986). The shelters measure about 1.5 m^2, and consist of a wooden frame and a flat or corrugated roof under which the spiny lobsters hide (Miller 1982). They are placed about 20 to 30 m apart and in some areas density may reach 10 000 casitas in 160 km^2. The yield from such an area may be 40 to 65 mt of spiny lobsters annually (Eggleston *et al.* 1990a).

The main effect of the shelters seems to be to reduce mortality due to predation by fish. The lobsters prefer shelters with small entrances so that larger predators are excluded. Survival of a range of sizes of lobsters may be enhanced therefore by the provision of a range of different sized shelters in otherwise unprotected areas where lobsters forage for food (Eggleston *et al.* 1990b). Small juvenile spiny lobsters can be collected in casitas made of close-fitting stacks of shelves and transplanted to areas where food and shelter are plentiful. While it is uncertain whether these activities augment the fishery or merely concentrate the stock, it does seem likely that the concept could be utilized to ranch those species of spiny lobster with low migratory instincts (Section 5.6.2).

7.9.9 References

Anon. (1987) Rock lobsters in New Zealand tests. *Fish Farming International*, **14** (5) 32.

Anon. (1989) Rock lobster culture – is it a serious proposition? *Austasia Aquaculture Magazine*, **3** (8) 5–8.

Chen L-C. (1990) *Aquaculture in Taiwan*. Fishing News Books, Blackwell Scientific Publications, Oxford.

Chittleborough R.G. (1976) Growth of juvenile *Panulirus longipes cygnus* George on coastal reefs compared with those reared under optimal environmental conditions. *Aust. J. Mar. Freshwater Res.*, **27**, 279–95.

Eggleston D.B., Lipscius R.N. & Miller D.L. (1990a) Stock enhancement of Caribbean spiny lobster. *The Lobster Newsletter*, **3** (1) 10–11.

Eggleston D.B., Lipscius R.N., Miller D.L. & Coba-Letina L. (1990b) Shelter scaling regulates survival of juvenile Caribbean spiny lobster *Panulirus argus*. *Mar. Ecol. Prog. Ser.*, **62**, 79–88.

Fee R. (1986) Artificial habitats could hike crab and lobster catches. *National Fisherman*, **67** (8) 10–12, 64.

Ingle R.M. & Witham R. (1968) Biological considerations in spiny lobster culture. *Carrib. Fish. Inst. Univ. Miami. Proc.*, **21**, 158–62.

Kittaka J. (1988) Culture of the palinurid *Jasus lalandii* from egg stage to puerulus. *Nippon Suisan Gakkaishi*, **54** (1) 87–93.

Kittaka J. & Kimura K. (1989) Culture of Japanese spiny lobster *Panulirus japonicus* from egg to juvenile stage. *Bull. Jap. Soc. Sci. Fish.*, **55** (6) 963–70.

Kittaka J., Iwai M. & Yoshimura M. (1988) Culture of a hybrid of spiny lobster genus *Jasus* from egg stage to puerulus. *Bull. Jap. Soc. Sci. Fish.*, **54** (3) 413–17.

Lellis W.A. (1989) Artificial food for post-larval *Panulirus argus*. Larviculture and Artemia Newsletter, (12) 9–10. State University of Ghent, Belgium.

Lovatelli A. (1990) *Regional seafarming resources atlas*. FAO/UNDP Regional seafarming development and demonstration project, RAS/86/024 Jan. 1990.

Marx J. & Herrenkind W. (1985) Factors regulating microhabitat use by young juvenile spiny lobsters, *Panulirus argus*: food and shelter. *J. Crust. Biol.*, **5** (4) 650–7.

Miller D.L. (1982) Construction of shallow water habitat to increase lobster production in Mexico. *Proc. Gulf Caribb. Fish. Inst.*, **34**, 168–79.

Miller D.L. (1983) Shallow water mariculture of spiny lobster (*Panulirus argus*) in the Western Atlantic. In *Proc. 1st Int. Conf. on Warm Water Aquaculture-Crustacea*, 9–11 Feb. 1983 (Ed. by G.L. Rogers, R. Day & A. Lim), pp. 238–45. Brigham Young University, Hawaii.

Oesterling M.J. & Provenzano A.J. (1985) Other crustacean species. In *Crustacean and mollusk aquaculture in the United States* (Ed. by J.V. Huner & E. Evan Brown), pp. 203–34. AVI Inc. Westport, Connecticut.

Phillips B.F., Joll L.M., Sandland R.L. & Wright D. (1983) Longevity, reproductive condition and growth of the western rock lobster, *Panulirus cygnus* George, reared in aquaria. *Aust. J. Mar. Freshwater Res.*, **34**, 419–29.

Serfling S.A. & Ford R.F. (1975a) Ecological studies of the puerulus larval stage of the California spiny lobster, *Panulirus interuptus*. *US Fish. Wild. Serv. Fish. Bull.*, **73** (2) 360–77.

Serfling S.A. & Ford R.F. (1975b) Laboratory culture of juvenile stages of the California spiny lobster, *Panulirus interuptus* (Randall) at elevated temperature. *Aquaculture*, **6**, 377–87.

Ting R.Y. (1973) Culture potential of spiny lobster. *Proc 4th Ann. Wkshop World. Maricult. Soc.*, 23–26 Jan., pp. 165–70, Monterrey, Mexico. Louisiana State University Press.

Van Olst J.C., Carlberg J.M. & Hughes J.T. (1980) Aquaculture. In *The biology and management of lobsters, Vol. 2, Ecology and management* (Ed. by J.S. Cobb, B.F. Phillips), pp. 333–84. Academic Press, London.

Yamakawa T., Nishimura M., Matsuda H., Tsujigado A. & Kamiya N. (1989) Complete larval rearing of the japanese spiny lobster *Panulirus japonicus*. *Bull. Jap. Soc. Sci. Fish.*, **55** (4) 745.

7.10 Crabs

7.10.1 Species of interest

Mud or mangrove crab (*Scylla serrata*); swimming crab (*Portunus trituberculatus*); blue crab (*Callinectes sapidus*); Caribbean king crab (*Mithrax spinosissimus*). Traditionally the culture of crabs for the table relies on wild-caught seed (e.g. in the Philippines); however some fish or shrimp hatcheries rear crab larvae for restocking (e.g. in Japan). Extensive polyculture as well as semi-intensive monoculture of *S. serrata* for the table is practised in Taiwan where the industry has become differentiated into hatchery, nursery, on-growing and fattening operations. A research hatchery for this species has also been established in Queensland, Australia (Anon. 1986; Garland 1988).

7.10.2 Broodstock and larvae culture

In 1981, production of *S. serrata* in Taiwan was based on wild-caught seed, but hatchery rearing is now possible. Eyestalk ablation of wild-caught broodstock females has increased the control over larval supplies, and survival rates of up to 60% are achieved at culture densities of six larvae per litre. The long spines typical of crab zoeae make these larvae particularly vulnerable to entanglement with filamentous material in the water especially at high culture densities (Figure 2.2d). Incubation takes 10 to 15 days at 25°–28°C and first stage zoeae are fed minced *Artemia* nauplii. Later stages are fed live *Artemia* nauplii at up to 30 ml^{-1}. Culture vessels vary in design but modified lobster 'kreisels' (Hughes *et al.* 1974) can be used in conjunction with recirculation systems (Heasman & Fielder 1983). At the megalopa stage minced fish and bivalve flesh are given, but survival frequently falls to around 20% at the first crab stage largely through cannibalism. Other crab species such as *Cancer irroratus* can be reared using very similar methods.

7.10.3 Nursery

Juvenile crabs are held at 2000–3000 m^{-2} in nursery ponds for two weeks. Taiwanese farms often have four to five earth-bottomed ponds, each of 15–20 m^2, containing a 10 cm layer of beach sand. Temperature is controlled by shading, salinity held between around 20 ppt and, if a flow is available, the water exchange rate is up to one pond volume per day. Blended trash fish (1 kg d^{-1} per 30 000 crabs) is given for the two weeks of culture after which the crabs have reached 1 cm carapace width (CW). Survival ranges from about 50 to 70%.

7.10.4 On-growing

Extensive polyculture of *S. serrata* is practised in the Philippines where it traditionally forms a secondary, low density crop in intertidal ponds with shrimp or milkfish (Table 5.6). Juveniles may enter with the tide or are purchased at a size of 2–3 cm CW from push-net fishermen. If crabs are stocked intentionally rather than incidentally, farmers may install an overhanging fence of bamboo stakes around the inside edge of the pond to prevent escapes. Berried females in particular try to escape to the sea to spawn. Stocking is usually done once a year (1000 crabs ha^{-1}) at the start of the growing season (May–August) although continuous stocking is sometimes attempted. Some farmers feed trash fish at 6–10% of the body weight per day, often on the rising tide to reduce pollution. Crabs grow to 200 g or 8 cm CW in 4–6 months and provide yields of around 340 kg ha^{-1} annually. It is likely that higher yields are possible with these methods. This simple technology is applied to edible crabs in other countries, for example in India, Thailand and Malaysia.

In Taiwan the crabs are often grown with *Gracilaria* (seaweed), shrimp or fish and pond management frequently gives priority to the conditions required by these other species. Crab monoculture is also practised (Chen 1990). Typically the ponds are 0.2–0.5 ha earth ponds with several sluice gates if they are tidal rather than pumped. Crab stocking density is 0.5–3 m^{-2} and feed includes live snails and trash fish (10–15 g m^{-2} daily). Crabs grow to 8–9 cm CW in three to four months (summer) or in five to six months in winter with 30–70% survival. Yields are estimated to be about 8000 to 9000 crabs ha^{-1}, or approximately 1800 kg ha^{-1} per crop.

Another form of crab monoculture, in Taiwan, is essentially a holding and fattening operation to supply 'red crabs' (females packed with an internal mass of orange eggs) for gourmet restaurants. Such specialist farms have five to 15 ponds of 50 to 600 m^2 each stocked at 2–4 crabs m^2 with 8–12 cm CW females. They are fed once each day with live snails, shrimp or trash fish (up to 200 g per crab). Survival over the one to three month holding period is around 70–90%. Similar entrepreneurial fattening operations are undertaken in Singapore and Hong Kong (Section 5.1.2).

7.10.5 Harvesting

Harvesting is done using baited lift nets, bamboo traps, cages, gill nets or when the pond is drained. During high tides crabs swim against the current and can be caught by dip nets.

7.10.6 Transportation

Crabs are commonly transported live, with the claws bound, and survive best if held in chilled, humid conditions (Section 3.3.5).

7.10.7 Processing

Most crabs are sold live, unprocessed.

7.10.8 Hatchery supported fisheries, ranching

Intensive culture of juvenile *Portunus trituberculatus* for restocking occurs in Japan, but there is very little culture to market size (Cowan 1983a). The hatcheries involved only rear crabs between three weeks and three months in a year and rear other species at other times.

7.10.8.1 Broodstock

Mated females are caught from March to August and transported in 5–10 ℓ of water (short 30 minute journeys) or in 1000 ℓ viviers (one to five hour journeys) at one crab per 20 ℓ of water.

7.10.8.2 Spawning and incubation

Hatchery tanks used for incubation and spawning typically contain 25 m^2 of sand 10 cm deep on a false floor. Water depth is 50 cm and the flow is around 200% per day or 25 000 ℓ per day. Shade covers are used to control algal growth and temperature, and the substrate is kept clean to reduce the risks of egg infections. The temperature is raised from 10°–20°C over eight to 12 days in March–April and then to 23°C. Stocking density is 1–3 females m^{-2} but to reduce fighting injuries some farmers remove the dactyl of the claws and increase the stocking density to 10 m^{-2}. Careful daily feeding with bivalves or fish (3–5 kg day^{-1} per 20–25 crabs) can result in 80% survival at the time of hatching. Females with a firm round orange egg mass are selected and moved carefully to 500–1000 ℓ hatching tanks in which water is gently aerated and continuously renewed. During the 20 to 25 day incubation period the eggs change colour from orange to black. The appearance of two red-purple spots in the eggs indicates hatching will occur in three days. Fecundity ranges from one to three million zoeae for 400 to 1000 g females. Rotifers (30 ml^{-1}) are added prior to hatching, which usually occurs between 8 pm and midnight. Small or non-phototactic batches of larvae are discarded and few females are retained to spawn a second time. Overall only a small proportion of the purchased broodstock are used.

7.10.8.3 Larvae culture

The product of one spawning is stocked to give a density of 10–50 larvae ℓ^{-1} in 75–300 m^3 capacity culture tanks. Many operators prefer circular 100 m^3 tanks. The tanks are supplied with sufficient water to give a daily exchange of 10%, but this may be adjusted to control algal growth and water quality. Salinity is 30–33 ppt, pH 8.0–8.5, but light levels below 3000 lux are inadequate. Some tanks are fitted with a slow (1 rpm) stirring bar to reduce bottom fouling. The larvae pass through four zoeal stages, each lasting three to four days, followed by one megalopa stage lasting five to seven days at 20°–25°C. Separately cultured *Chlorella* or a culture of bacteria and yeast together with rotifers (3–10 $m\ell^{-1}$) are fed to the first zoea but live *Artemia* are added for zoea two onwards (Cowan 1983b). Sieved clam and shrimp flesh (140 μ fragments) are given later five to six times a day. Hanging nets or plastic mesh are added for the megalopae to cling to, but a 30% mortality is common between zoea four and megalopa stage. The tanks are emptied through a siphon or drain and the post-larvae retained on 520 μ mesh netting.

7.10.8.4 Transport and release

Transportation to the release site is done in 1000 ℓ capacity vivier trucks at 15°–19°C. The crabs are stocked at 150 ℓ^{-1} and provided with frayed rope 'shelters'. The crabs are pumped to smaller containers and released at the sea surface from a boat or from the beach. Stage one crabs (C1) cling to seagrass and only burrow when they reach the C2 stage. The heavy losses following surface releases have led to the construction of specialized nursery facilities. These may be

(1) On shore tanks in which the crabs are stocked at 1–3 ℓ^{-1} and grow to C4 with 20–40% survival. After one to three weeks they are released at the sea surface;

(2) Netted inshore areas fitted with frayed rope shelters and stocked at 100–500 m^{-2}. The crabs are fed, survival is 30–40% over one to three weeks and they are released by opening the enclosure to the sea;
(3) Open inshore areas again fitted with frayed rope shelters but surrounded only by a low (40 cm) underwater fence to prevent early dispersal.

7.10.9 Soft shelled crab production

Production of newly moulted or soft-shell crabs is really a short-term holding operation in which the wild-caught adult crabs are generally not fed. These crabs are in demand by sport fishermen for bait but also form the basis of a small but expanding and lucrative gourmet food market, e.g. for the blue crab (*Callinectes sapidus*) in the USA. The same technology would be applicable with very little modification to other species for example *Cancer irroratus*, *C. magister* and *C. borealis* (Oesterling & Provenzano 1985). In North America two types of 'shedding' system are in use:

(1) Floating boxes (3.6 × 1.2 × 0.45 m deep) holding 200–300 crabs, and fitted with an external 20 cm wide lip to give stability. Problems arise from strong currents and predators;
(2) Onshore tables which may be provided with a through-flow of water or linked to a recirculation system. The tables may be stacked according to operational convenience and made of wood, fibreglass or other non-toxic materials. Typical dimensions are 2.4 × 1.2 × 0.25 cm deep with 10 cm water depth. Flow is designed to promote a self-cleaning action and may be set to three to four tray volumes per hour.

The stock are caught in traps or pots and inspected for the presence of 'peelers' and 'busters'. Peelers have a red line along the edge of the last two flattened sections of the 'paddle fins' (the last pair of pereopods), indicating that moulting will occur in one to three days. When a split develops along the posterior edge of the carapace, the crab becomes a 'buster' and will moult in two to three hours. The trays are examined every four to six hours and busters removed to other trays to moult. They are then observed from every 15 minutes to a few hours until they have expanded to full size, when they are removed for sale before the shell hardens and their value is lost. In some areas where peeler crabs are not abundant, inter-moult crabs are held and fed until they moult. The main constraint to the industry appears to be the supply of good quality peeler crabs.

7.10.9.1 Processing

Soft shelled crabs are chilled or packed and frozen for shipment to restaurants.

7.10.10 References

Anon. (1986) Barramundi, crabs and prawn hatchery established. *Austasia Aquaculture Magazine*, **1** (1) 7.

Cowan L. (1983a) Hatchery production of crabs in Japan. In *Proc. 1st Int. Conf. on Warm Water Aquaculture – Crustacea*, 9–11 Feb. 1983 (Ed. by G.L. Rogers, R. Day & A. Lim), pp. 215–20. Brigham Young University, Hawaii.

Cowan L. (1983b) *Crab farming in Japan, Taiwan and the Philippines*. Information series Q184009, Queensland Dept. of Primary Industries.

Chen L-C. (1990) *Aquaculture in Taiwan*. Fishing News Books, Blackwell Scientific Publications, Oxford.

Garland C.D. (1988) The hatchery basis of the Australian mariculture industry. In *Proc. 1st. Australian Shellfish Aquacult. Conf.* (Ed. by L.H. Evans & D. O'Sullivan), pp. 5–16, Perth, 1988. Curtin University of Technology.

Heasman M.P. & Fielder D.R. (1983) Laboratory spawning and mass rearing of the mangrove crab, *Scylla serrata* (Forskal), from first zoea to first crab stage. *Aquaculture*, **34** (3/4) 303–16.

Hughes J.T. Shleser R.A. & Tchobanoglous G. (1974) A rearing tank for lobster larvae and other aquatic species. *Progv. Fish Cult.*, **36** (3) 129–32.

Oesterling M.J. & Provenzano A.J. (1985) Other crustacean species. In *Crustacean and Mollusk Aquaculture in the United States* (Ed. by J.V. Huner & E. Evan Brown), pp. 203–34. AVI Inc. Westport, Connecticut.

Chapter 8
Techniques: general

8.1 Materials

Suitable materials for the construction of culture vessels and for use in pumps and plumbing are judged by two main criteria, their toxicity to the cultured species and their resistance to corrosion. Other attributes such as strength, weight, ease of working and cost are more easily recognised. As a general rule, toxicity is likely to be more of a problem in recirculation systems where dissolved substances can accumulate to harmful levels in the water (Portmann & Wilson 1971), while corrosion is a major problem in marine and brackish water systems. It is also worth remembering that many species accumulate dissolved substances to toxic levels in their bodies. One example is copper which is a vital component of crustacean respiratory pigment but which is also toxic when in excess. Wheaton (1977); Hawkins & Lloyd (1981); Dexter (1986); Muir (1988); and Huguenin & Colt (1989) provide useful reviews of materials suitable for aquaculture systems.

8.1.1 Concrete

Concrete is commonly used in aquaculture systems and storage reservoirs. Seawater resistant concretes are available (e.g. sulphate resistant Portland cement to British Standard BS 4027:1972) but two problems frequently arise. The first is the elevated pH and alkalinity levels that occur in water in contact with new concrete, the second is the break-up of the concrete caused by corrosion of the reinforcing steel inside. The risk of alkalinity problems are reduced if the concrete is thoroughly washed for several weeks with periodic renewals of water. In some cases it may be necessary to coat the concrete with an epoxy resin paint before the animals are introduced. Competent mixing and pouring is essential if corrosion is to be prevented.

8.1.2 Metals

As a general rule, and especially in recirculation systems, metals should not be allowed to come into contact with the water. In particular, metals to be avoided include copper, zinc and alloys containing these metals, such as brass, gun metal and bronze. Iron and most steels corrode readily in seawater, exceptions being titanium steel and, to a lesser extent, type 316 stainless steel. Newer, corrosion-resistant alloys with higher strength than 316 stainless steel have been developed, but may be too costly for general use. Protective coatings can be applied to most metals but in our experience are rarely satisfactory in the long term. The slightest fault or damage to the coat can lead to corrosion spreading beneath the coating and the unsuspected release of toxic materials into the water. Condensation dripping from galvanized bolts or cadmium-tipped masonry nails is another example of a potential source of contamination.

8.1.3 Plastics and other materials

A wide range of plastic, fibreglass and epoxy resin based materials are used in culture systems. The safest from a toxicity point of view are 'food grade' materials but it is nevertheless advisable to soak all materials in several changes of water for 10–14 days before use, to reduce levels of potentially toxic leachates (Carmignani & Bennett 1976). Additives such as colorants, plasticizers, antioxidants and stabilizers are often present in plastics and leach very readily from recycled plastics in particular. Mould release agents may be present on fibreglass tanks, which again should be well soaked before use. Wood preservatives are usually toxic and can be carried into the water, for example by drips from tanks held on wooden frames above reservoirs. Marine grade plywood sealed with epoxy resin paint is widely used in indoor installations.

8.1.4 Pond sealing materials

Several methods are used to line ponds built in areas where porosity, inward seepage or other soil inadequacies prevail (Wheaton 1977). All require proper prior pond preparation, removal of vegetation, sticks, stones etc. The methods include compaction of suitable soil (about 20% clay) to a depth of 20 cm with about six passes of a sheepsfoot roller or, more expensively, liners of butyl rubber, polyethylene, vinyl or similar materials. It is advisable to protect liners with about 15–25 cm of sand or coarse soil. The same recommendations about soaking for several weeks with frequent water changes apply here as to plastics.

8.2 Pond design and construction

We are concerned here with earthen or predominantly earthen ponds and we do not cover the design or construction of structures such as tanks or raceways. These aspects can be found in Wheaton (1977).

8.2.1 Layout and configuration

The layout and configuration of ponds within a crustacean farm will be largely determined by the type of culture to be performed (species and intensity) and the characteristics of the site, particularly the topography, soil type and the positioning of the water supply. Before design of the layout a comprehensive survey is vital to determine the soil type (to a depth at least 1 m below the intended base of the ponds) and detailed levels throughout the whole area involved (Sections 6.3.3 and 9.6.1)

If an extensive operation is planned, the general objective will be to perform a minimum of earthworks to create a maximum surface area of ponds. In the case of semi-intensive culture, attention should also be given to creating straight embankments and roughly rectangular ponds and, if intensive production is intended, ponds should be restricted in size to a maximum of about 0.5–1 ha. Whereas most sites can be suitable for building small ponds, large extensive culture ponds (5–100 ha) require land with a shallow gradient such as that found in coastal and estuarine margins or on alluvial plains.

If the water reaches the farm site by gravity (freshwater, river or stream) or tidal flow (brackish water or seawater), the elevation of the ponds will be restricted in accordance with the level of the water supply. A pumped water supply, on the other hand, enables land above the level of the water source to be exploited and thus allows more flexibility in the arrangement of the farm and its drainage. The water intake point must be located away from the discharge of the same and other farms.

A reservoir can be advantageous for tidally flushed shrimp farms since it allows for exchange to be performed for longer periods than would be possible just on high tides. Even if a pumped water supply is provided it may only be efficient to operate the pumps on high tides, so again a reservoir can be useful to extend the duration of water exchanges. With this in mind, the distribution canal in many Ecuadorian shrimp farms is often widened to 30–60 m so that it acts as a reservoir. If a water quality problem is encountered, for example low dissolved oxygen concentrations at night, the reserve of water can be used to rapidly flush a pond. A reservoir can however act as a large sediment trap, so ideally it should be drainable to enable accumulated silt to be removed. For some crayfish farms with a seasonal or intermittent water supply, a dam or reservoir may be essential.

Prevailing wind direction may need to be taken into account when a farm layout is designed, either to maximize wind induced circulation or, by orientating embankments to interrupt the fetch of the wind, to minimize possible wave damage.

Some examples of crayfish farm designs are given in Chapter 7 (Figure 7.1). Further arrangements for crustacean ponds are presented in Figure 8.1. Ponds with curved margins are feasible for extensive culture and can be arranged to take advantage of site contours. In the three-phase system (Figure 8.1d) the small triangular ponds act as nurseries, and juveniles are transferred by gravity to the larger adjacent ponds for two successive phases of on-growing. In theory this arrangement helps to maximise the productive use of the available surface area of the farm. In practice though, it is vulnerable to irregularities in the supply of juveniles and becomes inefficient when the three steps get out of phase. Rectangular or square shaped ponds are most convenient when they are of a uniform size since this allows standard size nets to be used (important for frequent *Macrobrachium* seine harvests (Section 7.3.6)) and allows straightforward calculation of drain and fill times. With inlet and outlet points positioned in opposing short embankments, rectangular ponds permit efficient water circulation and exchange. Modern square ponds often have central drain structures. Various other shapes, such as triangles and rhombuses, are often incorporated in farm designs to maximize use of the available land, and these too can be suitable provided no unproductive 'dead' spots are created by restricted water movement.

8.2.2 Construction

The first step in construction may need to be the formation of a perimeter drainage ditch to prevent waterlogging of the construction area. On the other hand, if the earth is too dry it may require moistening before embankment construction commences. The usual construction technique is by cutting and filling, in which material is skimmed from the pond beds and transferred to form embankments, usually with the aid of earth moving machinery. Alternatively, construction can be performed by excavators, although ponds created in this way are typically small and can be difficult to drain without the aid of pumps. If a series of ponds are to be constructed, it makes good sense to start with the water supply canal and then build the ponds around it, because in this way some ponds can be brought into production before the whole farm is completed.

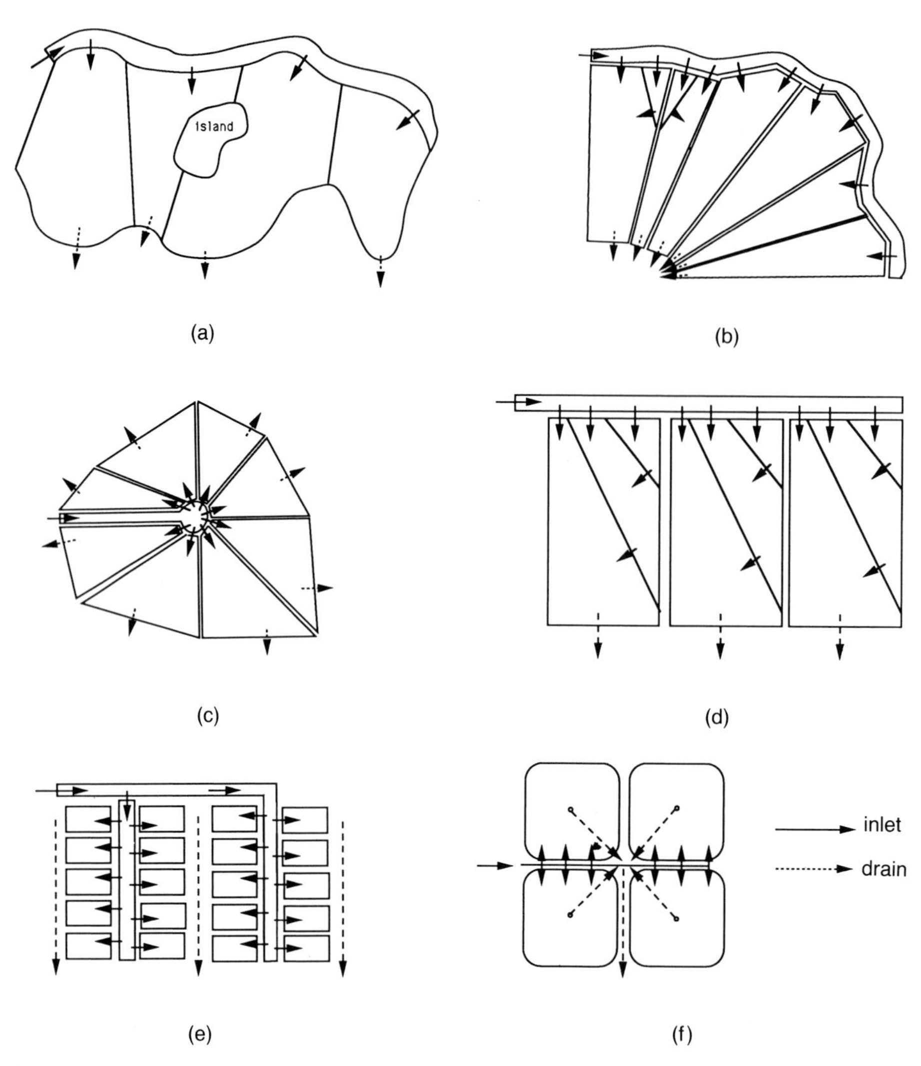

Fig. 8.1 Examples of pond layouts: (a) supply canal and ponds following natural contours; (b) supply canal following natural contours, two small nursery ponds are included; (c) daisywheel; (d) 3-phase pond system; (e) rectangular ponds; (f) centrally drained, concrete-walled ponds.

To ensure efficient drainage, pond beds should be sloped with a minimum gradient of 0.1% or preferably 0.2–0.5%. Good drainage can be essential for harvesting (Plate 7.2) and pond preparation (Section 8.3.5.1). Tree stumps, peat and other organic material should be removed from the pond beds and all holes or depressions must be filled because they can retain animals during harvesting and may harbour pests between crops. Channels cut into the bed can improve the drainage and provide deepened areas which serve as refuges from extremes of ambient temperatures. The channels may be arranged around the periphery of a pond, diagonally across the middle or in the shape of a fish bone, but to ensure efficient harvesting they must all be sloped to the outlet point. Deep peripheral channels, such as those found in many Thai and Ecuadorian farms, are created to provide material for adjacent embankments. However, these channels readily become fouled with black organic muds and often prove laborious and difficult to clean.

Sites for pond construction should be chosen with impermeable soils that are suited to embankment construction (Section 6.3.3). In some situations permeable soils can be sealed with clay blankets or with the addition of sealants such as bentonite, a fine-grained clay. When using sealants, laboratory analysis of soil will be necessary to establish what type of sealant is appropriate and what quantities will be needed (Wheaton 1977). Embankments can also be made impervious by using sealants or by incorporating a clay barrier. The latter may take the form of a clay layer on the embankment's inside surface, or a central clay core around 0.5 m thick and in contact with an impervious layer of the substrata.

Pond size and shape have an important bearing on the amount of earth that needs to be moved to construct a farm. The relationship between pond size and volume of earthworks for a 40 ha farm is shown in Figure 8.2a. Larger ponds require less embankment per hectare and consequently they are less expensive per hectare to build. On the other hand, they are unsuited to intensive culture methods because of the difficulty of effectively regulating water quality over such a large area. Eight hectares is considered an ideal size for extensive crayfish ponds in the USA (Avault & Huner 1985).

The relationship between pond shape and volume of earthworks is illustrated in Figure 8.2b. Clearly square-shaped ponds (length:width ratio 1:1) require the smallest volume of earth to be moved per hectare (only rectangular forms considered). All the same, despite their added cost, elongated ponds are often preferred since this shape facilitates feeding, harvesting and pond maintenance (Gibson & Wang 1977). Ponds of maximum width 30–50 m are favoured for freshwater prawn culture, especially where the frequency of seining is high, and narrow canal-type ponds, around 10 m wide, are popular for semi-intensive culture of crayfish because of the bank-burrowing habits of these crustaceans.

8.2.2.1 Embankments

In general, embankments should be over-built because they will always be subject to erosion, both from wave action at the edge and rainfall on the surface. Embankments made with sandy soils are particularly at risk and may need to be twice as wide as clay ones. The optimum height of embankments depends partly on the species in culture and the climate involved. Deep ponds can protect a crop against extremes of temperature but they also take longer to warm up than shallow ponds and this may not be desirable in some temperate climates. For extensive crayfish ponds a water depth of 0.3–0.5 m is adequate and enables low cost levees to be built (Avault & Huner 1985). Minimum water depth in other cases, however, should be 0.8 m, to reduce the growth of vegetation on the pond bed. In brackishwater ponds this minimum depth helps guard against salinity fluctuations caused by heavy rains. Average depths of 0.8–1.5 m are usually ideal for semi-intensive and intensive culture, although even deeper ponds may be desirable where plastic liners are used (Sections 7.2.6.5 and 12.2).

Embankment construction should allow for an additional freeboard of 0.3–0.7 m. The width of an embankment at its top depends primarily on access requirements but should at least equal the height and should never be less than around 1 m. Two to

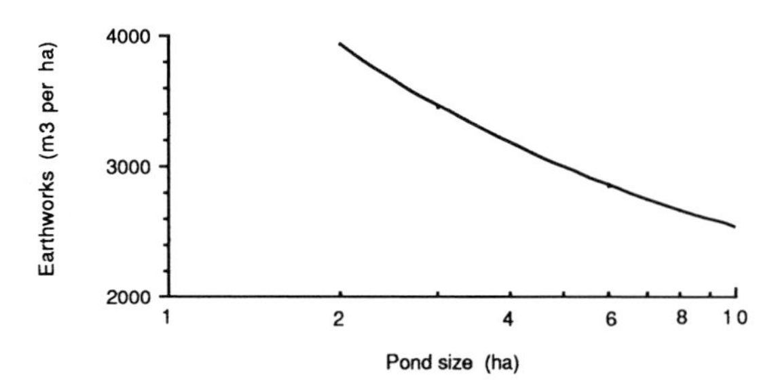

Fig. 8.2a Relationship between the amount of earth to be moved per ha and pond size for a 40 ha farm (based on Yates 1988).

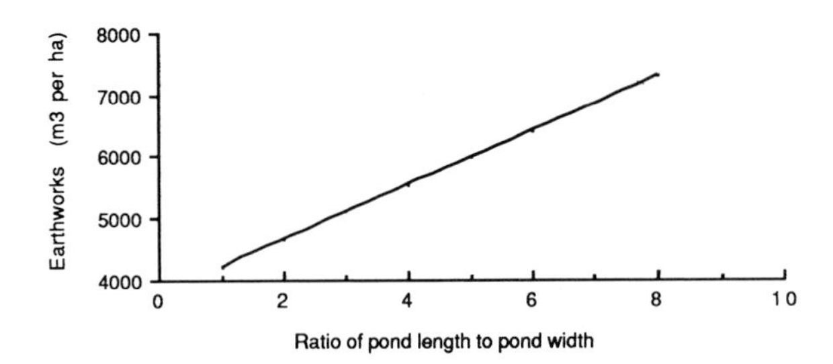

Fig. 8.2b Relationship between pond shape and amount of earth required to be moved for a 40 ha farm with ponds of 2 ha each (modified from Yates 1988).

three metres is more usual and 3.5–5.0 m may be required if the embankment is to carry vehicles safely. On exterior dry faces a slope of 1–2:1 (horizontal:vertical) is suitable, whereas inside faces in contact with water need shallower slopes with a minimum gradient of 2:1 and preferably 3–4:1. The shallow slope is especially important if embankments are made of light earth (non-cohesive) or will be subject to strong wave action.

It is important to construct embankments correctly at the outset because mistakes are very difficult to remedy and embankment failure can jeopardize the stock of a whole farm. The first step is to excavate down to a water-tight foundation. Embankments should then be constructed in a series of layers about 20 cm thick. Each layer must be thoroughly compacted, for which the soil may require moistening. Allowances should be made for embankment settlement which can be 10% of height, and for soil shrinkage which can be 10–20% of volume. Spaces can be left in the embankment for the installation of inlet and outlet gates, or alternatively the embankment can be constructed intact and later cut away in the relevant spots using a back-hoe. Baffle levees (small embankments) can be positioned in some extensive ponds to direct water flow over a greater area of a pond (Avault & Huner 1985).

Vegetation is valuable for embankment stabilization and it can be encouraged with a layer of topsoil. Excessive growth, however, is generally a hindrance to harvesting and should be controlled. Trees should not be planted because their roots will weaken the embankments. To reduce embankment erosion gravel or porous plastic sheeting can be placed on inside surfaces, and to limit further wave damage on downwind sections an array of vertical wooden stakes can be effective. If one corner of the embankment is widened with a ramp it will allow for vehicle access during pond cleaning or maintenance.

New ponds should be filled slowly over a period of several days to allow the embankments to become fully saturated before they are subjected to the full weight of water.

8.2.2.2 *Inlet and outlet structures*

In the simplest shallow ponds water may be controlled with plastic, metal or bamboo tubes fitted with valves or turn-down drains. The tubes are buried in the embankment and should ideally be fitted with collars to reduce water seepage. To prevent the passage of shrimp or other species a staked net may be positioned upstream of the inlet, or a strainer of netting or split bamboo located on the end of the tube.

Monks and sluices are more specialized water control devices (Figure 8.3). A sluice may be constructed of wood, brick or concrete and forms an opening in an embankment. Vertical grooves located in the sluice walls accept mesh screens and wooden

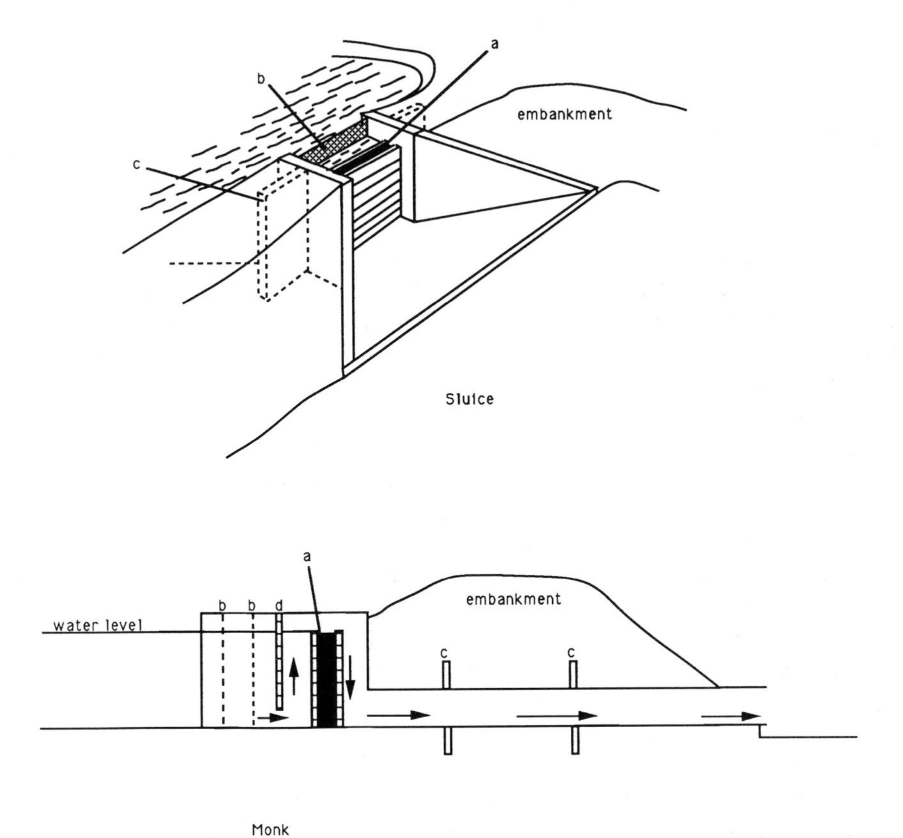

Fig. 8.3 Water control structures: (a) earth packed between two vertical stacks of wooden boards; (b) mesh screen; (c) brickwork or concrete keyed into embankment to stop seepage; (d) stack of boards with gap at base.

boards, the latter to control the water flow rate or pond level. A monk is located within the pond or water supply canal and is connected to a tube or channel which passes through the embankment. The advantage of this arrangement is that the embankment is left largely intact and vehicles can easily pass over it. Monks also incorporate grooves for mesh screens and for water control boards (unless water flow is regulated by a valve located in the tube). To increase the surface area of screens and thus reduce the frequency of blocking, the upstream end of a monk may be flared or formed in the shape of a Y. Centrally drained intensive ponds are sometimes equipped with a central monk structure or a vertical stand-pipe surrounded by a mesh screen.

If two vertical stacks of boards are used for water control they can be set 15–25 cm apart and the space between them packed with soil to make a watertight barrier. The monk depicted in Figure 8.3 is equipped with two sets of grooves for this purpose, another two sets for screens (coarse and fine) and a fifth set in which boards can be located to encourage the exchange of bottom water rather than surface water. To prevent the theft of a crop by draining, the water level control boards can be locked in position with a padlock or a secure lid. If drain harvesting is intended a harvesting basin can be located immediately behind the outlet gate. This will serve to keep the catch submerged as it is collected.

Pumping stations require solid foundations of concrete or wooden piles and the point where the pumps discharge their water must be protected from erosion with stones, rocks or a concrete spillway.

8.2.2.3 Construction in areas with acid sulphate soils

The special problems posed by acid sulphate soils are discussed in Section 6.3.3. Although affected sites should generally be avoided, some measures can be taken at the pond construction stage to limit the potential impact of soil acidity.

Some extensive hand-built fish ponds have been constructed in acid sulphate areas by minimizing the disturbance of existing soil layers (e.g. sometimes roots are left in place). Since embankments are responsible for large amounts of acidity, building larger ponds (less embankment per hectare) can also have some benefits. The construction of small ponds may only be feasible if plastic liners are used, if walls are made of concrete, or if embankments are covered with a deep layer of non-acidic topsoil.

Under normal usage it may take several years for the pH of a pond built in acid sulphate soil to reach acceptable levels. However, repeated applications of lime can be beneficial and a method for determining lime requirements is provided by Boyd (1982). One approach, described by Brinkman & Singh (1982), is designed to reclaim acid sulphate sites in a single dry season lasting only four months. It basically involves harrowing, drying and filling the ponds, and then allowing the pH of the water to stabilize before draining and repeating the procedure. Up to three or more treatment cycles may need to be performed to raise the water above pH 5. At the same time embankments are leached by repeatedly filling and draining shallow basins constructed along their tops. Before the pond is used agricultural limestone is applied at 500 kg ha^{-1} on the pond bottoms and at 0.5–1.0 kg m^{-2} on the embankments.

8.3 Pond management

8.3.1 Introduction

Every pond system has its own limitations to productivity, based on its physical and chemical characteristics and available inputs of water, feeds and juveniles. The pond manager is faced with the challenge of making the most efficient and profitable use of these resources, knowing that conditions will rarely be optimal and that for the most part the crop will remain hidden from view. Successful pond management depends largely on collecting reliable and regular data on the condition of the crop and the status of the culture environment, and deciding how this information may best be used in the application of fertilizers and feeds, and in the control of water exchange and aeration. Predator control is also of fundamental importance.

It is worth noting at this point that very significant variations in productivity can arise between different farms and ponds, due to differences in soil characteristics. This emphasizes the important role that site selection can have on project viability and the need to investigate soil quality before a site is chosen (Section 6.3.3).

8.3.2 Basic biology

An understanding of the basic biological processes at work in a pond, and the chemical changes they bring about, is very helpful to the farmer if he or she is to manage pond conditions effectively and keep water quality within acceptable ranges (Section 8.5).

Phytoplankton form the basis of the natural food chain in a pond ecosystem, and in the process of photosynthesis they use the energy of sunlight to synthesise organic molecules. This process can have a profound influence on water quality, largely because in the daytime it results in the liberation of oxygen and the consumption of carbon dioxide. Oxygen is essential for crustaceans and nearly all other organisms within a pond, since it is needed for respiration. However, it can rise to harmful levels (see below). Carbon dioxide concentration is important primarily because of its influence on pH. Carbon dioxide acts as an acid in water, so as it is removed during photosynthesis acidity declines and pH rises. On sunny days the pH in ponds rich in phytoplankton can rise to pH 9 or 10, at which levels the growth rates of the crustaceans can be impaired.

During darkness the metabolism of the phytoplankton changes from photosynthesis to respiration (the process by which organic molecules are oxidised to obtain energy); oxygen is consumed and carbon dioxide is released. The latter adds to the other respiration products (e.g. ammonia) of the crustaceans being farmed (and those of nearly all other non-plant organisms in the pond) with the result that oxygen concentrations and pH fall by night to reach a minimum around dawn. Figure 8.6a illustrates a typical pattern of variation in pH and dissolved oxygen levels within a diurnal cycle in an outdoor pond. Unfortunately an excess of oxygen at one time of day does not compensate for a deficit at another, and both high and low concentrations can be harmful to crustaceans. In fact, abrupt changes in any water quality factor are likely to be stressful and will adversely affect growth and susceptibility to disease (Section 8.9).

When circulation within a pond is poor, often as a result of calm weather, water can become strongly stratified with regard to temperature, pH and oxygen. In the absence of mixing, conditions deteriorate at the bottom of a pond as oxygen is consumed, pH declines and ammonia concentrations rise. Stratification can be especially bad in brackish water ponds following rainfall, when a layer of less dense freshwater may form on the surface. This severely impedes the process of gaseous exchange between the water and the atmosphere which normally helps to limit the fluctuations in carbon dioxide and oxygen levels that result from photosynthetic activity and respiration. Stratification can be counteracted by circulation and aeration (Sections 8.3.5.6 and 8.3.5.7).

Within the natural food web the phytoplankton are the primary producers that provide food for organisms at higher levels in the food chain. Zooplankton, for example, will graze on phytoplankton, and plankton productivity as a whole will support a community of micro-organisms (bacteria, fungi, protozoa) and invertebrates (worms, molluscs, crustaceans) on the pond bed, principally by providing a rain of nutrient organic material (faecal pellets, dead organisms, exuviae). The monitoring and control of algae populations are critical aspects of pond management (Section 8.3.4.2). If very dense phytoplankton populations develop they can rapidly exhaust supplies of inorganic nutrients and undergo a catastrophic mortality known as a 'crash'. The resulting mass of decomposing organic matter can consume much of the available oxygen and endanger the crop once more. Benthic algal mats and uncontrolled growth of macroalgae are also likely to jeopardise crustacean production.

Some cultured crustaceans feed directly on primary producers. Crayfish and prawns frequently feed on aquatic plants and most penaeid shrimp and crayfish will consume benthic micro-organisms. Hence the natural productivity of a pond represents a valuable source of nutrition for crustaceans and, in general, for this reason it should be encouraged. Since phytoplankton growth is often limited by the availability of inorganic nutrients it can be encouraged by the addition of fertilizers (Section 8.3.5.3).

Although natural productivity alone, or enhanced with fertilizer, can be adequate for extensive cultures, semi-intensive and intensive farming operations require the addition of supplementary feeds to maintain rapid growth rates. However, in addition to boosting growth, the application of feeds has important implications for water quality and must be controlled if the dangers of anaerobic conditions and ammonia toxicity are to be avoided (Section 8.3.5.4).

Feed is not only consumed by the crustacean crop; uneaten and partially digested fragments are also consumed (decomposed) by bacteria, other micro-organisms and invertebrates on the pond bed. Although this community often provides food for the crop, a major part of its impact relates to its heavy demand for oxygen. In one study on shrimp ponds receiving a daily average of 37 kg of feed ha^{-1}, oxygen consumption in the sediments accounted for 51% of the total oxygen demand of the whole pond system! The shrimp, by comparison, accounted for only 4%, with the remainder consumed by organisms within the water column (Madenjian 1990). Comparable results were obtained in freshwater prawn ponds, again emphasising the critical influence of sediment respiration on water quality in ponds receiving supplemental feeds (Moriarty & Pullin 1990).

8.3.3 Stocking

Although it is advantageous to standardize stocking densities, particularly beyond the level of extensive cultures, in practice variations may be necessary depending on

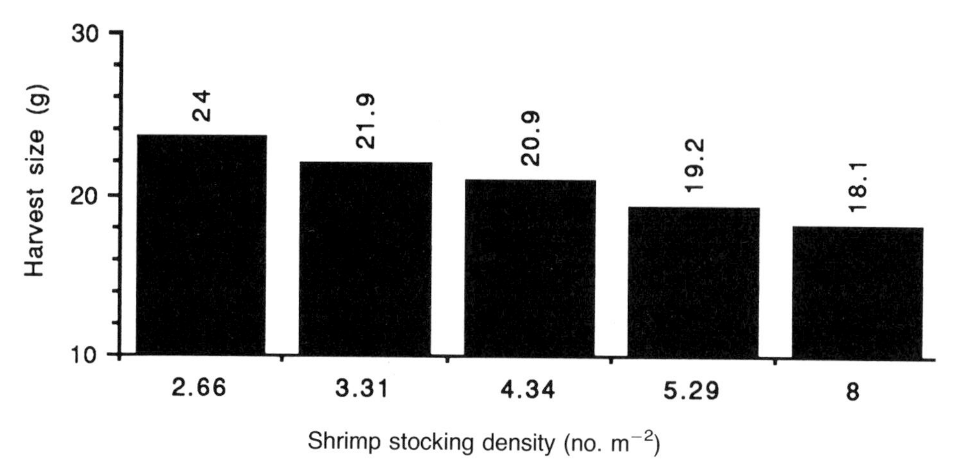

Fig. 8.4 Relationship between stocking density and harvest size for extensive/semi-intensive *P. vannamei* farming in Ecuador (based on 152 observations) (Hirono 1986).

season and the availability of juveniles. Periodic review of yield levels, feed conversion ratios and the size range of harvested animals will indicate the most profitable density at which to operate (Section 10.9). Since stocking density influences the size of animals at harvest, it can to a certain extent be adjusted so that the product meets the size requirements of the markets (Sections 3.2.3.1 and 7.3.7). Figure 8.4 illustrates the relationship between stocking density and harvest size for extensive/semi-intensive shrimp culture in Ecuador.

During the stocking process and the period directly afterwards, there is an enhanced risk of mortality due to predation and stress. When animals arrive for stocking they are usually weakened as a result of handling and transport and should be acclimatized gradually, minimising exposure to rapid changes in environmental conditions. Young post-larvae of hatchery origin are generally more delicate than larger nursery-reared or wild-caught juveniles. Requirements for acclimatization vary from one situation to the next, but in the tropics stocking during the heat of the day should be avoided.

During stocking it is useful to count a number of juveniles (usually 100 or more) into a test cage placed in the pond, where they can be fed if necessary and counted again two to four days later to estimate stocking mortality. The results are helpful to pond management since low survival can give an early warning of problems and may help in identifying the cause. Rather than waste time and effort on a pond where high stocking mortality is suspected, it may be better to re-stock. However, if high survival is obtained in a test cage this does not guarantee that other problems, notably predation, will not influence the outcome in the pond at large.

8.3.4 Monitoring

If ponds are to be managed successfully, regular and reliable information must be gathered on pond conditions and the status of the crop. Although certain management procedures can be standardized, many critical decisions must be based on the daily measurements made in the ponds rather than standard curves for growth, feeding and water exchange (Section 8.6).

Regularly reviewed, accurate records provide a basis for the understanding of performance trends and the effects of different management strategies. Sometimes as ponds mature over their first two to four years of operation they gradually become more productive. Eventually, however, a steady decline in yield may be detected indicating an overall deterioration in pond conditions. Assuming this is not related to general environmental degradation outside the farm (Section 6.3.1.4; 11.5.2), it can signify the need for improved pond bottom treatment between crops and possibly the need to overhaul the ponds and remove accumulated silt and organic sediment (Section 8.3.5.1).

Accurate records are also essential to the understanding of day-to-day problems and accounting for irregularities which only become apparent at harvest time (e.g. unexpected mortality). Any technicians involved in the collection of data should be aware of its importance and be well instructed in the use and callibration of their instruments (Section 8.7).

8.3.4.1 Crop biomass and growth

Obtaining good estimates of the crop biomass (or standing crop) present in a pond is essential for the efficient management of feeding rates (Section 8.3.5.4; Table 7.5). For the crustacean farmer this process is complicated by the fact that the crop remains largely invisible and survival rates cannot be determined with any precision until harvesting is completed. Crop biomass may be estimated by measuring the average size of the crustaceans and estimating the number of individuals present in the pond (population estimate).

The best methods of estimating population density before harvesting are only accurate to perhaps ±20%. They rely on laborious, repeated sampling and assume a relatively random distribution of animals within a pond. Accuracy can be improved if samples are taken at various points in a pond and results for several crops are compared with actual harvest numbers (Falguiere *et al.* 1989). In time a rough picture will emerge for each pond of how the animals distribute themselves, but this can vary with pond bottom configuration (Section 8.2.2), temperature, season, size of animals, stage of the tide (in coastal or estuarine ponds) and time of day. Once the most reliable procedure has been determined, sampling procedures can be standardized. Samples may be taken using cast-nets with a known catchment area or by seining specified areas. Sometimes it is possible to estimate numbers by snorkel-diving and counting the animals seen while swimming over a given distance. An added advantage of diving is that the amount of uneaten food remaining in a pond can be determined (Section 8.3.4.3).

Bearing in mind the general unreliability of sampling methods, the use of a standard mortality curve based on expected losses (low level cannibalism, predation, escapes) is often the simplest practical way of making a population estimate. This method is likely to be of real value only if it is continually updated in the light of farm yields (numbers and sizes) and takes account of differences in performance between groups of ponds. Large well-managed farms usually incorporate all these methods in order to minimize food wastage and pond fouling.

The unpredictability of initial post-stocking mortality can be overcome somewhat by using a mortality test cage (Section 8.3.3).

Regular samples, at seven to 14 day intervals, can be taken to estimate growth increments, and providing no large-scale mortality takes place within the crop, the growth rate can be taken as a fundamental indicator of the success of pond management. Growth which is close to predicted values indicates that pond conditions are adequate and that the crop is healthy and feeding well. Samples of individual weights of crustaceans provide an impression of size variability, which in the case of *Macrobrachium* may assist in the planning of partial harvesting (Section 7.3.6).

8.3.4.2 Water quality

Water quality measurements in a pond are best taken either at the water exit point or where good access can be obtained to deeper water. It may be useful to build a small jetty for this purpose, as long as its structure does not interfere with harvesting. The subject of water quality monitoring is of general importance to all phases and types of crustacean culture and is also discussed in Section 8.6. Desirable levels and ranges for some of the important water quality factors are given in Table 8.3.

Dissolved oxygen (DO) readings are usually taken at least twice a day for water close to the pond bed. Afternoon measurements can be used to monitor peak DO levels induced by photosynthetic activity. Since the most critical period for low oxygen levels is around dawn, readings can be taken in advance to give forewarning of likely problems. One method relies on plotting measurements taken at 8 pm and 11 pm, and extrapolating to obtain a predicted DO level at 6 am the next day (Boyd 1990). Remedial action (e.g. activating aerators) can be taken if the predicted level falls below a set minimum. Some shrimp farms aim to keep levels above 4–5 ppm and Boyd (1990) notes that feed conversion ratios in shrimp ponds increase drastically if DO levels fall below 2–3 ppm at night. The oxygen requirements of moulting crustacea are considerably higher than those of inter-moult animals and, since moulting usually occurs at night, it may be doubly important to increase aeration or exchange rates during this period.

Temperature measurements are simple to make and can be used to quantify the diurnal and seasonal variations which influence feeding and growth rates, and to observe any differences between surface and bottom waters which would indicate signs of stratification.

Turbidity can be conveniently measured using a Secchi disc at least once or twice a day, although in intensively managed ponds Wyban *et al.* (1989) recommend three daily observations to check that phytoplankton levels are stable (Section 8.3.5.3). If water samples are viewed under a microscope and phytoplankton counted using a haemacytometer, the concentration of microalgae can be compared to Secchi disc readings to establish what part of turbidity is due to suspended organic detritus and sediment and what part is due to the presence of the algae. Observations of water colour can suggest which type of algae is predominant. For example, in brackishwater ponds a green coloration is generally indicative of flagellates, while brown generally denotes diatoms. Unusual colours or shades can be useful indicators of the presence of toxic algae blooms similar to 'red tides' (see Glossary).

Salinity can fluctuate widely on estuarine sites and can rise due to evaporation. After heavy rains differences in salinity between surface and bottom water can be used to check for stratification. Salinity can be measured with a refractometer to a precision of ±1 or 2 ppt.

Commercial test kits can also be used to measure the concentrations of ammonia, nitrites, nitrates, phosphates and silicates, although when testing brackish or saltwater it should be remembered that certain kits are designed for use only in freshwater (Section 8.6). Usually it is not necessary to measure these concentrations on a regular basis, except for ammonia in intensive systems which is liable to build up unless adequate regular flushing is performed.

8.3.4.3 Other observations

Observations of the amount of uneaten food remaining every day should be made in more intensive pond systems because daily adjustments in feeding rates may be necessary, and the first sign of stress is often a cessation of feeding. Mesh trays with food on them can be lowered to the pond bed, left overnight and inspected the next morning. Such observations can help establish whether a problem with sluggish growth is related to underfeeding or not. If the farmer notices a decline in growth rate and responds by increasing feeding rates without checking that the food is actually being consumed, he or she risks severely polluting the pond (Section 8.3.5.4). Some observation of food remains and the condition of a pond bottom can also be made by snorkel-diving with a torch.

The softness of a pond bottom can be used to locate areas of accumulated organic sediments. Softness can be gauged using a pole from a boat and the information can be used to select the best sites for, or to re-orientate, aeration devices to prevent or disperse the sediments. Similarly, by wading into a pond, taking a sediment sample and observing odour, texture and colour, it is possible to make simple deductions about the organic load in a pond and the presence of the highly toxic hydrogen sulphide gas. Black sediments with a foul smell like bad eggs are indicative of the latter, which forms as the result of sulphide excretion by anaerobic bacteria. The toxic effects of hydrogen sulphide are felt at very low concentrations and are greatest when the pH is low (acidic conditions). The production of this dangerous gas can be minimised by maintaining aerobic conditions throughout the pond and the topmost sediments, by avoiding overfeeding and by allowing the drying and oxidisation of organic sediments between crops (Section 8.3.5.1). Problems with the build-up of hydrogen sulphide and ammonia in more intensive systems have been ameliorated by the addition of zeolite at 250 kg ha^{-1} (Chen 1990).

Observing the behaviour and condition of the crustaceans can provide a timely warning of existing and potential problems. Shrimp, for example, seen circling around the edge of a pond may be suffering from stress due to lack of oxygen, and crayfish under similarly low oxygen conditions may even start to migrate from a pond. The presence of dead animals in population samples or around the margin of a pond is an obvious cause for concern. Carcasses, however, are usually rapidly consumed and a steady mortality over a long period may go unnoticed. Recently dead animals or those showing signs of abnormality can be preserved for microscopic and histological examination if disease is suspected. Observed softness in shell texture indicates recent moulting or a problem with shell mineralization. Soft shelled or pre-moult crustaceans stop feeding, and if moulting is largely synchronous in a pond population (sometimes caused by the influx of new or freshwater) it may be necessary to reduce feeding rates or delay a planned harvest. In transparent shrimps and prawns the fullness of the alimentary tract can be observed, to establish whether the animals are feeding and (in support of observations of feeding trays and water quality) to assist in decisions on feeding rates.

8.3.5 Control

8.3.5.1 Pond preparation and rejuvenation

Since crustaceans dwell and forage on the pond bed, the condition of the substrate is vital to their well-being. Pond preparation, both initially and between cycles, has a considerable impact on the substrate and water quality, particularly at the early stages of the pond production cycle. Preparation basically involves draining, drying, ploughing and chemical treatment. Drying enables air to penetrate the sediments and assists in the breakdown and mineralization of organic matter and the release of hydrogen sulphide. The mineralization of organic matter produces inorganic nutrients (nitrate, phosphate, carbonate) which will improve the fertility of the pond, reduce the oxygen demand of the sediment and so reduce the impact of any previously formed anaerobic decomposition products.

Opinions on the best period of drying vary: seven days or until the top 1 cm of the soil has dried, has been recommended in brackish water shrimp ponds (ASEAN 1978).

Turning the top 10–15 cm of the pond bed with a plough exposes more of the sediments to the air, but may be prohibitively expensive between every crop. In addition, the pond bed must be firm enough to support the machinery and this may mean longer drying periods. After ploughing it may be necessary to recompact the soil to give a firm surface. With intensive ponds the deep rich organic sediments which accumulate may be pumped out after harvesting to avoid the excessively long drying-out and reoxidation period required to recondition the ponds.

Treating pond beds with lime (see Glossary) has several beneficial effects. It increases the pH of the mud, improves benthic productivity, buffers against large daily fluctuations in the pH of the water, boosts primary productivity by increasing the availability of carbon dioxide for photosynthesis, and improves the availability of nutrients, particularly phosphates. Its impact is most beneficial when accompanied by a program of fertilization. Methods for calculating lime requirements for ponds are detailed by Boyd (1979). Application rates of agricultural limestone (calcium carbonate) for shrimp ponds in Ecuador, recommended by Lucien-Brun (1988), range from 1840–410 kg ha^{-1} for soils with pH values from four to six. Limed ponds should be filled with water and left for at least a week, and the pH of the water checked before animals are introduced. Ponds are sometimes treated with other chemicals specifically aimed at eliminating predators (Section 8.3.5.2).

When a pond is refilled prior to restocking, a bloom of phytoplankton can be encouraged in a little water (10–30 cm deep) by the addition of fertilizers (Section 8.3.5.3). As the phytoplankton density increases, the water level can be raised in steps.

Pond preparation may include routine repairs to drainage channels, embankments and water control structures, and the filling of holes. In the case of extensive crayfish ponds it may be necessary to plant forage crops (Section 7.5.4).

8.3.5.2 Predators and competitors

Most predators and competitors in crustacean ponds can be eliminated by draining and drying between crops, provided pond beds are well constructed and no pools remain (Section 8.2.2). Chemical control may be required if complete draining cannot be achieved. While a pond is in production, mesh screens on inlet and outlet gates prevent the entry of most adult water-borne predators and competitors.

Teaseed cake, the residue after extracting oil from the seeds of *Camellia*, can be applied as a selective fish poison. It contains 10–15% saponin which is 50 times more toxic to fish than to shrimp, and biodegrades after a few days. The cake must be dried, ground and soaked in water for 24 hours. It is applied to shallow water and puddles at recommended rates of 12–20 g m^{-1} to give 1.2–3.0 ppm saponin (ASEAN 1978). The use of higher dosages, 2.5–10 ppm saponin, is reported by Chen (1990). Rotenone is another selective fish poison which is most effective in fresh or low salinity water and is often applied in the form of derris root which contains around 5% active ingredient. Recommended application rate is 4 ppm of dry root which gives 0.2 ppm rotenone. After poisoning, dead fish must be removed and destroyed and the treated pools left for several days for the chemicals to deactivate.

Some steps may be necessary to control birds. Scarers can be used, though these tend to loose their efficacy with time. Strings stretched across ponds can be an effective deterrent to diving birds but are only feasible for small units. Wading birds can be discouraged by regular attention to embankments to maintain their slopes, and by eliminating shallow areas in ponds. Baited jars sunk in pond banks can be effective traps for land crabs. Fences may be necessary around ponds to keep out various frogs, snakes and toads or to prevent non-native species of crayfish from escaping to natural waters. Night watchmen, guard dogs (or geese on small operations), lighting and perimeter fencing may be essential to guard against theft, and stakes can be positioned to interfere with cast nets (Plate 8.1). Total security on large farms, however, is very difficult to achieve.

8.3.5.3 Fertilization

Fertilizer is the principal method of promoting natural productivity in ponds where the concentration of inorganic nutrients is low. Fertilizers are of two types: organic and inorganic. The former represent much less concentrated sources of nutrients and are thus much more bulky to transport. For example, 37 kg of dry chicken manure supplies the same amount of nitrogen as 1 kg of urea. Although organic fertilizers provide additional material to boost benthic productivity, if used excessively their decomposition can create anaerobic conditions on the pond bed. Despite their disadvantages they are often readily available and cheap and can be ideal for small-scale and extensive aquaculture operations (Blakely & Hrusa 1989). The manure of geese and ducks is often preferred to others for its relatively high phosphate content. Chicken and cow manures, however, are often available in larger quantities. It should also be remembered that there is a danger of animal wastes being contaminated with pesticides and other chemical products such as antibiotics.

Fertilizers, particularly manures, are often applied to the pond bed in advance of filling and stocking. Once an initial phytoplankton population has been established, usually after 5–15 days, its maintenance usually requires further fertilizer applications

Plate 8.1 Bamboo stakes arranged around the margin of a Thai shrimp pond to deter cast-netting by poachers. Paddlewheel aerators and a hatchery facility are also visible.

unless a nutrient rich water supply is employed. Fertilizers may be slowly leached into the water from floating perforated plastic drums, sunken wicker silos or from porous sacks held in the inflowing water current or tied to stakes within the pond.

Alternatively they may be placed on a wooden platform 30 cm below the water surface. It is advisable not to broadcast solid fertilizers over the pond since the nutrients will be deposited on the pond bed rather than used to fuel primary productivity in the water column. This may cause a carpet of benthic algae to develop in ponds where light can reach the bottom, an effect which is undesirable and sometimes necessitates the application of algicides.

The control of phytoplankton productivity can be difficult, especially in large ponds. It usually requires manipulation of the rates of water exchange and feeding as well as fertilization. The objective is usually to keep turbidity levels (as measured with a Secchi disc) within set limits. Visibility to a depth of 25–40 cm is generally recommended for shrimp or prawn ponds, although some authorities recommend 15–35 cm for freshwater prawn culture (Malecha 1983). Supplementary feeds partly act as organic fertilizers, so fertilization rates usually need reducing as the standing crop and feeding rates build up in the later stages of a culture cycle.

The most important components of fertilizers for phytoplankton are nitrogen and phosphorus. To a certain extent fertilization regimes can be designed to favour particular types of phytoplankton. Diatoms, which are usually prevalent in moderate or high salinity water, are usually preferred in shrimp ponds and can be encouraged with fertilizer high in nitrogen. In contrast generally undesirable blue-green algae, which often bloom in lower salinity water, are able to fix dissolved atmospheric nitrogen and are thus likely to be favoured by fertilizers high in phosphates. Boyd (1990) recommends roughly a 20:1 ratio of nitrogen:phosphorus to maintain diatom dominated blooms in brackish water ponds. This equates to a 9:1 ratio of urea:triple superphosphate. Although diatoms require silicates, the value of using silicate fertilizer is unknown and likely to be site specific. Boyd (1990) notes that most tropical brackish waters have fairly high silicate concentrations. Interestingly, despite a general preference for diatoms in shrimp ponds, flagellates are preferred in some Taiwanese farms because their concentrations are more stable and easier to control (Chen 1990).

Although chemical analysis may reveal which nutrients, if any, are limiting to productivity, so assisting in the formulation of a suitable fertilization programme, the pattern of nutrient availability can be expected to change greatly with rainfall and season. Routines for fertilization should be established as experience is gained with each pond and at each site. Greater fluctuations in algal populations may occur in plastic-lined ponds compared to earthen ponds since the former may tend to develop less diverse and less stable populations of plankton. As a result, greater control over water exchange and the application of fertilizer becomes necessary.

Hard and fast rules for the frequency of application of inorganic fertilizers do not exist and recommendations vary from two or more times a week (Boyd 1990) to once every two to four weeks (Blakely & Hrusa 1989). Decisions should be based on changes in turbidity shown by Secchi disc readings. In response to excessive algae concentrations, applications should be reduced in size rather than eliminated, in order to reduce the risk of a population crash.

Examples of fertilization regimes (bi-weekly quantities ha^{-1}) include:

(1) For semi-intensive fish/*Macrobrachium* culture in Hawaii (Malecha 1983):
60 kg single superphosphate
60 kg ammonium sulphate or liquid ammonia;

(2) For shrimp ponds in Hawaii (Chamberlain 1987):
6.6 kg urea
2.7 kg triple superphosphate
10 kg calcium silicate
0.7 kg mineral mix (made from zinc oxide).

Details of 21 different fertilization schedules applied for aquaculture ponds are provided by Tacon (1990).

8.3.5.4 Feeding

The productivity of ponds relying on natural productivity alone or natural productivity boosted by fertilization rarely exceeds 500 kg ha^{-1} per crop. Feeding is essential if greater yields are required. Apud (1986) linked the need for feed in shrimp culture to the stocking density employed, placing it at densities greater than 2–5 shrimp m^{-2} for *Penaeus monodon* and above 5–10 shrimp m^{-2} for *P. indicus*.

The quantity of feed used should be based on regular estimates of the standing crop and should be reviewed daily in the light of water quality measurements and in some cases observations of food remains (Section 8.3.4.3). If weekly regular growth increments fall seriously below target, feeding rates should only be increased if it can be established that there is no water quality problem and that the great bulk of applied feed is being readily consumed. If adequate growth cannot be maintained it may be necessary to harvest all or part of the crop or switch to a diet with a higher protein content. Certainly there is no sense in adding feed to a pond where no growth is being registered.

The normal feeding rate can be expressed in terms of percent body weight per day and this percentage in most systems will decline as the average size of the animals increases. An example is provided in Table 7.5, where the daily ration for shrimp declines from an initial 50% to 3.2% during the culture cycle. The feeding rate for a whole pond will be the relevant percentage of the standing crop.

The utilization of feeds can be improved by increasing the number of applications per day. Two to four feeds per day are usual in semi-intensive shrimp and prawn cultures. Feeds are usually broadcast by hand either from the pond banks or from small boats. Mechanization is possible using feed blowers and the operators of one very large Ecuadorian shrimp farm (1600 ha) (Plate 7.1) found it convenient to distribute pellets using an adapted crop-spraying aircraft.

Standard commercial crustacean diets are usually incomplete in their nutritional profile (Section 8.8.2) and animals have to rely on natural productivity in a pond to make up for shortfalls in essential nutrients. As the standing crop in a pond increases, the supply of essential nutrients becomes limiting to growth unless a higher quality diet can be used. Improved diets usually contain greater proportions of protein and this is reflected in their higher price. It may be worth increasing the quality of a diet in the later stages of on-growing, provided that improved growth compensates for the extra expense.

The food conversion ratio (FCR) relates the weight of feed applied to the weight of crustaceans harvested. For dry diets low ratios (<2:1) are desirable and indicate that feeds are being converted efficiently into crustacean flesh and/or that natural productivity is making a significant contribution to growth. Higher food conversion ratios (>3:1) are suggestive of overfeeding, poor diet quality or slow growth. Efficient conversion of feeds is critical to the economics of crustacean culture.

8.3.5.5 Water exchange

Daily water exchange is important to maintain healthy phytoplankton blooms, to flush away toxic metabolites, to make good losses due to seepage, and in brackish water ponds to limit fluctuations in salinity caused by evaporation.

Water exchange rates in ponds are often expressed in terms of the inflowing volume as a percentage of the pond volume. This, however, does not represent the true exchange rate obtained (Table 8.1). (In a 1 ha pond 1 m deep, 5% per day = 347 ℓ min^{-1} and 10% per day = 694 ℓ min^{-1}.)

Water requirements differ widely. In some extensive crayfish culture in farm dams in Australia, water supply need only be sufficient to keep the reservoir full. A flow of 118 ℓ min^{-1} ha^{-1} from a well is considered adequate for a 32 ha crayfish farm in the USA (Avault & Huner 1985). Ideally in semi-intensive and intensive systems, the water supply should provide for a minimum exchange rate in each pond (usually about 5% daily) with additional capacity for emergency flushing. While average water use in freshwater prawn ponds in Hawaii has been recorded at 94–271 ℓ min^{-1} ha^{-1}, peak demand may be as high as 3700 ℓ min^{-1} ha^{-1} (Malecha 1983). In semi-intensive shrimp ponds a minimum flow of around 700 ℓ min^{-1} ha^{-1} may be needed, with the capacity to double this volume in an emergency (Lucien-Brun 1988). Calculations for overall water requirements on most modern shrimp farms are usually based on the need for a daily exchange of 10% to 20% of pond volume or up to 30% in intensive ponds. The amount of water normally used in each pond should be roughly in proportion to the amount of feed being added and should thus increase markedly during the later stages of the on-growing cycle. To keep water moving freely through the ponds all screens on water control structures must be regularly cleaned to prevent blocking with debris.

8.3.5.6 Circulation

The absence of mixing in a pond in calm conditions, and the stratification which often results, are barriers to the transfer of oxygen from the surface layers to deeper areas. A large part of the positive impact of mechanical aeration is due to physical mixing and destratification of the water layers. In addition to improving dissolved oxygen concentrations, mixing also evens out the distribution of phytoplankton and temperature.

Low energy water circulation devices based on large electrically driven propellers have shown potential to homogenize water quality within prawn ponds (Rogers & Fast 1988). The improved uniformity of prawn distribution that resulted reduced cannibalism and aggression and significantly decreased the heterogeneity of prawn sizes. In these tests, a 0.5 hp circulation unit was credited with moving 8300 litres of water per minute.

In US crayfish ponds boat and trapping lanes between forage crops are used to improve water circulation.

8.3.5.7 Aeration

Aeration devices have chiefly been used in small (<2 ha) intensive ponds to avoid water quality problems and permit high stocking densities. Their use in larger semi-intensive ponds for shrimp, for example in the USA, has accompanied trends towards intensification (Chamberlain 1989). The ability of aerators to destratify and ensure that oxygenated water reaches the depths of a pond is probably as important as water exchange in high density cultures. Boyd (1990) considers that mechanical aeration is most appropriate when large water exchange rates are impractical and moderate daily feeding rates of 50–60 kg feed ha^{-1} are employed. The value of aeration in reducing requirements for water renewal in shrimp ponds has also been noted by Chamberlain (1987). In trials in Hawaii, 0.4 ha shrimp ponds aerated with two × 1 hp paddlewheels required 62% less water exchange than non-aerated ponds and were also more productive. Nevertheless, at high feeding rates of 60–100 kg ha^{-1} per day or more, the build up of toxic metabolites necessitates significant water exchange as well as aeration (Boyd 1990).

The usual number of paddlewheel aerators (typically 1 hp each) used in intensive Taiwanese shrimp ponds has been put at four or more per hectare for ponds stocked at 10–30 animals m^{-2}, and eight or more per hectare for ponds stocked with more than 30 animals m^{-2} (Chiang & Liao 1985). Results in the USA quoted by Chamberlain (1987) are roughly in line with these levels of aeration: 10 hp ha^{-1} of paddlewheel aeration proved adequate for a shrimp density of 40 m^{-2}, and 1.5–3 hp ha^{-1}

avoided all but occasional oxygen problems at densities of 20–30 shrimp m^{-2}. Wyban *et al.* (1989) recommend aeration levels of 20 hp ha^{-1} in super-intensive round ponds and AQUACOP (1989) have used pairs of 2 hp propeller-aspirator pumps in super-intensive ponds of 0.1 ha (40 hp ha^{-1}).

The increasing demand for oxygen during the on-growing period can be met by bringing more aerators into operation. In Taiwan an initial 2.5 hp ha^{-1} may be increased to 10 hp ha^{-1} by the end of the culture period. Aerators can also be activated during particular critical periods of the day, when oxygen levels are at their lowest during the night or become excessive in the afternoon.

Floating paddlewheels and propeller-aspirator pumps have been the most popular aerators and have proved to be most cost-effective for regular use in small ponds (Engle 1989). Both types provide aeration and impart strong mixing currents. A 2 hp propeller-aspirator pump tested by Boyd & Martinson (1984) mixed salt throughout a 0.4 ha pond in 1.5 hours, whereas the same mixing by wind-driven surface currents alone took 60 hours.

8.3.5.8 Management of ponds with acid sulphate soils

The problems posed to pond culture by acid sulphate soils have been discussed earlier (Section 6.3.3.5). The successful management of affected ponds requires usual practices to be modified, particularly with regard to pond bottom preparation. Normal rejuvenation procedures involving drying and oxidation in air (Section 8.3.5.1) would serve to exacerbate acidity problems. Indeed, complete draining of ponds may need to be avoided to keep pond bottoms and embankments waterlogged and largely anaerobic. Between crops, repeatedly filling and leaching an affected pond can reduce acidity and will certainly be more beneficial than drying. Routine monitoring of pH will be necessary.

To reduce the leaching of acid from embankments into pond water, pond levels can be kept higher than surrounding waters. This is more easily achieved with a pumped rather than a tidal water supply. Another advantage of a pumped supply is that water can be exchanged each day to limit the build-up of acid in a pond. Tidal flushing, by comparison, may only be possible during spring tides.

8.4 Water treatment methods

This section summarises water treatment methods applicable to both fresh and sea-water, with any differencies indicated. The broad aims of water treatment are to:

(1) Provide a near optimal environment for maximum growth of the cultured crustacean;
(2) Economise on the quantity of water to be pumped or heated.

Unfortunately, the moment water is drawn from the natural environment changes occur in its physico-chemical and biological characteristics. Treatment attempts to slow or minimise these changes and to maintain conditions within limits tolerated by the animals (Section 8.5). A number of reviews cover the subject in more detail (Kinne 1976; Wheaton 1977; Boyd 1979; Spotte 1979; Wickins & Helm 1981; Romaire 1985); most authors agree that no amount of treatment will satisfactorily rectify problems arising from a poorly chosen site (Section 6.3.1).

8.4.1 Abstraction

Water from natural sources may be used untreated, for example in ponds, or if it is to be used in a hatchery or nursery it may be treated prior to entry (pre-treatment) or within the culture facility (treatment). Where discharge quality is to be controlled, treatment of site effluents may also be necessary.

Large scale, land-based on-growing units operated at well chosen sites generally have minimal pre-treatment needs. The water is drawn directly from the sea, river, lake or estuary by tide, gravity or pump with minimal mechanical filtration. Once on site, the water may be fed directly to the on-growing facility or, if destined for a hatchery or nursery, fed into a storage or pre-treatment reservoir where some settlement of solids occurs and phytoplankton growth may be encouraged by the application of fertilizers and inoculation with algae starter cultures. Shade netting is sometimes employed to control the amount of light reaching the algae in small reservoirs, while algal levels in ponds can be diluted by increased water exchange (Section 8.3.5.5).

Borehole water may be cascaded or aerated in a reservoir to eliminate carbon dioxide, hydrogen sulphide, ammonia or iron which frequently occur in ground waters. Smaller quantities of water suitable for hatcheries may be drawn through various types of filter before entry (Huguenin & Colt 1989). Where a suitable substrate exists or can be improvised, sub-sand extraction is a technique for drawing water down through layers of sand and gravel into a buried, screened intake pipe. Once the substrate around the intake point has stabilized, it acts both as a mechanical and a biological filter (see below) provided it is operated frequently (Cansdale 1981).

8.4.2 Primary treatment

Large land based filters used in pre-treatment take many forms (towers, sunken pits), contain any of a range of materials (gravel, rocks, coral, shells, plastic rings) and are operated in a variety of ways (up-flow, down-flow, submerged, trickling or under pressure). In circumstances where settlement or sedimentation is necessary, circular or long V-shaped, purpose-built tanks or compact plate separators can be used. Tilted plate or tube separators (also called particle interceptors) are static arrays of inclined parallel plates or tubes up which water passes in a laminar flow. Close packing of the elements permits the minimum of distance for particles to settle out on to the surfaces. Unfortunately these devices do not work efficiently with finely divided solids or 'sticky' organic materials which are reluctant to move off the inclined surfaces into the collection sump.

8.4.3 Secondary treatment

Adjustment of salinity, temperature, pH and oxygen levels in the water supply are most important but normally straightforward operations. In calculating the flow required to dilute metabolites, maintain temperature and in some cases add oxygen, it is necessary to compute the displacement time or exchange rate of water in the tank. In practice, supplying a given percentage of the pond volume each day does not actually exchange that amount of water. A better idea of exchange (assuming complete mixing) can be derived from the equation:

Table 8.1 The time (days) taken to achieve 50%, 75%, and 90% water exchange in a pond receiving 8%, 10%, 12% and 14% of its volume in new water per day (assuming complete mixing).

% exchanged	Flow rate (% pond volume day^{-1})			
	8	10	12	14
50	8.7	6.9	5.8	4.9
75	17.3	13.9	11.5	9.9
90	28.8	23.0	19.2	16.4

$$T = \ln(1 - F) \times V/R$$

where

T = days needed to get x% exchange;
F = fractional water replacement in time T;
V = pond volume (m^3);
R = inflow (m^3 day^{-1}).

Table 8.1 shows the approximate number of days taken to achieve partial replacement of 50%, 75% and 90% of water in culture tanks or ponds. By rearranging the above equation the daily water requirement needed to achieve, for example, 10% exchange per day in 40 ha of ponds of average water depth 1 m can be calculated:

$$40 \text{ ha} \times 10\,000 \text{ m}^2 \times 1 \text{ m depth} \times 0.1 \text{ (i.e. 10\%)} = 40\,000 \text{ m}^3 \text{ d}^{-1} \text{ or } 1.67 \text{ m}^3 \text{ hr}^{-1}.$$

Other than in extensive pond cultures, oxygen is added by means of specific aeration or oxygenation methods (Sowerbutts & Forster 1981; Engle 1989) rather than by the water flow alone, but it is worth noting that crustaceans can be adversely affected by total gas supersaturation, as can fish, and to avoid gas bubble disease, allowances have to be made for the presence of metabolic gasses already in the water when calculating the input from an aeration system (Colt 1986; EIFAC 1986).

It may sometimes be necessary to sterilize water destined for hatcheries or recirculation systems. A number of shrimp hatcheries, for example, add commercial sodium hypochlorite solution to a freshly filled indoor reservoir to give an initial concentration of around 5 to 20 mg free chlorine per ℓ. The water is then recycled through a rapid (pressure) sand filter for 24 hrs after which any remaining free chlorine is neutralized with sodium thiosulphate. The uses of chlorination and other methods including ozonation and ultra-violet irradiation were reviewed by Rosenthal (1981).

In specialised research hatcheries, additional treatment facilities for dark storage, ultra-fine filtration and activated charcoal treatment may be included (Wickins & Helm 1981). At certain times of the year phytoplankton blooms or elevated dissolved organic loads in the water may necessitate the use of air or air/ozone foam fractionation treatment (protein skimming) for their breakdown. The process typically involves the upward passage of fine air or air/ozone bubbles through a downward flowing column of water, and is often used in densely stocked recirculation systems for the breakdown of refractory organic molecules in solution that are not readily oxidized by biological filtration (Rosenthal 1981). Foaming in seawater invariably leads to an increase in suspended particulates and a filtration step generally follows. Ion exchange resins for the removal of dissolved substances (ammonia, metals and some organic compounds) may be appropriate in fresh water but not in seawater where chelation agents such as di-sodium EDTA, sodium metasilicate and Fuller's earth are widely used to deactivate a range of growth inhibiting substances (Wickins & Helm 1981).

8.4.4 Recirculation systems

Recirculation systems are frequently used in hatchery and pilot-scale battery operations to conserve heat and preserve water that has had expensive treatment or has cost a lot to prepare (e.g. artificial seawater (Bidwell & Spotte 1985). The proportion of water renewed varies from a continuous 'bleed-in' to almost closed systems where water is renewed only rarely. During recirculation the water is continuously treated to:

(1) Maintain the required temperature and salinity;
(2) Make good losses due to evaporation and leakage;
(3) Stabilize chemical changes by:
    (a) replacement of depleted components (oxygen, buffering capacity, calcium);
    (b) detoxification and dilution of substances which accumulate (ammonia, nitrite, nitrate, carbon dioxide, dissolved organic materials, suspended solids) (Wheaton 1977; Wickins 1982; Wickins 1985a,b; Hirayama *et al.* 1988).

The cost of maintaining temperature in a recirculation system is largely dependent on the temperature and amount of new water that has to be added to the system in order to maintain water quality. Higher heating costs arise in controlled-environment on-growing systems than in hatcheries, but in both the good control of chemical changes in the water combined with the use of heat exchangers or heat pumps can go some way towards minimizing these costs. Nevertheless, the high capital costs of providing adequate insulation and heat transfer equipment has so far prevented a successful demonstration of commercial viability in recirculation systems (Herdman 1988; Van Gorder 1990).

Control of salinity is normally by addition of artificial sea salts or fresh water and generally presents few technical problems. Mixing of seawater with some natural ground waters could cause problems of precipitation unless the latter are aerated first.

In any aquaculture system oxygen is consumed by the cultured animal, its live food (if any), other heterotrophic organisms in suspension and attached to surfaces, and by nitrifying bacteria. In cloudy, organically rich water considerably more oxygen may be consumed and ammonia and carbon dioxide may be produced by organisms other than those being cultured. This is particularly evident in intensive on-growing recirculation systems and at night in ponds after photosynthesis has stopped (Section 8.3.2). The ways in which the effects can be minimized in recirculation systems are, firstly, rigorous attention to feeding regimes and feeding husbandry (Section 8.7) and secondly, efficient mechanical filtration to remove much of the suspended matter. Aeration at this stage supplies oxygen and removes excess dissolved carbon dioxide, thereby tending to stabilize pH (Wickins 1984a).

During intensive culture, and particularly in recirculation systems, the mineral content of the water may change. For example, in marine recirculation systems both calcium (essential for shell formation after moulting) and magnesium may be lost by precipitation with phosphate and through uptake by the cultured species. In these circumstances the addition of new water or chemical restoration of the minerals becomes necessary (Wickins & Helm 1981).

8.4.5 Biological filtration

Biological filters have two primary functions: the oxidation of ammonia by autotrophic micro-organisms and the oxidation of dissolved and some fine suspended organic materials by populations of heterotrophic micro-organisms (simply, autotrophic = feeds on inorganic compounds; heterotrophic = feeds on organic compounds). By this definition crustaceans are heterotrophs and, like the microbes, also produce ammonia and carbon dioxide wastes. The autotrophs, on the other hand, feed on the ammonia and produce hydrogen ions and nitrate as waste products. The simplified reactions are:

$$55\ NH_4^+ + 5\ CO_2 + 76\ O_2 \xrightarrow{Nitrosomonas} C_5H_7O_2N + 54\ NO_2^- + 52\ H_2O + 109\ H^+$$

$$400\ NO_2^- + 5\ CO_2 + NH_4^+ + 195\ O_2 + 2\ H_2O \xrightarrow{Nitrobacter} C_5H_7O_2N + 400\ NO_3^- + H^+$$

The hydrogen ions (acid) produced by *Nitrosomonas* are normally neutralized or buffered by the alkaline reserve of the water, but in densely stocked recirculation systems they can result in a catastrophic loss of buffering capacity as the acid pushes the carbonate/bicarbonate equilibrium to the right (Wickins & Helm 1981):

$$4H^+ + 2\ CO_3^{2-} \rightleftarrows 2\ H^+ + 2\ HCO_3^- \rightleftarrows 2\ H_2CO_3 \rightleftarrows 2\ H_2O + CO_2$$

The loss of bicarbonate and associated rapid decline in pH is likely to prevent proper mineralization of the crustacean exoskeleton (Wickins 1984b). Ponds and lakes acidified by industrial wastes (e.g. acid rain) can produce similar moulting problems in freshwater crayfish (Appelberg 1989).

A biological filter provides a large surface area for colonization by these micro-organisms, by the material (the filter medium) with which it is packed. Commonly used materials include stone chips which may contribute (e.g. limestone), or may not contribute (e.g. granite), to the calcium content, alkalinity or buffering capacity of the water; plastic rings or sheets are also used, packed at random or coherently. The water to be treated may pass downwards or upwards through the filter, and down-flow filters may be submerged or percolating. In display aquaria and lightly loaded systems the filter functions may be combined with mechanical filtration (e.g. slow sand bed filters), but in heavily loaded systems more consistent and predictable performance is achieved with separate mechanical and biological filters because mechanical filters require regular backflushing and surface raking, processes that can disrupt the performance of the microbial populations. Modern biological filters contain plastic media designed to provide a large surface area per unit volume while at the same time containing a high percentage of voids so that the filter can never become blocked. Other types of biological filter include compact, rotating, biological contactors, which are discs or drum systems turning slowly in a sump tank, and fluidised bed filters in which finely divided filter particles are held in suspension by an upwelling flow of

Table 8.2 Examples of biological filter performance in marine and freshwater recirculation systems.

Input (mg N ℓ^{-1})	Filter type	Hydraulic load (day^{-1})	Temp. °C	pH	Daily ammonia removal	Time to establish nitrification (d)	Reference
0.5–2	Marine, plastic, percolating	—	20	8.2	—	37	Wickins & Helm 1981
1 + live lobsters	Marine, plastic, percolating	—	24	8.2	—	35	Wickins & Helm 1981
Live prawns	Marine, plastic, percolating	—	26	8.1	—	37	Wickins & Helm 1981
1–4 + 50 g mussels	Marine, 12–25 mm gravel, submerged	20.5(a)	26	—	501(e)	24–35	Forster 1974
	Marine, 12–25 mm gravel, submerged	82.0(a)	—	—	1112(c)	—	Muir 1982
	Marine, 12–25 mm gravel, submerged	246.0(a)	—	—	2178(c)	—	Muir 1982
0.1	Marine, plastic, percolating	95(a)	20	7.8–8.2	0.22(d)	—	Richards & Wickins 1979
0.2	Marine, plastic, percolating	182(a)	28	—	0.03–0.38(d)	—	Wickins 1982
0.28	Marine, plastic, percolating	153(a)	28	—	0.08–0.39(d)	—	Wickins 1982

0.39	Marine, gravel, 200 $m^2 m^{-3}$ sepecific surface area	26(a)	28	—	0.03–0.10(d)	—	Wickins 1982
—	Marine, gravel, 210 $m^2 m^{-3}$ specific surface area	360(a)	20	—	0.84(d)	—	Goldizen 1970
0.5	Freshwater, gravel submerged	—	6	—	0.25(d)	—	Wheaton 1977
	Freshwater, gravel submerged	—	12.5	—	0.64(d)	—	Wheaton 1977
	Freshwater, gravel submerged	—	20	—	1.03(d)	—	Wheaton 1977
0.08–9.3	Freshwater, plastic rotating biodrum	0.006–0.03 (b)	25–30.8	7.1–8.4	82–96(e)	—	Rogers & Klemetson 1985
	Rotating biological contactor	0.002–0.07 (b)	25–30.8	7.1–8.4	69–99(e) 2.83(d)	—	Rogers & Klemetson 1985
	Slag, percolating	0.003–0.03(b)	25–30.8	7.1–8.4	38–61 (e)	—	Rogers & Klemetson 1985

(a) $m^3 m^{-3}$; (b) $m^3 m^{-2}$; (c) $gN m^{-3}$; (d) $gN m^{-2}$; (e) %

waste water (Muir 1982). The most effective filters in terms of the ammonia removed per unit area at the low ammonia input concentrations typically found in culture systems seem to be the rotating disc types (Rogers & Klemetson 1985).

Calculation of the size and number of the biological filters required is one of the most uncertain elements of system design because the relative proportion of the available surface area occupied by heterotrophs and autotrophs in the filter alters as the cultured crustaceans grow, as the amount and composition of feed added to the system changes, and as changes occur in the quantity and composition of incoming make-up water. Calculations are simplified in systems where much of the particulate organic material in suspension is first filtered out mechanically, since heterotrophs generally grow faster and will colonise a biological filter more quickly than the autotrophs, leaving less capacity for ammonia oxidation (Wickins 1985a,b). Examples of filter performance under a range of culture conditions are shown in Table 8.2 and Wheaton (1977), Spotte (1979), Tiews (1981), Wickins & Helm (1981), Muir (1982), Wickins (1982; 1985a,b), and Rogers & Klemetson (1985). Rotating biodisc filters are among the most compact units available and are recommended for intensively stocked recirculation systems. Their use in seawater systems however requires special attention to the materials and engineering design to avoid corrosion and consequent mechanical breakdown of the moving parts.

The initial colonization of a filter begins as soon as food or animals are put into the system because the micro-organisms are ubiquitous in nearly all water supplies. The process may be hastened by adding chemical nutrients (e.g. ammonium citrate, ammonium chloride, sodium nitrite), some commercial filter seeding mixtures, or even a few freshly opened filter feeding bivalve molluscs to the system (Manthe & Malone 1987). The crustaceans to be cultured, however, should not be placed in the system until the filter populations have become established (about 3–9 weeks at 20–28°C). This is to avoid exposure to nitrite which invariably accumulates until the rate of consumption by *Nitrobacter* equals the rate of production by *Nitrosomonas* (Figure 8.5). Nitrate is much less toxic and levels are reduced by dilution as required.

8.4.6 Display, live storage and transportation

Biological filters and water recirculation are also widely used in display aquaria (Anon. 1988) and sometimes in live storage systems for clawed and spiny lobsters and crabs (Beard & McGregor in press). Vivier transport systems rely on refrigeration and aeration with or without recirculation to maintain live crustaceans during transportation by road, sea or air (Richards-Rajadurai 1989). Recent developments include cascade systems for road transportation of clawed and spiny lobsters and crabs (Whiteley & Taylor 1986) in which the crustaceans are held in vertically stacked trays continuously sprayed with recycled, chilled seawater. The gills of the animals are thus kept moist although the animals are not submerged in water. In this system the weight of water carried is considerably less and the space the trays can occupy considerably greater than in a conventional vivier truck.

At the other end of the scale, some crabs, crayfish and kuruma shrimp (*P. japonicus*) are typically transported to market out of water in baskets, damp sacks, and chilled sawdust respectively. A comprehensive review of live holding and transportation methods used for fish, molluscs and crustaceans in south-east Asia has been prepared by Macintosh (1987).

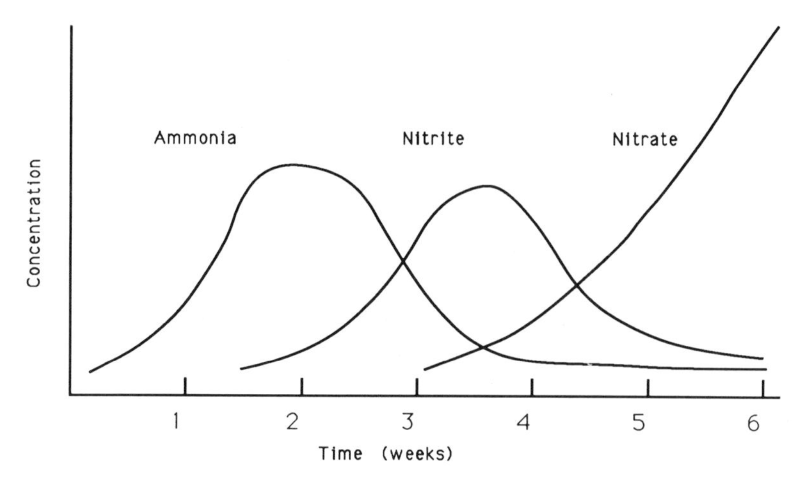

Fig. 8.5 Changes in levels of dissolved nitrogenous waste during maturation (start-up) of a recirculation system (NB. Time scale will change with temperature).

Broodstock shrimp and prawns, larvae and small post-larvae are commonly transported by road or air in double skinned polythene bags containing one third water and two thirds oxygen. The most important factors influencing survival under these conditions are temperature and oxygen. Attempts to improve survival by removing ammonia and buffering pH with chemical additives have not been successful (Robertson *et al.* 1987). Prior to shipment, animals are acclimatized to a suitably low temperature and starved to reduce their metabolism. The rostrum of adult prawns is sheathed or removed to prevent it puncturing the bags. Further details of transportation techniques are given under each species group in Chapter 7.

8.5 Water quality tolerance

In all aquaculture operations, but especially those where the water is treated, it is desirable to know the maximum and minimum acceptable levels of the changes that occur in water chemistry so that effort is not spent trying to achieve unnecessary goals. Unfortunately there is a shortage of such data based on long term growth studies with crustaceans and much reliance is necessarily placed on values extrapolated from short-term acute tests and from studies with fish. Unlike the crustaceans used in traditional laboratory tolerance tests, farmed crustaceans are exposed to mixtures of toxins or metabolites in the water which may act synergistically or antagonistically. In addition, the concentrations of each substance will undoubtedly vary throughout each day (see below). The examples shown in Table 8.3 of acceptable ranges for on-growing several major species or groups of crustacean can thus only provide a guide rather than definitive values for system design. Levels acceptable in a hatchery are generally more restrictive since larvae and small post-larvae are often more sensitive than juveniles and adults.

8.6 Monitoring water quality

The metabolic activity of all organisms in an aquaculture system varies throughout each 24 hr period, being generally less at night than in the daytime and reaching a

Table 8.3 Desirable ranges and levels of water quality factors.

Species/group.	Temperature (°C)	Salinity (ppt.)	Oxygen (mg ℓ^{-1})	pH	Un-ionised ammonia (NH_3-N mg ℓ^{-1})	Nitrite (NO_2-N mg ℓ^{-1})	Hardness ($CaCO_3$ mg ℓ^{-1})	Other
Penaeids	26–30(a)	15–30(a)	>5(b) 85–120%(i)	7.8–8.3	0.09–0.11(a) (<0.02 in presence of nitrite(e))	<0.2–0.25(b,e)	—	160–400 mg ℓ^{-1} Ca(i) 100–200 mg ℓ^{-1} NO_3-N <0.002 mg ℓ^{-1} H_2S(b) <10 mg ℓ^{-1} ferrous Fe(i) 2–14 mg ℓ^{-1} suspended solids(i)
Macrobrachium	26–30(a)	0, 12 for larvae(a)	>4.5(a) >75%	6.5–8.5	<0.1	<1.4(a)	50–200	3–8 mg ℓ^{-1} SO_4(c)
Crayfish								
Temperate	14–23	0, <5	>6(a) min 3	6.7–8.5(a) min 6.0	<0.1	<0.5	50–200 (min 40)	>5–16 mg ℓ^{-1} Ca <0.1 mg ℓ^{-1} H_2S(g) <0.1 mg ℓ^{-1} ferrous Fe(g)
Tropical	23–28(f) min 14	<1.5	>7.8(h) >80%(a)	7.0–8.0(f)	<0.1	<0.2(g)	60–100(f)	<0.1 mg ℓ^{-1} ferrous Fe(g)
Lobsters								
Clawed	18–22(a)	28–35(d)	6.4(d)	7.8–8.2(a)	<0.014(d)	—	—	—
Spiny	23–30	28–35		8.0–8.5	—	—	—	—
Crabs	23–30(j)	28–34(j) range 0–40(k)	>70% <130%	8.0–8.5(i,j)	—	—	—	—

(a) Wickins 1982; (b) Kuo 1988; (c) New & Singholka 1982; (d) Van Olst *et al.* 1980; (e) Chen & Chin 1988; (f) O'Sullivan 1988; (g) Culley & Duobinis-Gray 1989; (h) Sammy 1988; (i) Wickins 1981; (j) Cowan 1983; (k) Oesterling & Provenzano 1985.

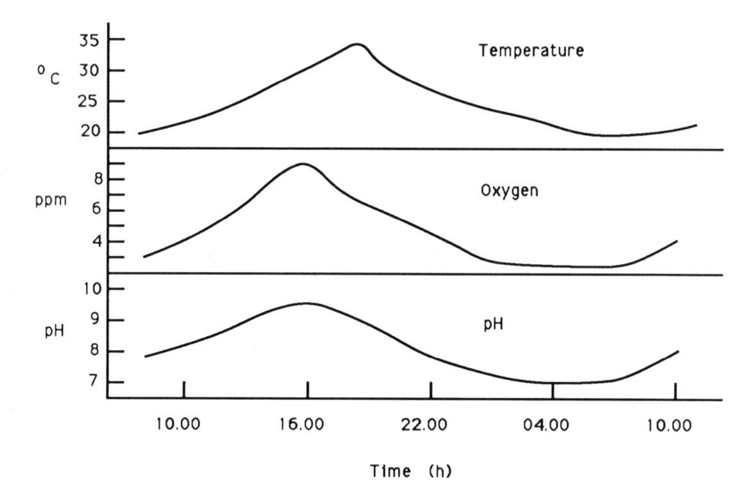

Fig. 8.6a Daily variations in vital water quality parameters in an outdoor earth pond.

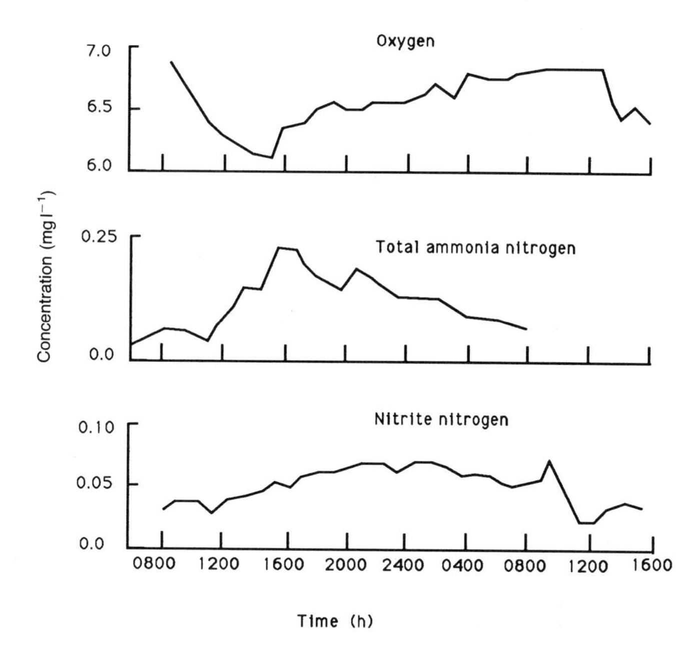

Fig. 8.6b Daily variations in selected water quality parameters in a marine recirculation system containing lobsters. Food was given between 0800 and 1000 hrs daily. Note the coincidence of adverse conditions just before 1600 hrs and the slow recovery of oxygen levels despite vigorous aeration in the system.

maximum during and just after feeding. It has been recommended (EIFAC 1986) that the daily cycle of metabolite levels in culture waters should be determined at least once in order to locate the periods of maximum and minimum levels of vital components and measure selected factors to obtain information relevant to water management. Monitoring changes in the levels of every factor likely to affect crustacean growth and survival is clearly impracticable on a commercial farm and it is worth considering which are the key factors. In the majority of situations they will include oxygen, temperature, pH, total ammonia nitrogen and turbidity, but priority will vary according to species, life cycle stage and the culture system used. Other factors often of less importance in established systems include salinity, alkalinity, nitrite nitrogen, carbon dioxide, mineral ions and dissolved organic materials. Figures 8.6a and b show typical changes in key factors over a 24 hr period in a) a shrimp pond and b) a hatchery recirculation system, and illustrate how samples taken at different times of day from the same system, or at the same time of day from different systems, will give totally different estimates of water quality.

Having considered when and what water quality factors to measure, it is appropriate to draw attention to the problems of making the measurements and analyses themselves. For example, ion sensitive electrodes used for pH and oxygen measurements should be calibrated daily, replaced as soon as performance declines (this can be as often as every 18 months for some pH electrodes) and the standard solutions and buffers used for calibration stored and renewed according to the maker's instructions. These and other examples associated with the calculation and expression of results are presented in detail by EIFAC (1986).

It is worth mentioning two points concerning samples sent to analytical laboratories. Firstly, proper collection and preservation of the sample is vital if changes are not to occur in transit, and secondly the measurements required must be specified in case the techniques normally used by the laboratory are unsuitable. For example, a laboratory routinely engaged in freshwater analysis may not be equipped to deal with ionic interferences from substances found in brackish or salt water. Several portable test kits have been found suitable for aquaculture use in fresh and seawater (Boyd 1980; Boyd 1981; Boyd & Daniels 1988).

Additional factors that arise when monitoring recycled water are discussed by Rosenthal *et al.* (1980), Wickins & Helm (1981) and Wickins (1985a,b). In brief, these concern short-term changes in the organic and inorganic load on the biological filters due to normal husbandry operations, and the cyclic water quality variations induced by this and by the natural, periodic sloughing of biological growths from the filter media. Long-term changes include the accumulation of refactory organic materials that cannot easily be broken down by biological filtration alone, a gradual decline in pH, and in some cases a loss of the buffering contribution normally expected from limestone or oyster shell filter media (Siddall 1974; Wickins 1985a). The latter may prevent normal mineralization of the crustacean exoskeleton after moulting (Wickins 1984b).

8.7 Husbandry and management practice

The planning and assessment of an aquaculture operation must allocate sufficient resources (manpower and equipment) for the implementation of good husbandry and staff management practices. Some general examples applicable to all species at all stages of the life cycle are given below.

8.7.1 Good husbandry practices

(1) The frequent, regular inspection of animals and water flow and aeration rates, and keeping a watch for unusual behaviour, animals not feeding, and signs of infestations and diseases;
(2) The adoption of a husbandry regime (stocking densities and handling methods)

sympathetic to the life cycle stage, especially for broodstock and incubating females;
(3) Frequent, regular feeding, monitoring of feeding rates and removal of uneaten food, detritus and sludge;
(4) Regular monitoring of food quality in both live and stored feeds;
(5) The frequent, regular inspection of water flow and aeration rates, and immediate attention to leaks;
(6) Rationalized, regular water quality monitoring (see Section 8.6);
(7) Attention to the control of temperature, lighting, weeds and predators, and minimization of disturbance and vibrations;
(8) Awareness of the precautions to be taken regarding use of chemicals, especially antibiotics;
(9) Planned reporting and documentation of all the above.

8.7.2 Good management practices

(1) The provision of clear instructions for and adequate supervision of staff;
(2) Good training with proper attention to the transfer of technological know-how from technical advisors/consultants to farm staff at all levels as appropriate;
(3) The installation of comfortable working practices (jobs involving a struggle do not get done properly);
(4) The definition of clear lines of communication and areas of responsibility;
(5) Consideration of staff welfare (health, cultural and social isolation, particularly on remote sites, and incentives);
(6) Protection of the site from poachers and vandals;
(7) Provision of clearly defined and regularly updated emergency procedures covering both staff welfare and culture operations. Shepherd and Morris (1987) review practical emergency procedures.

8.8 Diet

This section summarizes selected features of crustacean nutrition that are considered relevant to understanding project resource needs. It is beyond the scope of this book to provide a detailed review of crustacean nutrition; for this the reader is referred to the useful accounts of Cappuzo (1983), Conklin *et al.* (1983), Corbin *et al.* (1983), Pruder *et al.* 1983), Bordner *et al.* (1986), New (1987), Kuo (1988), Akiyama (1989a,b), Brown (1990), Tacon (1990). The special nutritional requirements of crustacean broodstock have been reviewed by Harrison (1990).

8.8.1 Larvae

The larvae of all farmed crustaceans grow and survive best on living foods. Examples include single-celled algae which are followed by rotifers or newly-hatched *Artemia* (Section 2.2) for penaeid shrimp and crabs with small larvae, and newly-hatched or partially grown *Artemia* for caridean prawns, other crabs and the clawed and spiny lobsters. Live foods are, however, expensive and their culture requires additional facilities and management costs. Much research effort has been applied to the development of micro-particulate and micro-encapsulated larval feeds during the past ten years, and today a wide range of proprietary non-living larvae diets exist. While it is claimed that many of these diets can completely replace living foods, experience indicates that the majority of hatchery operators use them as dietary supplements or as partial replacements, particularly at times when the quality of the cultured food organisms may be suspect. Information on the use of non-living foods for each species group is given in Chapter 7.

Live foods are cultured in a variety of ways. Traditional shrimp hatcheries encourage a 'bloom' of algae in outdoor or illuminated indoor tanks prior to spawning by the addition of fertilizers. Fish or chicken manures, or preferably clean agricultural fertilizers containing silicate as well as the usual phosphate and nitrogen compounds, are widely used (e.g. 50 μg at N ℓ^{-1}, 50 μg at Si ℓ^{-1} and 10 μg at P ℓ^{-1}; μg at = microgram atoms); the silicate encourages the growth of desirable diatom rather than flagellate species of algae. Sometimes it is necessary to inoculate the water from a culture of the preferred alga. Generally, however, a mixture of several endemic species develops. Under favourable conditions it is possible to obtain peak densities of three to four million diatom cells per ml after three to four days; however, lower densities of 50 000 to 150 000 cells per ml seem preferable in the larval tanks.

Advanced shrimp hatcheries culture algae at very high densities in nutrient enriched, sterilised seawater in illuminated culture vessels and feed controlled amounts at regular intervals during the protozoea and early mysis stages of culture (Section 7.2.4). Techniques and facilities describing the different approaches to live food culture are described by AQUACOP (1983), Fox (1983), Liao *et al.* (1983), Laing (1983a,b) Laing & Ayala 1990 and Laing *et al.* 1990.

Rotifers are sometimes cultured in penaeid hatcheries to provide an intermediate food during the transition from the filter feeding protozoea stage and the raptorial feeding mysis stage. Details of their culture methodology are described by Liao *et al.* (1983).

The brine shrimp *Artemia* is the most widely used food for crustacean larvae. It can be purchased in the form of cans of drought resistant cysts which have a shelf life of up to five years, or as frozen adults suitable for the large larvae of clawed and spiny lobsters. The cysts hatch within about 24 hours of being placed in aerated seawater to produce a nutritious, free swimming nauplius suitable for the mysis and post-larval stages of penaeids and for all stages of caridean larvae and immediate post-larvae. There are a variety of different strains of *Artemia* found throughout the world, some hatching larger or more nutritious nauplii than others. *Artemia* may be cultured and their food value enhanced by feeding them on micro-particulate foods or oil emulsions enriched with, for example, the specific lipids and fatty acids required by crustaceans (Section 8.8.2). Savings and improved hatchery and nursery performance may be achieved through the use of frozen nauplii, decapsulated cysts (but see Spotte & Anderson 1989) and partially grown *Artemia*. Details of culture, hatching, decapsulation and quality of different strains may be obtained from the Artemia Reference Centre at the State University of Ghent, Belgium (Sorgeloos *et al.* 1983; Léger *et al.* 1986; O'Sullivan 1986; Sorgeloos *et al.* 1986). Culture of other crustacean species suitable for live food is described by Ventura & Enderez (1980) and Ohno *et al.* (1990).

8.8.2 Juveniles and adults

All farmed crustaceans seem equally able to browse on detritus and benthic microorganisms, scavenge non-living material and become active predators during the on-growing phases. All species can also become cannibalistic when overcrowded or underfed and newly moulted individuals are particularly at risk. In nature, crayfish consume a higher proportion of vegetable material than lobsters, prawns and most shrimp, while crayfish, shrimp and juvenile lobsters are also adept at browsing on detritus and microscopic organisms in the substrate. In extensive and low density, semi-intensive on-growing systems, the natural production of these food sources is encouraged by the controlled addition of fertilizers (chicken, duck, cattle manures or agricultural chemicals) or in the case of crayfish by the addition of hay (lucern, sorghum) or the planting of forage grasses (low yield rice varieties). Specific examples are given in Chapter 7 for each species group and in the section on pond management (Section 8.3.5).

As the intensity of on-growing increases, so does the reliance on compounded feeds whose composition must be tailored to the nutritional demands of the species being cultured if rapid and economical growth is to be achieved. For example, it would not be economic to feed a high protein (60%) diet designed for *Penaeus japonicus* to *P. merguiensis* or *M. rosenbergii* which are able to thrive on low protein (25%) diets.

Although our understanding of crustacean nutrition is not as advanced as it is for chicken and trout, it is now clear that the dietary requirements of crustaceans differ in several important ways from those of farmed fish, birds and mammals. Research done since 1980 has, for example, highlighted the 'essential' nature of certain micronutrients, in particular cholesterol and certain fatty acids (see Glossary), which crustaceans have no or only limited ability to synthesize *de novo*. Four fatty acids are particularly important dietary ingredients for crustaceans: linoleic (18:2 n−6), linolenic (18:3 n−3), eicosapentaenoic (20:5 n−3) and docosahexaenoic (22:6 n−3). Dietary requirements also differ between crustaceans living in marine and freshwater environments. Again using lipids as an example, the freshwater prawn *Macrobrachium* contains more n−6 polyunsaturated fatty acids (PUFAs) than marine shrimp where n−3 PUFAs predominate. An important corollary of this relates to the vulnerability of lipids generally to degradation. Poor diet preparation and storage conditions rapidly lead to the development of rancidity which renders the diets useless.

The sources and fatty acid profiles of dietary lipids are especially critical in broodstock diets during the production of eggs. At present the majority of broodstock are fed a substantial proportion of fresh, natural bivalve, polychaete and crustacean flesh in order to ensure that adequate supplies of essential nutrients are present during the critical period of gonad maturation.

Many crustaceans feed to satisfy an energy need. If this is not met by lipid and to a lesser extent by carbohydrate in the diet, the expensive protein component is metabolized instead. The protein:energy (P:E) ratios and the total energy content of prepared diets are now recognized as critical to diet ingestion, utilization and subsequent growth performance. Good results with shrimp have been obtained with dry pelleted diets containing 3–4 K cals per gram and a P:E ratio of 0.07 to 0.10 at 25–35% protein content. Attempts to spare protein by further increasing dietary energy may be unsuccessful since energy in excess of requirements may prevent the intake of sufficient protein and other nutrients required for growth. High levels of carbohydrate in the food can be detrimental for other physiological reasons.

It is common practice for vitamin and mineral mixes to be added to diets. Early formulations were unsatisfactory since they were based on mixes used for poultry or mammals and contained soluble vitamins, especially vitamin C, which quickly leached out and were lost, together with high levels of iron and magnesium and imbalances in the calcium and phosphorous content. Indeed, considerable losses due to leaching of nutrients can occur within a few minutes of immersion and many of the more expensive commercial feeds now incorporate insoluble forms of critical nutritional components. Research over the past decade has certainly led to more critical formulations, though much remains to be done particularly in relation to aspects of moulting and shell mineralization.

Crustacean diets must be physically stable in water to prevent premature disintegration during repeated manipulation by the animal during feeding. Binding agents such as some glutens, starches or gums are commonly used, while manufacturing techniques which involve heat tend to gelatinize natural starches and increase the binding (New 1987).

It seems unlikely therefore that good performance could be obtained if crustaceans were fed solely on proprietary chicken or trout pellets. Such pellets, together with trash fish, are used as supplements on some extensive or semi-intensive shrimp and crayfish farms where they also provide food for small aquatic organisms on which the cultured species feeds. In other words, the pellets fertilize the pond water in the same way as additions of cheaper poultry or cattle manure (Section 8.3.5.3). Any thoughts of using chicken offal, vegetable or slaughterhouse wastes, particularly in semi-intensive and intensive operations, should be dismissed, as they are likely to cause considerable fouling, increased oxygen demand and contain unsuitable quantities or imbalances of micronutrients. In addition, some animal wastes can be contaminated with medicants, hormones or growth promoters that might be harmful or illegal if found in crustacean flesh.

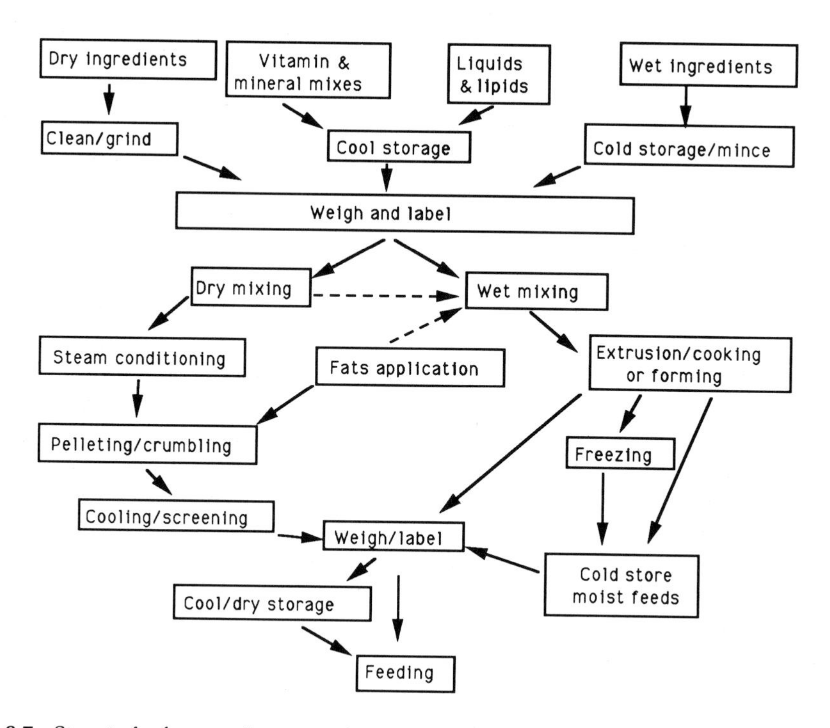

Fig. 8.7 Steps in feed preparation.

The best diets currently available are those manufactured for intensive and super-intensive penaeid shrimp farmers. It is not clear if the best of these would support good growth and survival during the prolonged on-growing period for clawed and spiny lobsters. At present research diets specifically formulated for lobsters allow only 80% of the growth achieved when natural diets are fed.

Compounded feeds are generally prepared using conventional plant but additional steps (finer grinding, activation of binding agents) may be required to control pellet stability (Figure 8.7). Significant losses or deterioration of feed can occur during storage because of theft or damage by insect, rodent and bird pests, fungal and mite infestations, and chemical changes in the feed due to enzymic action and oxidative rancidity. Dry, cool and well-ventilated storage areas out of direct sunlight are recommended and stock turnover should be carefully controlled to avoid prolonged storage. Moist diets and labile ingredients will require refrigeration.

The distribution of feed to the animals (e.g. Table 7.5) is done by hand on many farms with the advantage that the farmer can quickly detect any change in the habits or appearance of the stock. Since shrimp pond water is nearly always cloudy, some feed may be placed on a submerged lift net so that the condition of feeding shrimp can easily be observed. Some large farms use blower or mechanical feeders but, in contrast to fish farmers, few rely on automatic or demand feeders. The distribution of food to a large number of individually held crustaceans, for example battery reared lobsters or crayfish, requires special apparatus (Grimsen *et al.* 1987; Wickins *et al.* 1987).

8.9 Disease

It is generally accepted that animals kept in artificial (culture) conditions experience stress at some time or another and become more susceptible to disease. It is safe to assume that potentially pathogenic organisms are present in all culture systems and will create disease or infestation problems in weak and stressed animals. In pilot and newly established farms stressful periods may occur while operators gain experience with husbandry and water management. Almost inevitably some diseases and infestations will be encountered at this stage, but will not be a major threat later when more experience has been gained and sound working practices established.

Some pathogens are found naturally in many different host populations over a wide geographical range; others are more limited in their distribution. The widespread practice of shipping shrimp broodstock, nauplii and post-larvae and crayfish juveniles and adults between farms and countries has led to the introduction and re-introduction of serious diseases in a number of areas. The most serious outbreaks are frequently associated with pathogens found over a previously narrow range, which have been transferred with stock to new areas in which susceptible hosts exist. Primary pathogens often present more serious problems for crustacean farmers and are less easily prevented from entering outdoor culture operations than indoor controlled-environment systems where incoming water and (some) feeds may be sterilized prior to use. Several countries also conduct trade in live wild or farmed crustaceans for the table or aquaria, and quarantine measures must be enforced to reduce risks of disease transmission from escaped animals as well as from water and equipment with which they have been in contact (Turner 1988).

The first signs of stress or disease are often reduced appetite, abnormal swimming or postural behaviour and a continuous low level of mortality. More obvious signs include infestations of epibiotic growths on the cuticle or on the eggs of brooding females (Fisher 1986), increased cannibalism and moulting difficulties (Wickins 1972; Bowser & Rosemark 1981; Conklin 1990), discoloration, lesions, and finally, mass mortalities. Chronic diseases such as 'black spot' (El-Gamal *et al.* 1986) or soft and blue shell diseases (Baticados *et al.* 1987; Baticados 1988) while not necessarily fatal, reduce market acceptability and can result in considerable financial loss.

Management and containment of disease consists of careful site selection (clean water supplies, away from discharges from other farms, good quality soil) in the first instance followed by minimization of environmental and dietary stresses during culture. Most important are strict enforcement of recommended quarantine procedures whenever animals have to be imported (Turner 1988), control over quality and cleanliness of other farm or hatchery inputs such as water, feedstuffs and personnel, and, as a last resort, application of chemotherapeutic agents (Schnick *et al.* 1979, 1986). Disease control in recirculation systems presents special problems since many treatments designed to kill infectious or infesting organisms will also kill beneficial microbial populations resident in biological filters (Sandifer *et al.* 1974; Delves-Broughton & Poupard 1976; Bower & Turner 1982).

A vaccine is now available for an important blood disease (Gaffkaemia) which occurs in lobsters during live storage. The vaccine is administered by direct injection into adult lobsters (Anon. 1987). At least one other product has been developed which, it is claimed, will produce a broad immune response in some penaeid shrimp larvae. The demonstration that juvenile *Penaeus japonicus* can be protected against *Vibrio* spp. bacteria by spray and immersion vaccination is particularly encouraging (Itami *et al.* 1989). Viral diseases differ from bacterial, fungal and protozoan diseases in that they are not susceptible to therapy. At best, therapy might reduce the risks of secondary diseases attacking animals already weakened by viruses.

If an outbreak of disease cannot be controlled many operators prefer to kill any remaining stock, disinfect the facility and restart after a sanitary 'dry-out' period (Section 7.2.4). This is, of course, more readily practised in indoor hatcheries than in on-growing systems but even so there is no guarantee that the outbreak will not occur again. No published data are available on the costs and frequency of occurrence of disease outbreaks to farmers of particular species or groups of crustaceans, and it would thus seem prudent to make allowances for production losses at several levels during project appraisal.

The identification of diseases and their causative agents is a specialized subject beyond the scope of this book. For further information the reader is referred to the works of Johnson (1982), Brock (1983), Lightner (1983, 1985), Rosemark & Conklin (1983), Fisher (1986), Smith & Soderhall (1986), Vey (1986), Alderman & Polglase (1988) and Sindermann & Lightner (1988). A comprehensive, illustrated guide to the histology of healthy, disease-free tissues of penaeid shrimp has been prepared by Bell & Lightner (1988) and provides a reference base against which subtle changes in cell structure caused by inadequate diets, exposure to toxic materials as well as infection can be compared.

8.10 References

Akiyama D.M. (1989a) Shrimp feed requirements and feed management. In *Proceedings of the south-east Asia shrimp farm management workshop.* 26 Jul.–11 Aug. 1989, (Ed. by D.M. Akiyama), pp. 75–82. Philippines, Indonesia, Thailand. American Soybean Association, Singapore.

Akiyama D.M. (Ed.) (1989b) *Proc. People's Republic of China aquaculture and feed workshop.* 17–30 Sept. 1989. American Soybean Assoc., Singapore.

Alderman D.J. & Polglase, J.L. (1988) Pathogens, parasites and commensals. In *Freshwater crayfish: biology, management and exploitation* (Ed. by D.M. Holdich & R.S. Lowery), pp. 167–212. Croom Helm, London.

Anon. (1987) Vaccine can protect farmed lobsters. *Fish Farming International*, **14** (6) 25.

Anon. (1988). Selling live seafood. *Seafood International*, June 1988 pp. 37–43.

Appelberg M. (1989) Evaluating water quality criteria for freshwater crayfish: Examplified by the impact of acid-stress. In *Crayfish Culture in Europe* (Ed. by J. Skurdal, K. Westman & P.I. Bergen), pp. 140–51. Norwegian Directorate for Nature Management, Trondheim, Norway.

Apud F.D. (1986) Extensive and semi-intensive culture of prawn and shrimp in the Philippines. *Asian Aquaculture*, **8** (4) 3–5; **8** (5) 2–5.

AQUACOP (1983) Algal food cultures at the Centre Oceanologique du Pacifique. In *Handbook of Mariculture, Vol. 1, Crustacean aquaculture* (Ed. by J.P. McVey), pp. 3–14. CRC Press, Boca Raton, Florida.

AQUACOP (1989) Production results and operating costs of the first super-intensive shrimp farm in Tahiti. *Aquaculture '89 Abstracts*, from World Aquaculture Society Meeting 1989, Los Angeles.

ASEAN (1978) *Manual on pond culture of penaeid shrimp.* ASEAN National Coordinating Agency of the Philippines, Ministry of Foreign Affairs, Manila.

Avault J.W. Jr. & Huner J.V. (1985) Crawfish culture in the United States. In *Crustacean and mollusk aquaculture in the United States* (Ed. by J.V. Huner & E. Evan Brown), pp. 1–54. AVI Inc., Westport.

Baticados M.C.L. (1988) Diseases of prawns in The Philippines. *Asian Aquaculture* **10** (1) 1–8.

Baticados M.C.L., Coloso R.M. & Duremdez R.C. (1987) Studies on the chronic softshell syndrome in the tiger prawn, *Penaeus monodon* Fabricius, from brackishwater ponds. *Asian Aquaculture*, **9** (4) 3–9, 11–12.

Beard T.W. & McGregor D. (in press). *Care and storage of live lobsters.* Lab. Leafl., MAFF Direct. Fish. Res., Lowestoft.

Bell T.A. & Lightner D.V. (1988) *A handbook of normal penaeid histology.* World Aquaculture Society, Baton Rouge, Louisiana.

Bidwell J.P. & Spotte S. (1985) *Artificial seawaters: formulas and methods.* Jones and Bartlett Publishers, Woods Hole, Boston.

Blakely D.R. & Hrusa C.T. (1989) *Inland aquaculture development handbook.* Fishing News Books, Blackwell Scientific Publications, Oxford.

Bordner C.E., D'Abramo L.R., Conklin D.E. & Baum N.A. (1986) Development and evaluation of diets for crustacean aquaculture. *J. World Aquacult. Soc.*, **17** (1–4) 44–51.

Bower C.E. & Turner D.T. (1982) Effects of seven chemotherapeutic agents on nitrification in closed seawater culture systems. *Aquaculture*, **29** (3/4) 331–345.

Bowser P.R. & Rosemark R. (1981) Mortalities of cultured lobsters, *Homarus*, associated with a molt death syndrome. *Aquaculture*, **23** (1–4) 11–18.

Boyd C.E. (1979) *Water quality in warmwater fish ponds*. Auburn University, Alabama.

Boyd C.E. (1980) Reliability of water analysis kits. *Trans. Am. Fish. Soc.*, **109**: 239–43.

Boyd C.E. (1981) Comparisons of water analysis kits. *Proc. Ann. Conf south-eastern Ass. Fish Wildlf. Agencies*, **34**, 39–48.

Boyd C.E. (1982) *Water quality management for pond fish culture*. Elsevier Sci. Publ. Co., Amsterdam.

Boyd C.E. (1990) *Water quality management and aeration in shrimp farming*. American Soybean Association, Singapore.

Boyd C.E. & Daniels H.V. (1988) Evaluation of Hach fish farmer's water quality test kits for saline water. *J. World Aquacult. Soc.*, **19** (2) 21–6.

Boyd C.E. & Martinson D.J. (1984) Evaluation of propeller-aspirator-pump aerators. *Aquaculture*, **36** (3) 283–92.

Brinkman R. & Singh V.P. (1982) Rapid reclamation of brackish water fishponds in acid sulphate soils. In *Proc. Bangkok Symp. Acid Sulphate Soils* (Ed. by H. Dost & N. van Breeman), pp. 318–30. Int. Inst. Land Reclamation and Improvement, Publ. 18, Wageningen, The Netherlands.

Brock J.A. (1983) Diseases (infectious and non-infectious), metazoan parasites, predators, and public health considerations in *Macrobrachium* culture and fisheries. In *Handbook of Mariculture, Vol. 1 Crustacean aquaculture* (Ed. by J.P. McVey), pp. 329–70. CRC Press, Boca Raton, Florida.

Brown P.B. (1990) Review of crayfish nutrition. *Crustacean Nutrition Newsletter*, World Aquaculture Society, **6** (1) 18–19.

Cansdale G. (1981) Sea water abstraction. In *Aquarium Systems* (Ed. by A.D. Hawkins), pp. 47–62. Academic Press, London.

Cappuzo J.M. (1983) Crustacean bioenergetics: the role of environmental variables and dietary levels of macronutrients on energetic efficiencies. In *Proc. 2nd. Int. Conf. on Aquaculture Nutrition* (Ed. by G.D. Pruder, C. Langdon, & P. Conklin), pp. 71–86. Wid. Maricult. Soc. Spec. Publ.

Carmignani G.M. & Bennet J.P. (1976) Leaching of plastics used in closed aquaculture systems. *Aquaculture*, **7** (1) 89–91.

Chamberlain G. (1987) Status report: intensive pond management. *Coastal Aquaculture*, **4** (1) 1–7.

Chamberlain G.W. (1989) *Status of shrimp farming in Texas*. Presented at 20th. Meeting World Aquacult. Soc., Los Angeles, Feb. 12–16, 1989.

Chen J-C. & Chin T-S. (1988) Joint action of ammonia and nitrite on tiger prawn *Penaeus monodon* postlarvae. *J. World Aquacult. Soc.*, **19** (3) 143–8.

Chen L.-C. (1990) *Aquaculture in Taiwan*. Fishing News Books, Blackwell Scientific Publications, Oxford.

Chiang P. & Liao I. C. (1985) The practice of grass prawn (*Penaeus monodon*) culture in Taiwan from 1968 to 1984. *J. World Maricult. Soc.*, **16**, 297–315.

Colt J. (1986) Gas supersaturation – impact on the design and operation of aquatic systems. *Aquacultural Engineering*, **5** (1) 49–85.

Corbin J.S., Fujimoto M.M. & Iwai T.Y. Jr. (1983) Feeding practices and nutritional considerations for *Macrobrachium rosenbergii* culture in Hawaii. In *Handbook of Mariculture, Vol. 1, Crustacean aquaculture* (Ed. by J.P. McVey), pp. 391–412. CRC Press, Boca Raton, Florida.

Conklin D.E. (1990) Nutritional factors in 'molt death' of juvenile lobsters (*Homarus americanus*). *The Crustacean Nutrition Newsletter*, **6** (1) 71–2. World Aquaculture Society.

Conklin, D.E., D'Abramo, L.R. & Norman-Boudreau K. (1983) Lobster nutrition. In *Handbook of Mariculture, Vol. 1, Crustacean aquaculture* (Ed. by J.P. McVey), pp. 413–23. CRC Press, Boca Raton, Florida.

Cowan L. (1983) *Crab farming in Japan, Taiwan and the Philippines*. Information series Q184009, Queensland Dept. of Primary Industries.

Culley D.D. & Duobinis-Gray L. (1989) Soft-shell crawfish production technology. *J. Shellfish Res.*, **8** (1) 287–91.

Delves-Broughton J. & Poupard C.W. (1976) Disease problems of prawns in recirculation systems in the U.K. *Aquaculture*, **7** (3) 201–17.

Dexter S.C. (1986) Materials science in aquacultural engineering. *Aquacultural Engineering*, **5** (2/4) 333–45.

EIFAC (1986) Flow-through and recirculation systems. Report of the working group on terminology, format and units of measurement. *EIFAC Tech. Pap.* (49).

El-Gamal A.A., Alderman D.J., Rodgers C.J., Polglase J.L. & Macintosh D.J. (1986) A scanning electron microscope study of oxolinic acid treatment of burn spot lesions of *Macrobrachium rosenbergii. Aquaculture*, **52** (3) 157–71.

Engle C.R. (1989) An economic comparison of aeration devices for aquaculture ponds. *Aquacultural Engineering*, **8** (3) 193–207.

Falguiere J.C., Mer G., Gondouin P.H. & Defossez J. (1989) Evaluation of three population sampling methods for freshwater prawn *Macrobrachium rosenbergii* culture in earthen ponds. *Aquaculture '89 Abstracts*, from World Aquaculture Society Meeting 1989, Los Angeles.

Fisher W.S. (1986) Defences of brooding decapod embryos against aquatic bacteria and fungi. In *Pathology in Marine Aquaculture* (Ed. by C.P. Vivares, J.R. Bonami & E. Jaspers), pp. 357–63. European Aquacult. Soc. Spec. Publ. No. 9. Bredene, Belgium.

Forster J.R.M. (1974) Studies on nitrification in marine biological filters. *Aquaculture*, **4** (4) 387–97.

Fox J.M. (1983) Intensive algal culture techniques. In *Handbook of Mariculture, Vol. 1, Crustacean aquaculture*, (Ed. by J.P. McVey), pp. 15–41. CRC Press, Boca Raton, Florida.

Gibson R.T. & Wang J.-K. (1977) *An alternative prawn production systems design in Hawaii*. Journal series paper No. 2142, Hawaii Agricultural Experiment Station, Honolulu, Hawaii.

Goldizen V.C. (1970) Management of closed system marine aquariums. *Helgoländer Wiss. Meeresunters*, **20**, 637–641.

Grimsen S., Jaques R.N., Erenst V. & Balchen J.G. (1987) Aspects of automation in a lobster farming plant. *Modeling, Identification and Control*, **8** (1) 61–8.

Harrison K. (1990) The role of nutrition in maturation, reproduction and embryonic development of decapod crustaceans: a review. *J. Shellfish Res.*, **9** (1) 1–28.

Hawkins A.D. & Lloyd R. (1981) Materials for the aquarium. In *Aquarium Systems*

(Ed. by A.D. Hawkins), pp. 171–96. Academic Press, London.

Herdman A. (1988) Heating of hatchery water supplies. In *Aquaculture engineering technologies for the future.* Institution of Chemical Engineers Symposium series No. 111, pp. 343–56. EFCE Publication series No. 66, Hemisphere, London.

Hirayama K., Mizuma H. & Mizue Y. (1988) The accumulation of dissolved organic substances in closed recirculation culture systems. *Aquacultural Engineering,* **7** (2) 73–87.

Hirono Y. (1986) Shrimp pond management. In *Acuacultura del Ecuador* (in Spanish and English). Cámara de Productores de Camarón, Guayaquil, Ecuador.

Huguenin J.E. & Colt J. (1989) Design and operating guide for aquaculture seawater systems. *Dev. Aquacult. Fish. Sci.*, **20**.

Itami T., Takahashi Y. & Nakamura Y. (1989) Efficacy of vaccination against vibriosis in cultured kuruma prawns *Penaeus japonicus. J. Aquat. Animal Health,* **1** (3) 238–42.

Johnson S.K. (1982) Diseases of *Macrobrachium.* In Giant prawn farming (Ed. by M.B. New). *Dev. Aquacult. Fish. Sci.*, **10**, 269–77.

Kinne O. (1976) Cultivation of marine organisms: water quality management and technology. In *Marine Ecology* 3 (1) (Ed. by O. Kinne), pp. 19–300. Wiley, London.

Kuo J. C-M. (1988) Shrimp farming management aspects. In *Shrimp '88, Conference proceedings,* 26–28 Jan. 1988, pp. 161–74. Bangkok, Thailand, Infofish, Kuala Lumpur, Malaysia.

Laing I. (1983a) Large scale turbidostat culture of marine microalgae. *Aquacultural Engineering,* **2** (3) 203–10.

Laing I. (1983b) *A simple method for the production of marine algae in polythene bags.* MAFF Direct. Fish. Not. 73. Fish. Res. Lowestoft.

Laing I. & Ayala F. (1990) Commercial mass culture techniques for producing micro-algae. In *Introduction to applied phycology* (Ed. by I. Akatsuka), pp. 447–77. SPB Academic Publishing, The Hague, Netherlands.

Laing I., Child A.R. & Janke A. (1990) Nutritional value of dried algae diets for larvae of Manila clam (*Tapes philippinarum*). *J. Mar. Biol. Ass. UK,* **70**, 1–12.

Léger Ph., Bengtson D.A., Simpson K.L. & Sorgeloos P. (1986) The use and nutritional value of *Artemia* as a food source. In *Oceanogr. Mar. Biol. Ann. Rev.* (Ed. by H. Barnes) **26**, 521–623.

Liao I.C., Su H-M. & Lin J-H. (1983) Larval foods for penaeid prawns. In *Handbook of Mariculture, Vol. 1 Crustacean aquaculture* (Ed. by J.P. McVey), pp. 43–69. CRC Press, Boca Raton, Florida.

Lightner D.V. (1983) Diseases of cultured penaeid shrimp. In *Handbook of Mariculture, Vol 1, Crustacean aquaculture* (Ed. by J.P. McVey), pp. 289–320. CRC Press, Boca Raton, Florida.

Lightner D.V. (1985) A review of the diseases of cultured penaeid shrimps and prawns with emphasis on recent discoveries and developments. *Proc. 1st. Int. Conf. cult. penaeid prawns/shrimps,* pp. 79–103. Iloilo City, Philippines.

Lucien-Brun H. (1988) *Guía para la producción de camarón en el Ecuador* (in Spanish). Expalsa, Guayaquil, Ecuador.

Macintosh D.J. (1987) *An overview on the handling of aquaculture products in ASEAN (Thailand, Malaysia, Singapore, Indonesia and the Philippines).* ASEAN Food Handling Bureau, Kuala Lumpur, Malaysia.

Madenjian C.P. (1990) Patterns of oxygen production and consumption in intensively managed marine shrimp ponds. *Aquaculture and Fisheries Management*, **21** (4) 407–17.

Malecha S.R. (1983) Commercial pond production of the freshwater prawn, *Macrobrachium rosenbergii*, in Hawaii. In *Handbook of Mariculture, Vol. 1 Crustacean aquaculture* (Ed. by J.P. McVey), pp. 231–60. CRC Press, Boca Raton, Florida.

Manthe D.P. & Malone R.F. (1987) Chemical addition for accelerated biological filter acclimation in closed blue crab shedding systems. *Aquacultural Engineering*, **6** (3) 227–36.

Moriarty D.J.W. & Pullin R.S.V. (1990) Detritus and microbial ecology in aquaculture. *Proceedings of the conference on detrital food systems for aquaculture in 1985*, Bellagio, Como, Italy.

Muir J.F. (1982) Recirculated water systems in aquaculture. In *Recent advances in aquaculture* (Ed. by J.F. Muir & R.J. Roberts), pp. 357–446. Croom Helm, London.

Muir J.F. (1988) Shrimp farming: engineering and equipment. In *Shrimp '88, Conference proceedings*, 26–28 Jan. 1988, pp. 136–60. Bangkok, Thailand. Infofish, Kuala Lumpur, Malaysia.

New M.B. (1987) Feed and feeding of fish and shrimp. *A manual on the preparation and presentation of compound feeds for shrimp and fish in aquaculture.* FAO Aquaculture Development and Coordination Programme ADCP/REP/87/26. FAO Rome.

New M.B. & Singholka S. (1982) Freshwater prawn farming. *A manual for the culture of* Macobrachium rosenbergii. FAO Fish. Tech. Pap. 225.

Oesterling M.J. & Provenzano A.J. (1985) Other crustacean species. In *Crustacean and Mollusk Aquaculture in the United States* (Ed. by J.V. Huner & E. Evan Brown), pp. 203–34. AVI Inc. Westport, Connecticut.

Ohno A., Takahashi T. & Taki Y. (1990) Dynamics of exploited populations of the calanoid copepod, *Acartia tsuensis. Aquaculture*, **84** (1) 27–39.

O'Sullivan D. (1986) Growing brine shrimp – Australian companies see big potential. *Australian Fisheries*, Nov. 1986 pp. 43–6.

O'Sullivan D. (1988) Queensland cray farmers opt for local species. *Aquaculture Magazine*, **14** (5) 46–9.

Portmann J.E. & Wilson K.W. (1971) The toxicity of 140 substances to the brown shrimp and other marine animals. *MAFF Shellfish Information Leaflet No. 22.* MAFF Direct. Fish. Res., Lowestoft.

Pruder G.D., Langdon C. & Conklin P. (Eds.) (1983) *Proc. 2nd. Int. Conf. on Aquaculture Nutrition: Biochemical and physiological approaches to shellfish nutrition.* 27–29 Oct 1981, Lewes/Rehoboth Beach, Delaware. World Maricult. Soc. Spec. Publ. 2, Louisiana State University Press, Baton Rouge, Louisiana.

Richards P.R. & Wickins J.F. (1979) Ministry of Agriculture, Fisheries and Food. *Lobster culture research.* Lab. Leafl. (47) MAFF Direct. Fish. Res., Lowestoft.

Richards-Rajadurai P.N. (1989) Live storage and distribution of fish and shellfish. *Infofish International*, (3) 23–8.

Robertson L., Bray W.B. & Lawrence A.L. (1987) Shipping of penaeid broodstock: water quality limitations and control during 24 hour shipments. *J. World Aquacult. Soc.*, **18** (2) 45–56.

Rogers G.L. & Fast A.W. (1988) Potential benefits of low energy water circulation in

Hawaiian prawn ponds. *Aquacultural Engineering*, **7** (3) 155–65.

Rogers G.L. & Klemetson S.L. (1985) Ammonia removal in selected aquaculture water reuse biofilters. *Aquacultural Engineering*, **4** (2) 135–54.

Romaire R.P. (1985) Water quality. In *Crustacean and Mollusk Aquaculture in the United States* (Ed. by J.V. Huner & E. Evan Brown), pp. 415–55. AVI Inc. Westport, Connecticut.

Rosemark R. & Conklin D.E. (1983) Lobster pathology and treatments. In *Handbook of Mariculture, Vol. 1, Crustacean aquaculture* (Ed. by J.P. McVey), pp. 371–7. CRC Press, Boca Raton, Florida.

Rosenthal H. (1981) Ozonation and sterilization. In *Aquaculture in heated effluents and recirculation systems* (Ed. by K. Tiews), **1**, 219–74. Heenemann Verlagsgesellschaft, Berlin.

Rosenthal H., Anjus R. & Kruner G. (1980) Daily variations of water quality parameters under intensive culture conditions in a recycling system. In *Aquaculture in heated effluents and recirculation systems* (Ed. by K. Tiews), **1**, 113–20. Heenemann Verlagsgesellschaft, Berlin.

Sammy N. (1988) Breeding biology of *Cherax quadricarinatus* in the Northern Territory. In *Proc. 1st. Australian Shellfish Aquacult. Conf.* (Ed. by L.H. Evans & D. O'Sullivan), pp. 79–88. Perth, 1988, Curtin University of Technology.

Sandifer P.A., Smith T.I.J. & Calder D.R. (1974) Hydrozoans as pests in closed-system culture of larval decapod crustaceans. *Aquaculture*, **4** (1) 55–9.

Schnick R.A., Meyer F.P. & Gray D.L. (1986) *A guide to approved chemicals in fish production and fishery resource management.* US Fish and Wildlife Service, National Fisheries Research Laboratory, LaCrosse, Wisconsin.

Schnick R.A., Meyer F.P., Marking L.L., Bills T.D. & Chandler J.H. Jr. (1979) Candidate chemicals for crustacean culture. In *Proc. 2nd biennial crustacean health workshop* (Ed. by D.H. Lewis & J.K. Leong), TAMU-SG-79-114 pp. 245–94. Sea Grant Program, Texas A&M University, College Station, Texas.

Shepherd B.G. & Morris J.G. (1987) A review of practical emergency procedures for fish culturists. *Aquacultural Engineering*, **6** (3) 155–69.

Siddall S.E. (1974) Studies of closed marine culture systems. *Progv. Fish Cult.*, **36**, 8–15.

Sindermann C.J. & Lightner D.V. (Eds.). (1988) Disease diagnosis and control in North American marine aquaculture. *Dev. Aquacult. Fish. Sci.*, **17**, 1–431.

Smith V.J. & Soderhall K. (1986) Crayfish pathology: an overview. In *Freshwater crayfish 6* (Ed. by P. Brinck), pp. 199–211. Int. Assoc. Astacology, Lund, Sweden.

Sorgeloos P., Lavens P., Léger Ph., Tackaert W. & Versichele D. (1986) *Manual for the culture and use of brine shrimp* Artemia *in aquaculture.* Artemia Reference Centre, State University of Ghent, Belgium.

Sorgeloos P., Bossuyt E., Lavens P., Léger P., Vanhaecke P. & Versichele, D. (1983) The use of *Artemia* in crustacean hatcheries and nurseries. In *Handbook of Mariculture, Vol. 1, Crustacean aquaculture* (Ed. by J.P. McVey), pp. 71–96. CRC Press, Boca Raton, Florida.

Sowerbutts B.J. & Forster J.R.M. (1981) Gases exchange and reoxygenation. In *Aquaculture in heated effluents and recirculation systems* (Ed. by K. Tiews) **1**, 199–217. Heenemann Verlagsgesellschaft, Berlin.

Spotte S. (1979) *Seawater aquariums: the captive environment.* Wiley-Interscience, New York.

Spotte S. & Anderson G. (1989) Chemical decapsulation of *Artemia franciscana* resting cysts does not necessarily produce more nauplii. *J. World Aquacult. Soc.*, **20** (3) 127–33.

Tacon A.G.J. (1990) *Standard methods for the nutrition and feeding of farmed fish and shrimp*, **1**, 1–117; **2**, 1–129; **3**, 1–208. Argent Laboratories Press, Redmond.

Tiews K. (Ed.) (1981) *Aquaculture in heated effluents and recirculation systems*, **1**, 1–513; **2**, 1–666. Heenemann Verlagsgesellschaft, Berlin.

Turner G.E. (Ed.) (1988) *Codes of practice and manual of procedures for consideration of introductions and transfers of marine and freshwater organisms.* Int. Counc. Explor. Sea, Cooperative Research Report No. 159. Copenhagen, Denmark.

Van Gorder S. (1990) Closed systems: a status report. *Aquaculture Magazine*, **16** (5) 40–47.

Van Olst J., Carlberg J.M. & Hughes J.T. (1980) Aquaculture. In *The biology and management of lobsters, Vol. 2, Ecology and management* (Ed. by J.S. Cobb & B.F. Phillips), pp. 333–84. Academic Press, London.

Ventura R.F. & Enderez E.M. (1980) Preliminary studies on *Moina* sp. production in freshwater tanks. *Aquaculture*, **21** (1) 93–6.

Vey A. (1986) Disease problems during aquaculture of freshwater crustacea. In *Freshwater crayfish 6* (Ed. by P. Brinck), pp. 212–22. Int. Assoc. Astacology, Lund, Sweden.

Wheaton F.W. (1977) *Aquacultural engineering.* Wiley-Interscience, New York.

Whiteley N. & Taylor E.W. (1986) *Handling, transport and storage of live crabs and lobsters.* Sea Fish Industry Authority open learning module, Seafish open tech. project. HMSO.

Wickins J.F. (1972) The food value of brine shrimp, *Artemia salina* L., to larvae of the prawn, *Palaemon serratus* Pennant. *J. exp. mar. Biol. Ecol.*, **10**, 151–70.

Wickins J.F. (1981) Water quality requirements for intensive aquaculture: a review. In *Aquaculture in heated effluents and recirculation systems* (Ed. by K. Tiews), **1**, 17–37. Heenemann Verlagsgesellschaft, Berlin.

Wickins J.F. (1982) Opportunities for farming crustaceans in western temperate regions. In *Recent advances in aquaculture* (Ed. by J.F. Muir & R.J. Roberts), pp. 87–177. Croom Helm, London.

Wickins J.F. (1984a) The effect of hypercapnic seawater on growth and mineralization in penaeid prawns. *Aquaculture*, **41** (1) 37–48.

Wickins J.F. (1984b) The effect of reduced pH on carapace calcium, strontium and magnesium levels in rapidly growing prawns (*Penaeus monodon* Fabricius). *Aquaculture*, **41** (1) 49–60.

Wickins J.F. (1985a) Organic and inorganic carbon levels in recycled seawater during the culture of tropical prawns *Penaeus* sp. *Aquacultural Engineering*, **4** (1) 59–84.

Wickins J.F. (1985b) Ammonia production and oxidation during the production of marine prawns and lobsters in laboratory recirculation systems. *Aquacultural Engineering*, **4** (3) 155–74.

Wickins J.F. & Helm M.M. (1981) Sea water treatment. In *Aquarium Systems* (Ed. by A.D. Hawkins), pp. 63–128. Academic Press, London.

Wickins J.F., Jones E., Beard T.W. & Edwards D.B. (1987) Food distribution equipment for individually held juvenile lobsters, *Homarus* sp. *Aquacultural Engineering*, **6** (4) 277–88.

Wyban J.A., Sweeney J.N., Kanna R.A., Kalagayan G., Godin D., Hernandez H. & Hagino G. (1989) Intensive shrimp culture management in round ponds. In *Proceedings of the south-east Asia shrimp farm management workshop* (Ed. by D.M. Akiyama) pp. 42–7. Jul. 26–Aug. 11, 1989, Philippines, Indonesia, Thailand, American Soybean Association, Singapore.

Yates M.E. (1988) The relationship between engineering design and construction costs of aquaculture ponds. MSc Thesis, Texas A&M University. Extract in *Texas A&M University shrimp farming short course materials*, article 46 (1990) (Ed. by G.D. Treece). Texas A&M Sea Grant College Program, College Station, Texas.

Chapter 9
Project implementation

9.1 Introduction

Translating an initial idea into a viable project can be a complex and often lengthy operation which requires careful planning if the desired objectives are to be achieved and costly mistakes avoided. The aim of this chapter is to describe a logical approach to the planning and implementing of a project, while giving an account of the key factors and options involved in:

(1) Getting the best value for money;
(2) Identifying the many compromises that must be made;
(3) Evaluating and minimising the risks.

The role of consultants and the value of technical and government assistance are discussed, along with ways to make the best use of these resources. Figure 9.1 shows the basic steps in the planning and implementation of a crustacean farming project.

It may be necessary to allocate between 5% and 15% of the total capital cost of a project to the planning and design process, sometimes even more (FAO 1988), but despite the time and cost involved in adopting a systematic approach to project planning, the benefits are considerable. Critical decisions to abandon, modify or proceed with a project can be taken in an orderly progression, and the accuracy of technical specifications and cost estimates can be steadily improved through each stage of analysis. A basic planning framework enables full accounts of the proposed operations to be agreed between the participants, and at the same time clearly expressed and attainable objectives at each planning stage promote efficient implementation of the project. Well documented planning provides continuity in the event of staff changes or delays in the implementation process, and in addition the accumulated information can be very valuable to a financing agency or to the development of the industry as a whole (FAO 1988).

9.2 Objectives

As a first step in the planning process clear and attainable objectives need to be formulated in the light of the original reasons for interest in the project. If interest arises from a national development strategy, specific objectives may include plans for social and economic progress, regional development, alternative and future land usage, and the need to attract aid or foreign investment, or to earn foreign exchange.

Commercially motivated interest in crustacean farming is usually based on generating profits, although the objectives may include gaining tax advantage, making best use of land or a facility already owned, or diversification into new areas of activity. Choices will have to be made on the form of involvement in a project, which may be

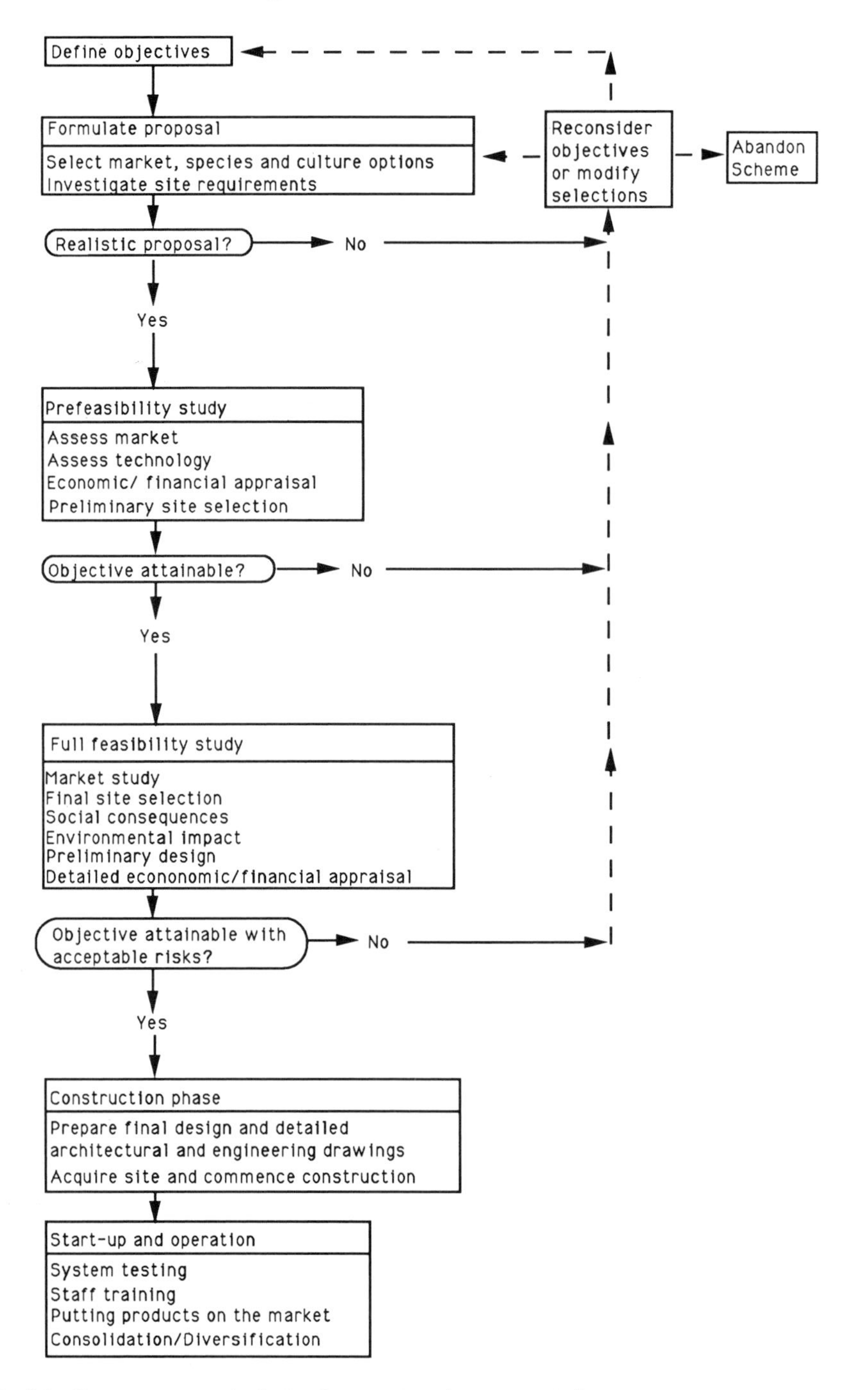

Fig. 9.1 Progressive steps in the implementation of a crustacean farming project.

on the basis of complete ownership, a joint venture agreement or as a participant in a contract growing scheme.

9.3 Project proposal

Once the objectives have been clarified, a proposal for a project of an appropriate type and scale can be formulated based on an understanding of the relative merits of the various options available. At this stage knowledge of the subjects covered earlier in this book is essential, i.e. markets (Chapter 3), candidates for cultivation (Chapter 4), culture options (Chapter 5), and site selection (Chapter 6). More detailed and more specific knowledge of these topics will be required later in the implementaton process.

Figure 9.2 presents a range of possible culture options arranged on the basis of locality, scale of operation, motive for involvement and climate. Apart from operations centred exclusively on on-growing, there are the phases of hatchery, nursery, processing and marketing which can also be considered, either on their own or combined in various ways to provide vertical integration within a project. There are also related activities, including feed production (for larvae rearing, nursery and on-growing operations); collection and sale of wild juveniles or broodstock; production of shrimp nauplii; and provision of various services such as consulting, technical assistance, harvesting and pond cleaning. Alternative and specialized options which may also present significant opportunities include the production of soft-shell crustaceans and the rearing of juveniles for wild stock enhancement. Development project proposals may incorporate one or more of the following components: government or aid backed hatcheries; feed mills and marketing programmes; or facilities for demonstration, training, research and extension services.

A proposal is generally prepared with the objective of convincing backers or landowners to part with money or land and to invest in a project. It must therefore be well-prepared, unambiguous, accurate, and, above all, realistic in its claims. A well-prepared proposal will usually contain significant elements of the prefeasibility and feasibility studies outlined below.

9.4 Prefeasibility

The prefeasibility analysis is generally conducted by the potential investor or backer in order to verify claims and assess markets. It represents the first judgement of a project proposal and may be based on purely technical factors or on a combination of both technical and financial/economic aspects. In all cases the technology to be employed must be assessed for its reliability and any possible constraints identified. It may be necessary to consider the potential of alternative species and culture options to ensure that the best choice has been made in the project proposal.

As a first step the existence of a market for the final products must be confirmed and an outline of marketing policy established in order to answer a range of basic questions. These will relate to where and in what form the products will be marketed; whether the product can be sold to a processor or directly to consumers or buyers at the farm gate; and whether any special arrangements or facilities will be necessary for handling or distribution. Important aspects of crustacean markets are discussed in Chapter 3.

If the culture technology appears to be appropriate and reliable, the emphasis of

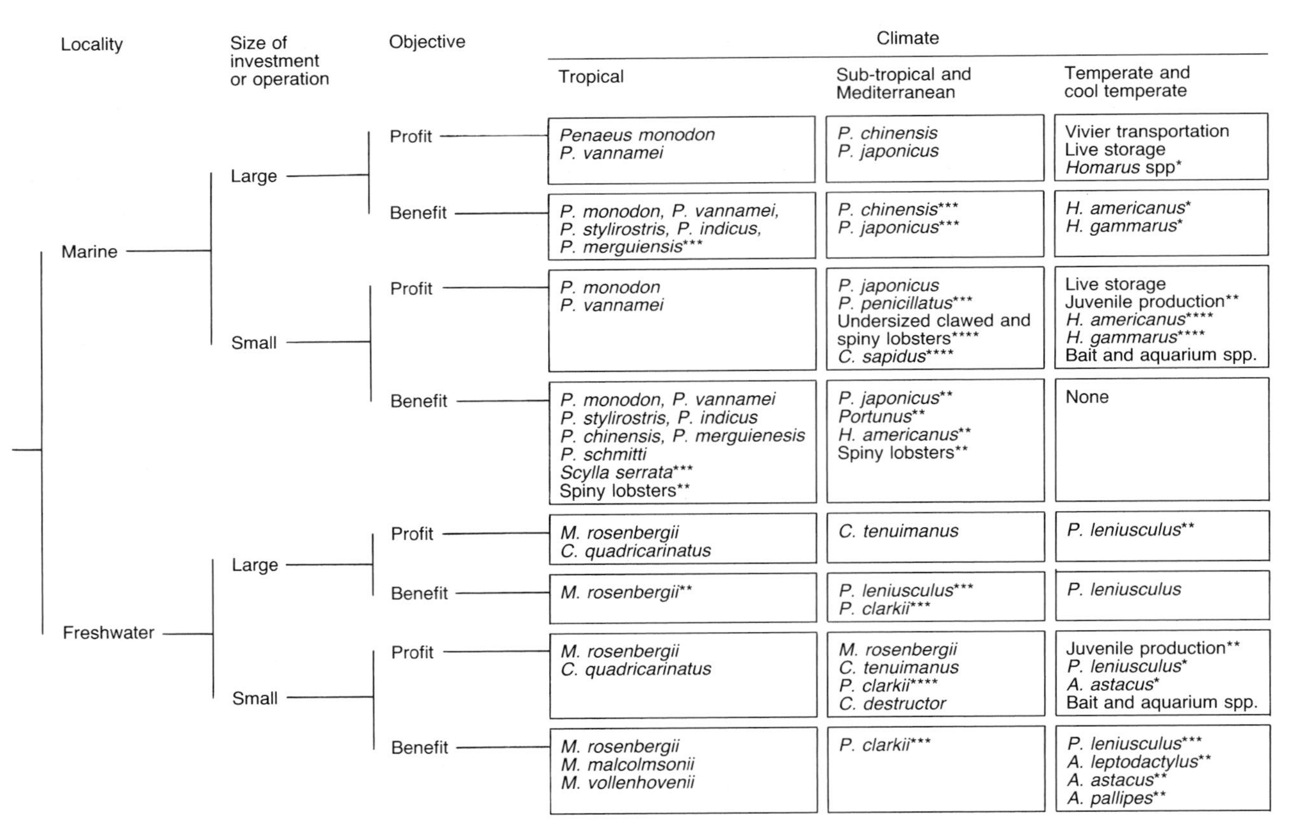

Fig. 9.2 Possible culture and business options: decision making chart Profit = private enterprise/commercial venture; Benefit = can be government/public restocking programmes, supply of seed to artisanal farmers, rural aid and development programmes. * Ranching on artificial reefs, ** Stock enhancement, restocking, *** Catch crop or diversification interest; **** Soft-shell production.

the prefeasibility study can be switched to financial and economic considerations. Proposed sources and types of funding, and the availability of grants and aid need to be investigated, bearing in mind that at this early stage only rough costings for a project will be available. Some idea of likely investment and operating costs for a range of different crustacean farming projects is provided in Chapter 10. If specific rates of return on capital are expected it should be determined whether the project has a reasonable capability of achieving them. It may be necessary to eliminate some high risk or low profit schemes at the prefeasibility stage unless these characteristics are compatible with the stated objectives.

Perhaps the most important pre-investment decision is the selection of a suitable site. At the prefeasibility stage it should be established whether the area being considered is appropriate with regard to climate, availability of broodstock/seedstock and infrastructure. Although the final site selection, including surveying and soil and water analysis, will usually take place later as an initial part of the feasibility study, the suitability of sites can be assessed at the prefeasibility stage to see if they conform to basic requirements of topography, soil characteristics and water supply. Some of the most important topics to be considered during site selection are described in Chapter 6. Additional sources of information that will assist prefeasibility studies are considered in Section 9.5.1.

If the prefeasibility study concludes that the project shows good potential to achieve the desired yields and objectives, the decision can be taken to proceed with the more thorough and costly full feasibility study. In such cases the prefeasibility study should be used to highlight the most important factors influencing the projects viability, and should indicate the areas (e.g. marketing, water supplies) requiring more detailed investigation in the full feasibility analysis. Identification of potential problems and weaknesses in underlying assumptions concerning productivity is a primary objective of a prefeasibility study and, if achieved satisfactorily, will greatly enhance the effectiveness and value of the full feasibility study. If on the other hand the outcome of the prefeasibility study is negative and the proposed project shows little promise, the findings may facilitate the redefinition of more realistic objectives, the selection of different culture options, or indicate that interest in crustacean farming should be abandoned altogether (or at least until more favourable conditions prevail).

Prefeasibilty studies generally take only two to three weeks and can eliminate unworkable schemes at an early stage, minimising wastage of time and money later. Although many small-scale projects might be approved and started at this stage, large projects involving significant resources would normally proceed to a full feasibility study.

9.5 Feasibility

In a full feasibility study a detailed analysis and assessment of a project is carried out to enable the levels of risks and rewards to be more fully quantified. This phase is of critical importance because it is the last analysis to be performed before the decision is taken to proceed with construction and full-scale investment.

The study will contain full technical details of processes involved in the project, and all assumptions underlying anticipated markets, yield predictions and cost estimates. To achieve this all available information on the site, water, soil and infrastructure must be collected and assessed. Any gaps in the data must be exposed to establish if they are likely to undermine confidence in the technical assumptions. In a comprehensive study it may be possible to include comparisons with the performance of similar, viable projects elsewhere. In addition, it is usually at this stage that possible social and environmental implications are considered in depth (Chapter 11). The question of how to go about implementing the construction phase of the project is then addressed and choices made regarding the possible purchase of turnkey packages; if and how consultants, technical assistance and government assistance will be used; and what safeguards (risk management) will be sought (Sections 9.5.2 to 9.5.5).

Part of the study will consider how to finance the total operation and will include the proposed arrangements for loans and equity, taking into account the overall investment costs, the period for start-up, and the expected running costs. The study should convince not only those who prepare it but also the project's backers (bankers, landowners and government officials) of the overall viability of the operation. If it is prepared well it should be easy to recognise and identify *bona fide* objectives and purposes. However, it should be remembered that even with an accurate and thorough feasibility analysis, success can never be guaranteed. It is difficult to be certain about all the elements required to make a project function as planned. Weather and longer term climatic change can be hard to predict and water quality can deteriorate, especially if aquaculture activities or other developments are expanding in the chosen area.

9.5.1 Sources of information

Market information at the prefeasibility stage will usually rely on second-hand sources rather than performing actual market studies. Information can be sought in trade journals and obtained from existing producers, processors, retailers and restaurants. From these sources a basic idea of sales potential and the best product forms can be established. A market study, involving a market survey or the test marketing of sample products, may be required as part of the full feasibility analysis, in which case specialized help may be needed from a company or consultant with relevant expertise.

Information about culture techniques can be obtained from scientific and technical literature including Chapters 5 and 7 of this book and the references therein. Consulting with industry sources, university departments, government departments and extension services in the chosen area can be a valuable aid to assessing the 'state of the art', and to establishing if any special technical adaptations may be necessary for local conditions.

Preliminary information on possible sites can be obtained from maps, preferably ones which indicate relief and vegetation type as well as access points (roads, tracks). Large-scale maps can be most useful once suitable general areas have been identified; however, in some countries they may either not exist or may only be available to military personnel. Marine charts sometimes provide the most accurate information on the form of coastal margins and indicate the extent of areas such as salt flats, mangroves or swamps. Oceanographic institutes, national government resource and fishery departments or universities can be good sources of information on prevailing sea currents and temperatures. Climatic information, particularly rainfall and temperature data, may be more easily found through government or university departments concerned with agriculture, and these sources can also be useful for information on dominant soil types. Mineral resource surveys can provide valuable information on soil types and may be available from commercial resource surveying reports, government agencies or departments concerned with mining, and as a result of gas and oil exploration and geographical surveys. Land registry details are usually available through local government and district offices.

Light aircraft can be chartered to obtain an overview of large areas of land or coastline. Aerial photographs can be taken and analysed. Some satellite images are ideal for identifying zones with different types of vegetation and land usage and can provide up-to-date information on the extent and location of existing aquaculture operations such as coastal shrimp farms.

Economic and financial information can be obtained through foreign trade missions and government departments concerned with industrial, agricultural, fisheries and aquaculture development. Trade associations may also be of some assistance, along with aid agencies and various types of commercial, co-operative and development bank.

9.5.2 Consulting services

9.5.2.1 Types of organisations

Consulting services are available from individuals or from groups of consultants in many different types of organisation. Individual consultants may offer particular specialist help, for example with marketing, economics, biology or engineering, or they may provide more general expertise covering a range of such disciplines. Aquaculture consulting groups often consist of in-house or sub-contracted specialists and may be organised as partnerships or private companies. Some groups, however, are government enterprises or receive government backing and others are attached to universities. The manufacturers and distributors of aquaculture feeds, equipment or other supplies often provide consulting services in support of their main business. Some shrimp processing companies provide free technical advice to farmers, generally on condition that all harvested product is sent to their operation. Assistance with the technical aspects of crustacean farming may be provided as an integral part of a turnkey package (Section 9.5.4).

9.5.2.2 How and where to locate consultants

Many consultants and consulting companies advertise in aquaculture trade publications; others can be contacted through universities, government research stations, trade associations, professional institutions or people already engaged in crustacean farming. Recommendations can also be obtained from aquaculture supply companies but if they offer their own consulting services their advice may tend to be centred on the use of their own products. The advantage of such companies is that they are often in routine contact with many operators and are in a good position to compare results at different operations and recommend consultants accordingly.

9.5.2.3 How to use consultants

Consultants can be valuable for their specialized skills and general advice right from project prefeasibility analysis though to construction and operation. Their importance generally increases with the scale of the project. In the planning stages they can be employed to put together complete prefeasibility and feasibility analyses or to tackle particular tasks such as site surveys, market surveys or environmental impact assessments. The special case of turnkey operations, in which comprehensive design and management services are provided, is discussed separately in Section 9.5.4.

When it comes to the construction phase and the start-up period of operations, it is generally beneficial for supervision of the work to be undertaken by the same people who performed the feasibility study. However, some tasks may require independent and specialized help, such as pond construction and site management, staff recruitment and training, accounting and financial planning.

Technical advice can be essential during the start-up period when inexperienced staff require instruction and training. Advisors may need to be retained until smooth running is established with all routines defined and documented, including technical operations, sanitation, quality control, data collection and report formats. Assistance may be needed on a long-term basis, either continuously or at regular intervals, although by emphasising the importance of the training role and by encouraging the delegation of responsibilities to other technical employees, reliance on individual outside experts can gradually be reduced.

During the operational phase consultants may be brought in as 'troubleshooters' to identify and resolve production problems. However, Muir (1988) warns against taking the advice of consultants who offer 'quick-fix' solutions because these solutions tend to be very short-term in their effectiveness. Valuable advice about production problems can be obtained from aquacultural suppliers, especially if they have encountered similar problems in other operations. Nevertheless, it should be ensured that they do not simply take advantage of the opportunity to promote their products, such as therapeutants, in ways that treat the symptoms of problems rather than the causes. Some of the larger processing companies may offer practical assistance and advice to client farms because of mutual interest in maintaining productivity levels.

At any stage of a project's development, the use of a second technical opinion to provide a detached viewpoint can be very valuable, especially in critical areas of production. During the operational phase consultants can assist technical staff in identifying problems and communicating with the project owner. In some situations it may even be worth considering a confrontational approach: Mock (1987) recommends that 'you (a project owner) bring someone in once in a while unannounced, that can review and challenge your so-called experts'.

9.5.2.4 Choosing a consultant

Larger aquaculture projects often require substantial and complex engineering design work, and it may be advantageous to employ an engineering design or consulting engineer group to prepare proper design specifications, engineering drawings and tender documents. Since the interpretation from biology to engineering must be done with great care, the engineering group must have aquaculture experience, otherwise a bio-engineer should be hired to shape the design to the needs of the species to be cultured.

Small aquaculture consulting companies may be competent with regard to biology but may not possess the skills needed for all aspects of planning and implementation.

Some recruit additional experts as and when they need them for a specific project, in effect acting like recruiting agencies and providing back-up to the experts and a link with the clients. In all cases it is advisable to verify the suitability of the experts that are being hired.

Consulting roles can be broadly divided into specialist and general. The value of a general aquaculture consultant may be greatest in the early phases of planning when project options must be chosen to suit desired objectives, and a prefeasibility study is required to highlight the areas where more specialized inputs may be needed for full feasibility analysis and implementation work. The general consultant may also play a key role in compiling the feasibility study and co-ordinating and supervising the work of the more specialized consultants. Specialist consultants are most valuable in the later stages of planning after the needs for specific expertise have been clearly defined and when the implementation process has advanced towards the construction phase. When specialists are given precise roles it is usually a straightforward matter to maintain control, monitor progress and obtain value for money.

During the operational phase, both specialists and general consultants can be useful. If a production problem can be identified as relating to a particular subject, such as nutrition, disease or water quality, specialist help may be appropriate. However, production problems are often caused by a series of interrelated factors and an aquaculture consultant with broad experience may be able to identify the most likely causes, suggest general improvements in areas such as husbandry, and recommend specialist help only if it is needed. Alternatively, if a production system is in general disarray and a rapid resolution of problems is required, it may be advisable to hire a team of experts each with different specializations. This may be costly but can result in considerable time savings.

9.5.2.5 How to get the best from consulting services

Since the project backers pay for the services of consultants in order to obtain guidance in areas outside their direct experience, it is of fundamental importance that they are convinced of the consultants' expertise and are fully prepared to take the advice in the end. There is no sense in paying for advice simply to ignore it. If at any stage overall confidence in consultants is lost, they should be replaced rather than allowed to carry on and reach unreliable conclusions. The danger of expensive mistakes is at its greatest when over-optimistic or over-ambitious project developers combine with incompetent or inexperienced advisors. A good consultant will always be prepared to recommend that a project be stopped or redirected if it looks unfavourable.

To be sure of the competence of consultants, investigate and discuss with them their past activities. People with whom they have worked previously should also be questioned to establish if the consultant or consulting group do in fact perform the services they claim, or if their skills in the relevant field are only limited. Experience should cover the chosen species and culture system and also, ideally, its application in the relevant location or at least the same type of location. Entrepreneurs and investors should be aware that, as in any industry, there are individuals and companies which may make false claims and others which lack sufficient experience to be credible.

Muir (1988) recommends studying the attitude of the consultants and considering whether they are too slick and fast in their approach; i.e. very efficient at promoting and selling their services but possibly less proficient at performing them. Muir also considers that consultants should be prepared to admit that they have made mistakes in the past because this is an essential aspect of learning. Their methods, style of work and personal approach must be acceptable to the employer if efficient communications and working relationships are to be established and maintained, particularly over long-term contracts.

An initial statement of the work that is required from the consultant, the terms of reference (TOR), should be drawn up, including a timescale and if possible a budget framework for the project. This can be especially critical for pre-investment studies where the scope and form of the assignment are open to different interpretations (World Bank 1981). The TOR should specify the information that the consultant is required to produce, along with relevant outline formats for reports, plans, design drawings, budgets and recommendations. Details of training assignments may need to be included, along with requests for background documentation in support of reports.

More than one consultant or consulting group should be approached; the World Bank suggests that offers be sought from three to six. Price comparisons can then be made, making sure to take into account any hidden extras above the specified daily or monthly rates. However, experience and the quality of the services offered should be the overriding factors used in the selection of consultants, rather than the cost (World Bank 1981).

Although most internationally active consultants are from developed countries, there are many situations where suitably qualified people in developing countries can provide valuable advice and good value for money. Such experts can take advantage of cultural affinities and may be particularly familiar with local climatic or physical conditions, design practices, and legal and bureaucratic complexities.

Contracts can be drawn up on several bases. Time-based contracts, relating to a specified number of man months, are commonly used and are favoured, for example, by the World Bank for general planning and feasibility studies, for design, detailed engineering and construction, and for technical assistance assignments. The time-based rate includes salary, social costs, the firm's overheads, fee or profit and usually an overseas allowance. Other types of contract include lump sum (fixed fee), percentage (based on a percentage of construction costs), and cost-plus-fixed-fee; these are discussed in a World Bank booklet (World Bank 1981). It may be necessary to include a contingency allowance in the consultant's contract to allow for the cost of any unforeseen work.

Production targets and incentives can be used to encourage commitment from technical advisers as well as production workers. Incentives should not, however, be allowed to promote imbalance in the production schedule by encouraging 'storming' – i.e. reaching a production target by a set time without regard to the quality of the work, its impact on the infrastructure or environment, or the quality of the final product. If an incentive scheme extends to the whole workforce bonuses should be paid in addition to set salaries, and salary levels should not be restricted because of the earning potential of the bonus scheme. Performance-related clauses can also be included in the contracts of technical advisers to allow for termination in the event of continually bad results. However, the root causes of problems are not always the fault of the technical assistants and initially, during start-up operations, some legitimate delays in reaching production targets may need to be accommodated.

9.5.2.6 Indemnity insurance

It is worth establishing whether, and up to what level, a consultant or consulting company is insured against problems that arise from their negligence and incorrect advice or information. If they have indemnity insurance from a reputable source it usually means they have an established track record and are recognised as an authority in their field. Indemnity insurance is more commonly possessed by consulting companies than by individuals.

9.5.3 Contract growing (nucleus/plasma) schemes

In contract growing (nucleus/plasma) schemes, a nucleus operation is responsible for providing individual contract farmers (the plasma) with technical assistance. Farmers also usually receive pond construction services, feed, seed and preliminary training, and in return are required to provide the land for their ponds, manage the on-growing phase and then sell their product back to the central processing plant. In one such scheme for shrimp, Aquastar in Thailand, the nucleus operates a hatchery, nursery, feed mill and processing plant (Brown 1989).

In theory, a nucleus/plasma scheme provides a more even distribution of wealth than the establishment of a single farm on the same scale, because it involves local people in the role of small-scale farmers rather than simply as farm labourers. The technical back-up from the nucleus is usually of a far higher quality than that normally available to small-scale operators, and the improved dissemination of good advice is beneficial to the industry in general. The long-term prospects for nucleus/plasma systems, however, remain to be seen (Brown 1989).

Success with small-scale farms relies greatly on the diligence, motivation and innovation of the farmers (Chien & Liao 1987) and a nucleus/plasma scheme must encourage the development and expression of these same positive attributes. Good results are likely to rely on very skilful management of the nucleus operation. The system must be flexible enough to meet the requirements of a large number of individual farmers and maintain an atmosphere of mutual co-operation with equal priority given to each grower, even when complications arise as a result of such problems as a shortage of seed or water. Contract growers, for their part, must learn to co-ordinate stocking and harvesting operations with the nucleus operation. One potential difficulty arises because of the need for protection against incompetent contract growers. Some form of dismissal may be a reasonable sanction, but it would not be proper to expel a farmer from his land if he had previously owned it, or to deprive his pond of water from the company supply system.

9.5.4 Turnkey projects

When a project developer acquires a turnkey package he usually receives comprehensive design and management services and in return provides the funding and sometimes the land and water. Many consulting companies, particularly those whose forte is engineering, offer turnkey options. The package may be arranged to cover the operation from initial prefeasibility analysis right through the planning stages to the provision of technical assistance during the operational phase. Sometimes, however, the decision to use a turnkey package is taken after independent prefeasibility or feasibility studies have concluded that this is the best approach.

The turnkey approach is only advantageous if very high quality services are being offered. Great trust will be placed in the consulting company and much of the control over the project will be relinquished by the owner. The results of buying a bad or inappropriate package may not become apparent until the project is installed and then fails to perform to specifications, by which time it may be extremely expensive or impossible to take corrective measures. To establish the quality of the services that are offered and to judge the likelihood of success, the track record of the consulting company should be investigated. Ideally, more than one company should be approached and the final choice based on results obtained previously with similar projects.

The advantage of employing a single company in the turnkey approach is that continuity is provided in the precise translation of an idea into a viable operation. Responsibility for good site selection, functional design and operational success rests with a single group which has unique control. The approach can be especially useful for particular specialized projects or newly emerging technologies where expertise is not widely available. However, as crustacean farming technology becomes more widespread, the turnkey approach may become increasingly less attractive because it is often far less expensive to hire the services of individuals with wide experience in the industry.

The use of a performance related contract can help obtain full commitment but disputes can arise over the payment of royalties on production. In one Ecuadorian hatchery project the owners refused to pay agreed royalties to the company which provided the technology, maintaining that it was only the development of their own management and technical inputs which had secured a viable production level. Because of such problems, it is often in the interest of the technology provider to receive the bulk of remuneration on completion of the project rather than as royalties on production. Once an operation is finished and running successfully the owner has effectively acquired the product he paid for and the technology provider is in a weak bargaining position since technical developments and improvements are a feature of present day crustacean farming. Contracts may need to be carefully formulated to prevent disputes and ensure mutual interest in the success of a project. If contractual disputes do arise they can be very expensive, time consuming and, in some countries, impractical to resolve.

Some large aquaculture engineering companies (with access to financial resources) acquire equity in the projects they are setting up, either by injecting cash or exchanging technical input for a percentage share of the project ownership. This joint venture approach encourages commitment from both partners.

9.5.5 Government assistance

Governments often offer technical and financial assistance for crustacean farming in relation to national interests. Support goes to training, research and extension services associated with universities, agricultural colleges and research institutes. Demonstration hatcheries and production units may be established commercially with national government grants in order to extend new technologies to traditional farmers and to train new technicians and farm managers (Wickins 1986).

Plate 9.1 Bundles of plastic pipes supplying water to intensive shrimp ponds in southern Taiwan. Each pipe carries seawater from a separate electric pump situated on the beach (out of sight behind the trees). A line of vertical power supply poles, each bearing an electricity meter, is visible in the background.

Governments may also back promotional campaigns for crustacean products on home and export markets, and encourage auxiliary industries such as those producing seed and feed. In Hawaii during the mid 1980s some prawn farmers relied on a supply of juveniles from a government sponsored *Macrobrachium* hatchery and would otherwise probably not have remained in business. When government involvement is sought in the development of a new project, the process of obtaining land concessions, leases and other permits can be greatly simplified.

Extension services, research stations and universities can be very useful for providing scientific and technical information on crustacean farming. Experienced extension workers, available to make routine and emergency visits, can provide a valuable service, particularly for small-scale farmers with limited access to other types of assistance (Section 11.6.2.1).

The Malaysian government promotes its national shrimp culture industry by providing advice and technical assistance through the Department of Fisheries (USDC 1989). It also offers tax rebates and tax relief to this and other developing industries. In an investment tax credit scheme, half of the first five years' qualifying capital expenditure may be deducted from taxable income. Qualifying expenditure covers land clearing, pond construction, the purchase of plant and machinery, and construction of buildings. Private investors are also able to borrow from a special fund at reduced rates or benefit from a 'reduced interest rate export credit re-financing scheme'. Also in Malaysia, state-backed economic development corporations are active in shrimp farming in conjunction with foreign (joint venture) partners.

As part of the Shrimp Farming Industry Development Plan in Taiwan, low interest loans and infrastructure grants were made available and water resources re-allocated (Lee 1988). However, these measures, together with a paddy conversion plan and subsidies on electricity, helped to boost shrimp farming to levels that resulted in severe environmental damage (Section 11.5).

9.6 Construction and start-up

The final implementation phases of a project usually begin with the production of the final design drawings and construction documents and proceed through construction and start-up to full operation.

9.6.1 Acquisition of site and construction

Before a site is acquired, and as a final part of the site selection process, detailed surveying is usually necessary in order to prepare a topographic map on a scale of 1:1 000 or 1:2 000. In addition to marking elevations, buildings, access points, paths, channels, bodies of water, vegetation types and the positions of soil sample cores, the map will need to define the precise location of boundaries. Once these have been established, arrangements can be made to take possession of the land, either by purchase or a lease agreement.

For remote sites road access may need to be constructed and accommodation provided for a construction workforce. If freshwater is not available tanks may be needed to provide a reservoir for drinking water and for mixing concrete and mortar. Water may also be needed for compacting pond embankments (Section 8.2.2).

Construction work can be managed by a site engineer but higher level technical advisers with responsibility for the whole project should supervise and monitor progress. It is important to proceed with constructions as swiftly as possible because any delays can severely affect the economic viability of a project (Section 10.7). Careful planning and good logistical support are essential and may be aided by employing a quantity surveyor.

Cheap locally-available materials should be used for buildings, providing the overall quality of construction suits the planned life of the project (Section 8.1). Whereas a back-yard hatchery may be planned to withstand just a few seasons, a large hatchery may be expected to last a minimum of ten years. After construction, time must be allocated for the installation of equipment such as pumps and generators; suitably qualified personnel may be required for supervision and to ensure all electrical and water systems function correctly.

9.6.2 Operational phase

Start-up operations usually begin after construction has been completed and all equipment has been installed and tested. However, to generate early income in some farms it may be possible to start production before all the ponds have been built. During the start-up of hatcheries, defects in plumbing and electrical installations often come to light and need to be resolved before production can commence. It usually takes some months or even years before a reliable production routine can be established, and investors and project backers must allow for the learning curve as new staff are trained and gradually gain practical experience.

If an integrated operation is planned, to incorporate a hatchery, nursery, farm, processing unit and possibly feed mill, then it may be beneficial initially to construct only the farm and rely on purchasing juveniles and feed from outside sources. This will enable production assumptions to be tested, site suitability to be confirmed, and income to be generated before investment in the additional operations. Ideally, each operation in a vertically integrated set-up should be economically viable in its own right and be able to compete in outside markets for its products or services. However, if one operation such as a hatchery is newly-built and unique in a particular area, its value as a vital link in an integrated process may justify subsidization.

In order to monitor and control operational costs and maintain profitability in an increasingly competitive industry, strict accounting procedures may need to be introduced. Costs for materials, energy and manpower can be collated for each centre of operations in order to more easily identify areas where economies can be made. For example, when assessing a hatchery operation it may be worth splitting costs between different operational units, e.g. broodstock and maturation, larvae rearing, and live feed production. Problems with very detailed cost accounting systems can arise, however, due to unpredictability in the performance of aquaculture systems.

On a farm it can be advantageous to keep records of the amounts and costs of all inputs, such as labour, feed, seed, energy and fertilizer, destined for each pond. After a harvest the revenue can then be compared to total production cost, and results interpreted in terms of the management practice applied to that pond. This approach may be particularly useful in nucleus/plasma schemes where individual operators or individual ponds may vary in performance.

At some stage the decision to scale-up operations may need to be taken and a choice faced between expanding an existing operation or building new facilities elsewhere. For an on-growing operation, the limitations of the site and its water supply will strongly influence the choice. For a hatchery, overall size has implications for the flexibility of operations as well as the cost of production (Section 10.9.1.1) and rather than continually increase the capacity of an original facility it is sometimes preferable to establish an independent operation at a new site. Most hatcheries suspend production for cleaning and sterilization, so independent operations may be able to provide a more consistent supply of seed.

9.6.3 Consolidation

From an operational base there may be many opportunities to sell alternative products and skills and provide advice to other people involved or interested in the crustacean farming business. The demand may be especially great in a pioneer industry. Indeed, some early marron growers in Australia made more money selling juvenile crayfish than retaining them for on-growing (O'Sullivan 1988).

Despite the existence of opportunities to sell technology and the need to restrict the number of visitors to a busy installation, there are dangers in maintaining 'secrecy' in order to retain some perceived commercial advantage. If technical staff and other members of the workforce are convinced that they are engaged on a very special project they will expect special salaries in recognition or otherwise will be encouraged to go elsewhere and sell their 'secrets'. Mayo (1988) mentions people who are new to aquaculture and who keep building fences around their projects and research, and notes that fences work both ways, so that in trying to retain some advantage through secrecy the influx of ideas and solutions from outside is stifled. In the long-run free exchange of information and ideas between operators and researchers alike will have a much greater impact on success. Good results in aquaculture are much more closely related to hard work, diligence and trial and error than to the application of 'secret' or 'magical' formulae.

9.7 References

Brown J.H. (1989) A 'franchising' scheme for shrimp farmers. *Fish Farming International*, **16** (10) 8–9.

Chien Y-H. & Liao I-C. (1987) *Bioeconomic consideration of prawn farming*. Dept. of Aquaculture, National Taiwan College of Marine Science and Technology, Keelung, Taiwan.

FAO (1988) *Planning an aquaculture facility – Guidelines for bioprogramming and design*. ADCP/REP/87/24, UNDP, FAO, Rome.

Lee J.-C. (1988) Government policies and support for shrimp farming. In *Shrimp '88, Conference proceedings*, 26–28 Jan. 1988, pp. 175–81, Bangkok, Thailand. Infofish, Kuala Lumpur, Malaysia.

Mayo R.D. (1988) *The Birdsong Junction handbook*. JM Montgomery Consulting Engineers, Bellevue, WA, USA.

Mock C.R. (1987) A penaeid shrimp farm in Ecuador. *Aquaculture*, **13**, (3) 36–43.

Muir J. F. (1988) Aquaculture consultancy services – the realities (Parts 1 and 2). *Infofish International* (5) 19–20, (6) 48–50.

O'Sullivan D. (1988) *The culture of the marron (*Cherax tenuimanus*) in Australia: A review*. Presented at the World Aquaculture Society Meeting, Hawaii, 1988 (mimeo).

USDC (1989) *Malaysian shrimp culture*. National Marine Fisheries Service, National Oceanographic and Atmospheric Administration, US Dept. of Commerce, F/IA23: KK, PN, IFR-89/50.

Wickins J.F. (1986) Prawn farming today: opportunities, techniques and developments. *Outlook on Agriculture*, **15** (2) 52–60.

World Bank (1981) *Guidelines for the use of consultants by world bank borrowers and by the World Bank as executing agency*. World Bank, Washington, D.C.

Chapter 10
Economics

10.1 Introduction

Since the setting up of a crustacean farming project is likely to require considerable investment, it is worth considering fully the financial and economic implications before making any commitments. Too often financial disaster strikes as newcomers are captivated by the euphoric press accounts of the industry and fail to appreciate the costs and risks involved.

This chapter, by reference to published examples of budgets and feasibility studies, provides an indication of the likely costs of projects and gives an idea of which factors are most critical in the determination of economic viability. Established financial and economic appraisal methods are commonly used to analyse project proposals, although numerous species specific assumptions underly the calculations and must be fully understood in advance. If the assumptions for key factors such as productivity levels are based on false or misleading estimates, incorrect conclusions will be drawn. It is thus essential to evaluate the technical aspects of a project (Chapter 7) to gain an understanding of its likely performance before commissioning an economic appraisal. The technique of sensitivity analysis, which is applied to investigate the effect of variations in prices and yields, is outlined and some examples of its application are provided. Even though prices and price structures are constantly changing, important general conclusions are drawn concerning the various opportunities crustacean farming has to offer.

10.2 Objectives

The results of the appraisal of any scheme or project can only be interpreted by referring to specified objectives and these should be clearly established as a first step in project planning (Chapter 9).

10.2.1 Strategic or policy objectives

In the case of governments or aid agencies, the policy objectives for developing aquaculture schemes will usually be based on broad issues of social and economic development. They may be described in terms of benefits to the national economy (e.g. attracting foreign investment, or improving the balance of trade through exports or import substitution) or with reference to rural development schemes (benefiting the local economy, providing employment, improving local incomes, avoiding population drift to urban centres). The possible socio-economic benefits of an aquaculture project are summarized in Figure 10.1. In general, the objective of economic development can be defined as providing the greatest possible increase in the standard of living of a population. However, the economic benefits of crustcean culture projects must be

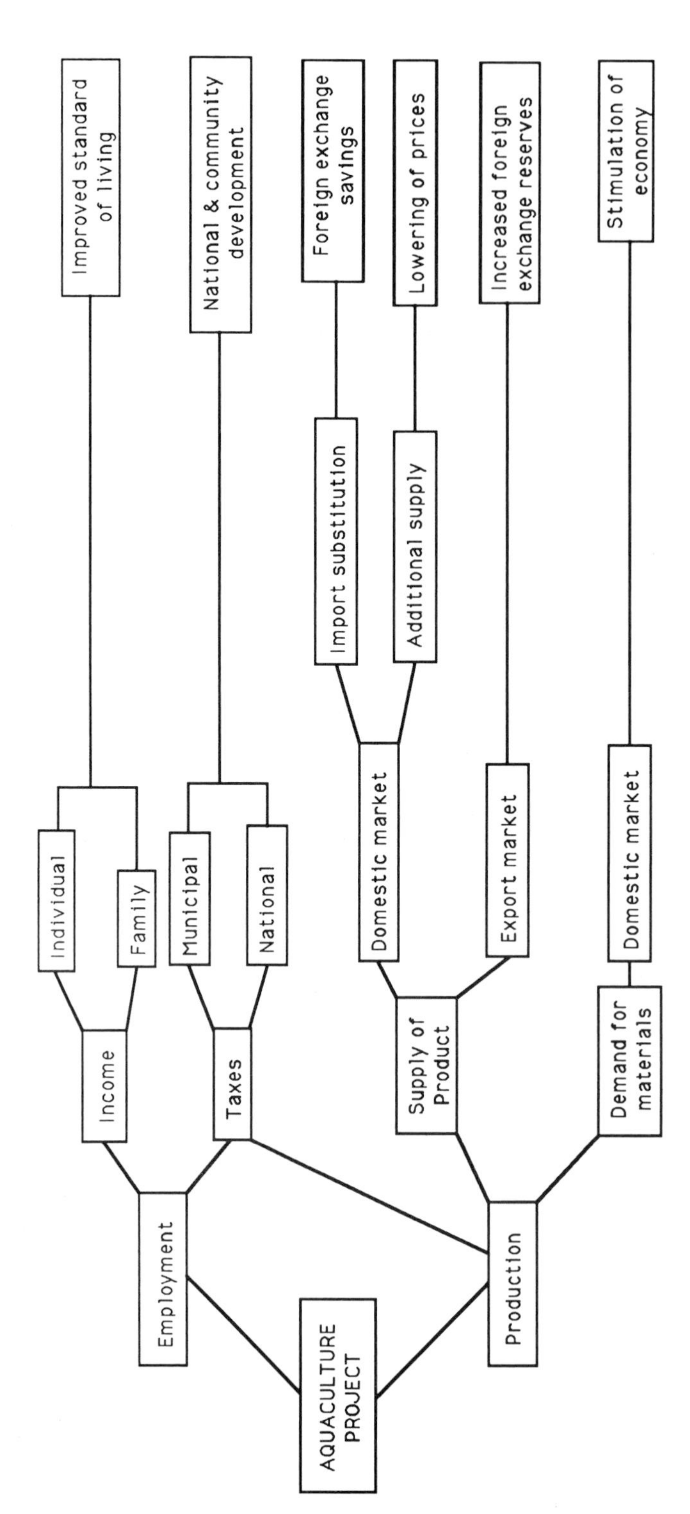

Fig. 10.1 Socio-economic benefits derived from an aquaculture project (modified from Israel 1987a).

weighed against a range of possible negative impacts: social (Section 11.2); ecological (Section 11.4); and environmental (Section 11.5).

In countries which lack financial resources, the need to attract foreign investment is usually a crucial consideration in aquaculture development plans. Several methods are applied by government to achieve this objective, including:

- Relaxing exchange controls;
- Offering flexibility in regulations regarding national involvement in joint ventures;
- Simplifying legal requirements and bureaucratic 'red-tape';
- Providing training, information, extension and other support services;
- Promoting auxiliary industries, e.g. feed production;
- Cutting duties on imports of equipment and raw materials;
- Providing 'tax holidays' during start-up;
- Permitting repatriation of profits and assets;
- Providing infrastructure (e.g. designated areas for aquaculture, 'aquaparks').

The introduction or rehabilitation of crustacean culture on the scale of smallholders or artisanal farmers has been an objective of some governments in developing nations (e.g. Indonesia (Sidarto 1977)). If supplies of wild or hatchery juveniles are adequate and markets for crustacean products are accessible, this approach has obvious potential for raising incomes in rural communities. However, as a means of producing food to supplement local diets, crustacean culture does not provide the more immediate benefits of finfish farming, which can generate greater yields with lower input costs using simpler and more easily introduced techniques (Section 11.2). In addition, the lower value of fish does not tempt the producer to export his crop and it thus becomes available locally.

Though private crustacean farming ventures can range in size from vertically integrated multi-national concerns down to smallholdings and backyard hatcheries, each will ultimately be interested in generating profits. Objectives and investment decisions, however, will be based on different criteria at different scales of operation. For example, the largest companies may view involvement in crustacean culture as part of a long-term strategy to exploit trends in increasing seafood consumption. As such, significant investment in unprofitable new technology and production on a pilot or research scale is sometimes judged to be justifiable in the light of potential long-term profits and a leading position in the industry.

Smallholders, on the other hand, may be attracted to crustacean culture as an alternative to other activities such as agriculture (e.g. rice growing) or the production of fish or salt. In such cases, the potentially greater profits and risks of crustacean farming must be weighed against the immediate loss of income from an established crop. Proven 'low-risk' culture practices are sometimes employed to keep the impact of the critical start-up and learning period to a bare minimum. 'Backyard' hatchery operators, who produce juveniles only in response to the seasonal availability of broodstock and demand for juveniles, are primarily concerned with the profitability of each production run. This type of operation allows great flexibility but alternative occupations or sources of income are needed to cover periods of inactivity.

The considerable areas of overlap between government and private sector objectives allow for co-operation and co-ordination of effort. Examples of this include the establishment of government hatcheries to supply juveniles, and the promotion of contract growing 'nucleus/plasma' schemes in which a privately owned, vertically integrated 'nucleus' operation supplies numerous small holders ('the plasma') with materials, technical assistance and a guaranteed market (Section 9.5.3). Further aspects of government assistance to the aquaculture sector are discussed in Sections 9.5.5 and 11.6.3.

10.2.2 Tactical objectives

In many cases the objective of a particular project can be specified in terms of a desired rate of return. When overall economic benefits are considered, this may be expressed in the form of the economic internal rate of return (EIRR) (Section 10.5). Individual farmers and private investors can usually express their objectives in terms of desirable rates of personal income or levels of financial return on investments.

The returns that are required by an investor reflect the sources and cost of the capital employed. The cost of risk capital (equity) will depend on the return expected by the shareholders, while the cost of funds borrowed from a financial institution will vary depending on interest rates and the duration and size of the loan. Since the cost of loans is usually lower than the cost of risk equity, the average cost of capital for a particular project will depend on the proportions (the gearing ratio) of the two different types of capital employed. Achieving the correct balance between equity and loan capital can be critical to the financial success of a project. Though heavy reliance on loans may be an attractive option when interest rates are low, when they are high a highly geared company can be placed in an impossible situation due to the excessive cost of servicing their debt. In such situations, a venture that relies mainly on equity capital can be better placed to survive.

When deciding what rate of return is suitable for a particular type of project it is also important to take account of the risk involved. Investment in crustacean culture is very hazardous when compared, for example, to the purchase of secure treasury bonds, and in order for it to be a worthwhile proposition it should demonstrate a potential for generating higher returns. Just what rates of return might be acceptable is a subjective matter, but one method of evaluation involves adding a risk premium to the return possible from an alternative low-risk investment. For example, in a financial analysis of shrimp culture in the USA, Johns *et al.* (1983) considered that the minimum rate of return required (on equity) would be 19.75%, which represented a risk premium of 5% added to the return from US corporate bonds of 14.75%. Risk premiums of between 5% and 15% are appropriate for most crustacean culture projects, although the highest premiums should be used when intensive or untried technology is to be employed.

10.3 Finance

The sources of funding for aquaculture projects can be divided into two categories: capital assistance and private investment.

Capital assistance, in the case of developing nations, usually comes from external sources and is provided in the form of loans from institutions such as the World Bank, and grant aid packages from donor countries and organizations. In the case of developed countries capital assistance is made available in the form of government enterprise grants and regional development subsidies, sometimes linked to providing employment.

Private investment, which can be foreign and/or domestic, may involve private or public sale of equity, and is usually backed-up by commercial bank financing.

The bulk of external capital assistance is provided in the form of 'soft' concessionary loans with low interest, extended repayment terms. The Asian Development Bank (ADB) has played a leading role in this area, promoting aquaculture, particularly of shrimp, in countries such as Indonesia, the Philippines, Pakistan and Sri Lanka. Some World Bank (WB) loans are provided for schemes in which aquaculture is one of many components. For example, within a total WB package of US$200 million, US$5–30 million may be aimed at aquaculture (Nash 1988). Other organizations providing 'soft' loans to aquaculture include the Inter-American Development Bank (IDB), the International Development Agency (IDA) and United Nations organizations such as the International Fund for Agriculture Development (IFAD). Governments and various organisations including the EC and the United Nations provide assistance in the form of grants which, of course, do not require repayment. While part of this grant aid is aimed directly at the construction of research facilities, hatcheries and farms, part is usually provided in the form of technical assistance to help maximize the beneficial impact of the whole aid package. In Cuba, for example, the Food and Agriculture Organisation of the United Nations currently provides both financial backing and technical assistance for the farming of shrimp and prawns.

External capital assistance is chanelled to individual projects via government departments, commercial or development banks and specialised credit institutions. Domestic government input will usually involve the provision of infrastructure and the funding of support activities such as extension services, hatcheries, training and research facilities or grants. Domestic and international commercial or development banks such as The International Finance Corporation (IFC), a USA based development bank, will typically provide credit for large scale projects over US$1–4 million and thus tend to favour large vertically integrated projects incorporating hatchery, farm, processing plant and sometimes a feed mill (Leeds 1986).

Specialized credit institutions, often in the form of agricultural, fishery, or co-operative banks, are geared to supplying loans to numerous scattered smallholders because, unlike commercial banks, they usually have a wide network of branches in rural areas. As such their role can be central to the development of community-based, small-scale aquaculture. If the specialized needs of small-scale operators are to be met, however, loans need to be provided promptly, with a minimum of bureacracy and preferably with the back-up of technical extension and marketing services (FAO 1977). In addition, as part of a co-ordinated effort, briefing also needs to be provided on the purpose of credit and how to make best use of it. Development banks have been urged to follow this approach to the promotion of aquaculture and the argument that they cannot bear the added cost of supervising numerous small loans is considered spurious bearing in mind that they are ready to provide subsidised credit for large-scale projects (Smith 1984). Certainly, in the parallel case of fisheries development, success in the past has relied on the provision of appropriate, flexible and timely credit (Tietze 1989).

In developed nations, governments often provide support for aquaculture in the form of enterprise grants and subsidies, particularly if a proposed project is likely to stimulate the economy of a depressed area and provide jobs. Examples of this include

Spanish government and EC grants towards the capital expense of constructing shrimp hatcheries and farms in Andalusia. Other schemes to assist businesses in rural areas are aimed primarily at agriculture but can equally be applied to crustacean culture.

Private investment in the crustacean farming business can be backed by venture capital through the private or public sale of equity. Venture capital is available from specialized companies but they expect high rates of return and usually consider the capital needs of aquaculture firms too small to be of interest. McCoy (1986) lists typical criteria for venture capitalists:

- Annual growth rates of at least 25%, preferably 50%;
- Pre-tax margins of at least 35%, preferably 50–60%;
- Minimum return on investment (ROI) of 30%;
- Management sufficiently 'attractive' to make a public offering within five to eight years;
- Unique patentable technology is preferred;
- Up to 30% ownership of the project;
- Minimum investment level around US$1 million.

Funds for aquaculture operations are rarely generated directly through the public sale of equity although some companies, usually ones involved in salmon production, have made public offerings on the Vancouver Stock Exchange in British Colombia, Canada. One, Aquarius Seafarms Ltd of Vancouver, raised C$4 770 000 on an initial public offering of 1.8 million shares at C$2.65 each. Virtually all publicly traded firms with interests in crustacean aquaculture (e.g. British Petroleum, Unilever, SANOFI and the International Proteins Corporation) count aquaculture as a diversification interest rather than their prime activity.

Loans for crustacean culture operations are difficult to obtain in some countries unless the activity has a proven track record of success. Main & Deupree (1986) note that there have been problems in Hawaii in obtaining capital for expansion and operating expenses in the wake of large short-lived projects; and in Africa, Skabo (1988) complains of a lack of confidence from commercial banks, foreign aid organizations and international finance institutions, due to general ignorance of the aquaculture industry.

It is worth remembering that most companies go bankrupt for a lack of cash rather than a lack of profits, and that aquaculture-based operations are no exception to this general rule. A financial crisis will lead to abrupt failure even while most problems of a biological or technical nature can, in time at least, be solved. Despite the obvious need for careful financial planning, many if not most companies in the aquaculture industry are under-capitalized and unable to fulfil their potential (McCoy 1988). Problems are particularly common during the start-up phase when insufficient cash is available to cover operating expenses. For most projects positive cash flows should not be expected for a minimum of three to five years after initial investments have been made.

10.4 Joint ventures

In many developing tropical nations the profit potential of crustacean culture and its support industries has attracted foreign investment. The attraction is based primarily on the presence of favourable climatic conditions and the availability of relatively cheap land and labour. Activity so far has been centred on shrimp culture in south-east Asia and Latin America, with production aimed at export markets because of the limited purchasing power of domestic consumers.

Regulations regarding the establishment of joint ventures vary from country to country. Sometimes a certain minimum level of local participation is required on the basis of equal or majority domestic shareholding. In some cases the foreign party is required to finance local stock acquisition while in others, if the requirements for foreign capital are high, the foreign partners often take the majority interest.

Joint venture agreements often involve the provision of technical expertise by the foreign partner. In such instances it is usually worth establishing pilot-scale operations before expanding to full production, unless the foreign partner has an excellent record of technical success and the proposed site for the project is considered to be completely satisfactory. The subject of technical assistance and how to get the best from it is discussed in Chapter 9.

Many opportunities for joint venture agreements exist, and often interested parties advertise in trade publications such as Infofish International. Great care needs to be taken in choosing a partner. Ideally, selection should be made from a number of candidates and preference given to parties with reliable histories and appropriate experience. Some of the most successful ventures in shrimp culture involve large established companies with experience in the export of fishery or agricultural products.

10.5 Investment appraisal methods

In order to evaluate and compare the potential benefits of a series of different investment opportunities, various analysis techniques can be employed. To evaluate purely financial returns appraisal methods include pay-back period (PB), return on investment (ROI), net present value (NPV) and internal rate of return (IRR). To measure the likely economic benefits of a project it is possible to calculate the economic internal rate of return (EIRR). Although appraisal methods are often employed as a separate and final stage in the development of a project, the principles should be applied at all stages of project design, selection and analysis (ODA 1977).

The pay-back (PB) method simply calculates the time it will take to recover the cost of the original investment. Its advantages are its simplicity and the fact that it can reflect the risk level of a project, i.e. short pay-back periods usually imply lower risk. It does not, however, take account of cashflows beyond the pay-back period and thus does not calculate the overall profitability of a project over its expected lifetime. Because of this basic limitation the PB method is best used only in conjunction with other appraisal methods.

The rate of return on investment (ROI) method calculates the average yearly return over the whole life of the project but does not adjust future revenues to take into account yearly incidence (the 'time value' of money) and as such it is inferior to the NPV and IRR methods.

The net present value technique (NPV) provides a figure for the overall value of a project by totalling all the expected annual cash flows over its intended life. However, in order to account for the decline in the value of money over time, each yearly cash flow is adjusted by a discount factor calculated on the basis of the discount rate – the minimum rate of return desired from the project. A project which yields a positive NPV will in theory generate sufficient returns to pay interest at or above the discount rate as well as repay the original investment.

The internal rate of return method (IRR) (also known as the discounted cash flow yield) takes the NPV technique one stage further and calculates the discount rate at which discounted revenues, totalled over the life of the project, just balance the original investment, i.e. the discount rate at which the NPV is zero. The IRR represents the true rate of return from the investment.

The appraisal of a project by the economic internal rate of return method (EIRR) is closely related to the cash flow analysis method of IRR, but in order to estimate economic returns to society as a whole (as opposed to solely financial benefits) the prices of goods, services and labour are adjusted so that they more closely represent their true economic value rather than simply their financial cost. Distortions in a country's domestic price structures usually result from the imposition of trade barriers and are often most marked in less developed nations where there is an acute shortage of foreign exchange. Wage rates can also become distorted as a result of a large pool of unemployed labour and rigidities in labour markets, and so inadequately reflect the cost to the economy of employing additional labour. In economic appraisal, 'shadow' wage rates compensate for such distortions and conversion factors adjust domestic prices to more representative world-equivalent prices. Conversion factors vary between countries and between different categories of goods, and together with shadow wage rates they can be obtained from the World Bank. Although it is difficult to establish the true economic values of certain goods and services such as power, construction and internal transport, economic appraisal provides a valuable indication to governments and aid agencies of a project's overall worth and exposes schemes which are only viable on a financial basis because of protectionist policies.

10.5.1 Project life

For investment appraisal the life of a project is usually set at ten years. Although this is somewhat arbitrary, if profitability is not foreseen within ten years or less, the project is unlikely to be an attractive investment proposition. Some assessments, for example for development projects, are made on the basis of longer project lives of 12, 15 or 20 years.

10.5.2 Inflation

It is usually assumed that the returns generated by a project will keep pace with inflation, i.e. the inflation of production costs will be compensated by inflation in the prices obtained for the final products. While it is advisable to test possible variations in the market prices for products using sensitivity analysis (Section 10.7), the above assumption is generally valid and means that the IRR technique yields a value for the true rate of return (exclusive of inflation). However, because the interest rates of loans and the cost of equity are expressed as nominal rates (inclusive of inflation) rather than true rates, it is sometimes useful to also express the rate of return from a project as a nominal rate. This can be done most simply by adding the expected rate of inflation to the estimated true rate of return. Thus if the inflation rate is expected to be 5%, a project with an estimated IRR value of 20% has a nominal rate of return of 25% and should be able to pay interest at this higher rate.

10.6 Assumptions

In an assessment of the future viability of any project, a series of assumptions must be made. For financial appraisal some of these assumptions are made to simplify calculations and permit general conclusions to be drawn. For example, it is often assumed that unlimited capital is available and that tax payments and allowances do not apply. Clearly in any specific situation allowances must be made for these factors. Taxation effects in particular can have a large influence on profitability.

Other assumptions relate to the market prices obtained for products, the dimensions and operational parameters of the project, and the cost and amounts of materials, labour and utilities required for construction and operation. As none of these can be estimated with precision, feasibility studies should take into account the likely accuracy of underlying assumptions.

The market prices for the final products will have the greatest and most direct effect on profitability. They are variable and estimates can only be based on knowledge of existing markets and market trends (Chapter 3).

The projected output level of any operation rests on numerous assumptions, some of which are under the control of the designer or operator and some of which can only be guessed from experience. Some key design and operational parameters for a shrimp nursery, on-growing and processing operation are given in Table 10.1. In actual farms, rates for survival, growth and feed conversion will depend on manage-

Table 10.1 Example of assumptions affecting productivity, associated with the design and operation of a shrimp nursery, on-growing unit and processing plant (based on IFC 1987).

Phase	Factor	Assumed value	Units
Nursery			
	number of ponds	11	—
	average pond size	1.4	ha
	nursery pond area (total)	15.7	ha
	stocking density	400 000	Post-larvae ha^{-1}
	duration of cycle	35	days
	*survival rate	70	%
On-growing			
	on-growing pond area (total)	300	ha
	average pond size	10	ha
	number of ponds	30	—
	stocking density	40 000	juveniles ha^{-1}
	duration of on-growing period	125	days
	cycles per year	2.6	—
	number of harvests per year	78	—
	*feed conversion ratio	2.5:1	wt. feed:wt. shrimp harvested
	*survival rate	70	%
	*average harvest weight	28	g
	*average harvest	784	kg ha^{-1} $crop^{-1}$
Processing			
	capacity	5	mt day^{-1}
	processing yield	65	%

* Expected outcome rather than design or operational choice.

Table 10.2 On-growing results for *Penaeus chinensis*.

Location	Temp. (°C)	Stocking density (no. m^{-2})	Salinity (ppt.)	Survival rate (%)	Size at stocking (g)	Size at harvest (g)	Yield per crop (kg ha^{-1})	Crop duration (days)	Reference
China, north	—	—	25	25	0.005*	18–22	—	137	Mock 1990
UK, indoor tanks	26–30	25	28–30	67	0.15	22.8	3819E	112	Forster & Beard 1974
UK, indoor tanks	26–30	166	28–30	42	0.13	14.6	9930E	112	Forster & Beard 1974
Taiwan	22–34	4	30–38	55	0.01	14.3	314	120	Liao & Chien 1990
Taiwan	22–34	15	30–38	55	0.01	11.7	965	120	Liao & Chien 1990
Taiwan	22–34	20	30–38	55	0.01	9.4	1034	120	Liao & Chien 1990
Korea	14–22	—	26–31	—	0.01	25.0	—	168	Kim 1990
China, south	18–34	7.8	11–34	—	0.05*	16.5*	—	146	Hu 1990
China, south	18–34	7.8	11–34	—	0.05*	22.4*	—	190	Hu 1990
China, north	19–30	(5.4) h	—	—	0.007*	20.8*	1123	141	Zhang 1990
China, north	19–28	(11.6) h	—	—	0.007*	19.9*	2308	141	Zhang 1990
China, north	22–30	—	—	—	0.005*	41.3*	—	131	Zhang 1990
China, north	—	—	—	—	—	40.8	1263	—	Zhang 1990
China, north	—	(2.9) h	—	—	0.01	43.5	1275	126	Zhang 1990
USA assumptions	15–29	30	—	60	1.00	25	4500	112	Main & Fulks 1990

h = Harvest density. * Weight (g) calculated based on length (cm) using $W = 0.01412 \times L^{2.9465}$ (Zhang *et al.* 1983). E = Extrapolation based on tanks of 0.62 m^2.

ment skills and the environmental conditions. They can be predicted most reliably when proven species and culture techniques are applied in established culture areas. When the output level of a proposed project is calculated from a string of assumed values any errors can have a multiplicative effect. Table 10.1 shows that the annual production per hectare from a crustacean farm can be calculated as the product of stocking density (no. ha^{-1}), survival rate, average harvest weight, and number of crops per year. A 20% over-optimistic estimate for one of these factors will distort the annual production per hectare (2038 kg) by a factor of 1.2 (2446 kg), but if all four are over-estimated by 20% the productivity will be exaggerated by a factor of 2.074 (4227 kg).

To correctly interpret the results of a financial appraisal, the significance of the underlying assumptions must be understood. This is illustrated by the appraisal of proposed *P. chinensis* culture in mainland USA, as shown in Table 10.2 (Main & Fulks 1990). The most effective scenario presented involves the production of two crops per year of *P. chinensis* and predicts IRR values of 31% and 34% for operations based in South Carolina and Texas respectively. These levels of profitability are significantly greater than the 1% and 5% IRR figures given for current *P. vannamei* culture in the same states. The improvement possible with *P. chinensis* is attributed to its faster growth rate and the extended growing season possible because of the cold-tolerant nature of the species. Assumptions for the on-growing phase which underly the financial appraisal include:

• Stocking density (1 g nursed juveniles)		300 000 ha^{-1}
• Survival rate		60%
• Crops per year		2
• Shrimp size at harvest	(first crop)	23.5 g
	(second crop)	25 g

From these figures it is possible to calculate that a first crop of 4230 kg ha^{-1} will be followed by a second of 4500 kg ha^{-1} to give a total projected annual yield of 8730 kg ha^{-1}. This level of productivity greatly exceeds the current US average of 3000 kg ha^{-1} yr^{-1} which is based mainly on the semi-intensive culture of *P. vannamei* (Rosenberry 1990). Though not specified in the text of the appraisal, it is this greatly improved level of productivity which is at the root of the greatly enhanced financial returns. However, it is probable that the regular production of 4000–4500 kg ha^{-1} $crop^{-1}$ of any shrimp species would require the use of more intensive culture techniques than those currently applied in the USA. (In fact a trend towards intensification in the USA has been noted (Chamberlain 1989).).Clearly, the predicted improvement in profitability possible with *P. chinensis* has as much to do with underlying assumptions regarding the use of more intensive culture techniques as it has to do with the use of a new species with improved growth potential. This has important implications for US shrimp farmers attracted by the idea of generating greater profits through the farming of *P. chinensis* instead of *P. vannamei*. Table 10.2 presents actual on-growing results for *P. chinensis* and provides the type of information necessary to evaluate the culture potential of this species and more fully appreciate the significance of the assumptions underlying a financial appraisal.

The analysis of the feasibility of on-growing, nursery and hatchery operations is generally more complex than the appraisal of projects centred on processing, marketing or feed production because it is harder to predict output levels and estimate the duration of the delay between investment in construction and attaining profitable production levels. By comparison, a processing plant receiving material from many aquaculture operations as well as from wild fisheries, or a feed mill with reliable supplies of raw materials, can achieve a more stable output within a more predictable period.

10.7 Sensitivity analysis

If the assumptions which underly a particular project appraisal are fixed, the conclusions that can be drawn are necessarily limited. However, if the assumptions are changed to allow for uncertainties and to take account of possible alternative scenarios, a more complete picture of viability can be obtained. The technique is known as sensitivity analysis and relies on altering critical variables such as product market prices and output levels and observing the effects on production costs, profitability or financial indicators such as the pay-back period (PB), the NPV, or the IRR (Section 10.5). Critical factors are varied by a particular percentage, e.g. ±10%, or within a range of likely values established from experience, case studies or experiments.

In the most complex analyses computers are programmed to generate values for selected factors using probability functions, and then to use these values to run simulations. By repeatedly running the program the results for a particular financial indicator can be expressed in the form of a probability distribution or a mean value with its standard deviation. In an example of this, Staniford (1988) ran 500 simulations of the profitability of a hypothetical 10 ha integrated crayfish (*Cherax destructor*) farming operation in which market prices varied between A$8 and 12 per kg, yield varied between 2000 and 3800 kg ha^{-1}, and periodic crop failures occurred due to mismanagement. Results for the IRR were presented in the form of a probability distribution and showed that an IRR of around 15% was most probable and that there was only an 18% chance of exceeding an IRR of 20% (Table 10.3). Simulations of this type are useful for quantifying levels of risk and can be used to study the impact of random events such as hurricanes. In the studies by Staniford (1988; 1989) and another based on shrimp culture by Hanson *et al.* (1985) it was observed that a much more realistic picture of likely profitability could be built up using this repetitive technique rather than straightforward sensitivity analysis.

Table 10.3 Probabilities of different levels of profitability for a simulated yabbie culture operation (from Staniford 1988).

Internal rate of return (IRR) %	Probability of project exceeding IRR value %
5	96
10	87
15	56
20	18
23.7	3

Table 10.4 Factors typically used in sensitivity analyses.

(1) Market price	(4) Design features
(2) Productivity level	size of operation
(3) Factors affecting productivity	size of ponds/tanks/containers
species or combination of species cultured	water exchange rate
availability of juveniles for stocking	water re-circulation rate
stocking density	(5) Major cost components
survival rate	Investment
growth rate	land
harvest weight	construction
length of growing season	duration of start-up period
crops produced per year	Operation
environmental temperature	stocking/broodstock
operating temperature	feed
salinity	labour
dissolved oxygen levels	energy
random events e.g. hurricanes, crop failures	interest rates

Table 10.4 lists various factors that are typically included in sensitivity analyses to test their impact on profitability. All analyses should at the very least test the assumptions regarding market prices and productivity levels. The selection of other factors would ideally relate to scenarios that are considered probable or possible in any particular situation. For example, a shrimp on-growing operation in a temperate climate may investigate the effect of variations in the length of the growing season, and an intensive operation may need to test assumptions regarding energy costs, particularly if based in heated indoor facilities.

Some general conclusions can be drawn from published cost and sensitivity analyses:

(1) *Economy of scale* As the size of a system increases, production costs are reduced because fixed costs are spread over greater output levels (Shrimp: Hanson *et al.* 1985. Prawns: Sandifer & Smith 1985. Crayfish: Dellenbarger *et al.* 1988; Staniford & Kuznecovs 1988.)

(2) *Pond size* Smaller ponds are more controllable and more productive per hectare but are more expensive to build and maintain and hence have greater investment and operating costs. In the study by Hanson *et al.* (1985) large shrimp farms and large ponds gave higher rates of return, although increasing the number of ponds in a facility resulted in less variability in IRR values.

(3) *Start-up time* Ideally this should be kept to a minimum. In yabbie culture Staniford & Kuznecovs (1988) found it was very important that output in a new operation reached 100% in year two, and Hanson *et al.* (1985) found that a 100 ha shrimp farm built in two years was more profitable than a 200 ha farm built in three years.

(4) *Site selection* Factors affecting yield and annual production potential have the most significant effects on capital and operating costs. In an assessment of brackish water pond systems Muir & Kapetsky (1988) found these factors to include water temperature, salinity and silt content, and soil acidity – characteristics which also proved to be especially critical in more extensive pond systems (Sections 6.3 and 8.2).

10.8 Intensity level

The attraction of intensive and super-intensive crustacean culture is its apparent ability to maximize financial returns per hectare of operation. However, though the output of culture systems can certainly be boosted through intensification, resulting changes in the economics of production do not necessarily improve overall profitability. This is because, with the exception of very extensive systems, operating costs per kilogram rise with increasing intensity and can result in reduced profit margins. The commercial viability of intensive crustacean culture is limited to three specific situations, where:

(1) A premium price market can be exploited;
(2) High land prices or a shortage of suitable sites preclude the construction of extensive or semi-intensive ponds;
(3) The culture of a particular species requires controlled-environment facilities.

In Japan, high land costs and the high price market for live *Penaeus japonicus* combine to support high intensity culture operations. Table 10.5 compares three Japanese systems, the production costs of which rise with increasing intensity and are greatest in super-intensive Shigueno-type tanks (Shigueno 1975). Also reliant on a premium price market are the intensive systems for producing soft-shell crabs and crayfish, though strictly speaking these are holding operations rather than culture operations.

The high production and overhead costs of intensive crustacean culture relate to the high cost of establishing and operating the necessary systems. In intensive shrimp farming increased investment in reinforced pond embankments, extra water control structures and aerators may be involved, together with greater feed and juvenile requirements. Taken together these inputs are rarely fully offset by increased production levels. In the case of extensive pond culture, however, adopting some elements of intensification can have a dramatic effect on productivity and can reduce overall production costs per kilogram. For example, by raising the stocking density and using compound feeds (i.e. by adopting semi-intensive practices) increased productivity will usually more than compensate for the extra costs incurred. This strategy has been widely adopted in extensive shrimp farms except in situations where seed or feed or the cash to buy them are in short supply, or where the farm operator is concerned with minimizing risk rather than maximizing profits.

Table 10.6 presents data based on *P. vannamei* culture in Ecuador. Whereas

Table 10.5 Comparison of production cost and culture intensity for *P. japonicus* in Japan (based on Yamaha 1989).

Pond type	Density (no. m^{-2})	Yield (kg m^{-2} $crop^{-1}$)	Production cost (US\$ kg^{-1})
Partial embankment(a)	20–50	0.4–0.5	34
Full embankment(b)	40–60	0.6–1.0	38
'Shigueno' tank(c)	70–80	1.5–2.0	45

(a) Impoundment wall submerged at high water and topped with a net to prevent escapes. (b) Normal pond. (c) Circular tank with sand-covered false floor.

Table 10.6 Estimated direct operating cost kg^{-1} and net revenue ha^{-1} for an Ecuadorian shrimp farm operated at different stocking densities (based on Shang 1983).

Stocking density (no. m^{-2})	Feed applied	Production (kg ha^{-1} $crop^{-1}$)	Direct operating cost (US\$ kg^{-1})	Net revenue (US\$ ha^{-1})
1.8	–	500	1.93	3354
4	+	1100	1.80	8535
7	+	2000	2.04	14 580
10	+	2500	2.45	16 129

output rises with each increase in intensity, the direct operating costs per kg are at a minimum at a stocking density of four shrimp m^{-2}. The figures for profitability per hectare rise sharply with stocking density because a selling price of US\$5.67 kg^{-1} is assumed, which is well in excess of the production costs in each scenario. When profit margins are good, as in this example, operating at greater levels of intensity will boost profitability per hectare. When profit margins are slim, however, the best strategy for maximizing profitability is to operate with the lowest production costs per kilogram.

In the case of shrimp production, if prices for the farmed product continue to drop globally, viable culture operations will increasingly be found where less intensive techniques are applied and shrimp is produced at the lowest cost per kilogram. The economics of intensive and super-intensive culture are highly sensitive to increases in the cost of feed, seed, labour and energy. Semi-intensive culture will dominate except where poorly financed operators with limited technical skills or support continue to rely on traditional methods.

10.9 Review of cost analyses

In the sections that follow, published details of the costs of culturing shrimp, prawn, crayfish, lobster and crab are reviewed and where possible compiled to illustrate the major components of investment and operating costs. Although most individual studies necessarily relate to a particular situation and are based on a specific set of underlying assumptions, taken together they serve as useful reference material for the financial appraisal of crustacean farming proposals and may help project sponsors gauge whether they are getting value for money.

10.9.1 Shrimp and prawns

10.9.1.1 Hatcheries

The cost of setting up a simple backyard *Macrobrachium* or penaeid hatchery with the capacity to produce 0.5–5 million post-larvae per year can be less than US\$10 000. SEAFDEC (1985) gave a figure equivalent to US\$6000 for a hatchery in the Philippines designed to produce 0.6 million *P. monodon* PL_{35} annually, and New (1990) quoted a figure as low as US\$1000 to establish a backyard *Macrobrachium* hatchery in Thailand. These low prices are possible if cheap or locally available materials, such as bamboo, wood, cement blocks and plastic sheeting, are used in construction. Inexpensive tanks can be made from up-ended sections of concrete drainage pipes or plastic pool liners within wood or bamboo frames. Such structures, however, may only be temporary.

In backyard hatcheries equipment is kept to a minimum; where possible, submersible pumps and hoses are used in preference to fixed pumps and rigid pipework, and stand-pipes are fitted in place of tank valves. All the same, if the hatchery is dependent on mains electricity, provision must be made for emergency power to maintain aeration in case of power cuts. This may take the form of a small generator or a petrol motor that can directly operate an air compressor.

Since *Macrobrachium* hatcheries can usually rely on supplies of berried female broodstock from local farms, they do not require separate facilities for broodstock production or maturation. Generally speaking, penaeid hatcheries that receive supplies of broodstock or nauplii from elsewhere are also able to concentrate resources on rearing larvae and post-larvae.

Large penaeid hatcheries with the capacity to produce more than 200 million post-larvae per year can easily represent investments of more than US$2 million. The expense can often be attributed to the construction of facilities with a planned life of 10–20 years, and investment in multiple back-up systems. The latter may include broodstock ponds and a controlled-environment maturation unit (Section 10.9.1.2) to provide a supplementary supply of nauplii, as well as spare generators, pumps and compressors to provide security for the electricity, water and air supplies respectively. Plumbing is usually fixed, and some hatcheries incorporate dual seawater distribution lines which enable pipework to be alternated between normal usage and disinfection. A controlled indoor environment is usually provided for the maintenance of algae stock cultures and the early stages of algae culture. The largest projects often include buildings for the accommodation of staff and as much as 20% of the total investment may be spent on acquiring a design and technical assistance package (Section 9.5.2).

If a hatchery is to operate at above ambient temperatures, investment costs are increased by the need to provide a heating system and sometimes insulation. A *Macrobrachium* hatchery designed by AQUACOP (1982) included a solar heating system which represented 20.5% of the total investment. Many hatcheries in Ecuador have boilers and heat exchangers to raise water temperatures during the cool season.

Table 10.7a lists examples of operating and investment costs for a range of different shrimp and prawn hatcheries. The proportionate components of the major costs are presented in Tables 10.7b and 10.7c, with summaries in Figures 10.2a and b. In some cases, values in these tables and figures are somewhat subjective since there are gaps in the data, and individual studies do not divide their costs precisely between the categories presented. However, it is clear that construction costs (buildings) are the major outlay in establishing a hatchery and that labour is the greatest operational cost. Table 10.8 lists the works, equipment and services needed to set up a representative medium-sized operation with a production capacity of four to eight million penaeid post-larvae (PL_{7-10}) per month. If broodstock ponds are required, their inclusion will constitute a major additional expense. If complete reliance is to be placed on pond reared stock, IFC (1987) estimate that 1 ha of broodstock ponds are needed for each 10 million post-larvae produced annually.

The relationships between production capacity and investment and operational costs are shown in Figures 10.3a and 10.3b, respectively. For both costs the data

Table 10.7a Examples of operating and investment costs for shrimp and prawn hatcheries.

Species	Location	Output (millions per month)	Post-larvae age(a)	Operating cost (US$ yr^{-1}) (b)(c)	Investment (US$)(b)	Facilities included	Production cost (US$ per 1000 PLs)	Reference/source
P. monodon	Philippines	0.05	35	11 000	6 000	0.1 ha land, nursery	18.33	SEAFDEC 1985
P. monodon	Indonesia	0.25	16	7 700	1 400(e)	—	2.57	Yap 1990
M. rosenbergii	Dominica	0.26	—	—	215 000(e)	—	—	New *et al.* 1978
P. monodon	Vietnam	0.50	10–15	—	2 900	—	—	Quynh 1990
P. monodon	Philippines	0.83	21	112 000	113 000	—	11.24	NACA 1986
M. rosenbergii	Tahiti	0.83	—	189 000	336 000	0.3 ha land, broodstock ponds	18.98	AQUACOP 1979
M. rosenbergii	unspecified	1.25	—	196 000	412 000(e)	solar heating system	13.07	AQUACOP 1982
P. semisulcatus	Kuwait	1.67	—	108 000	394 000	land free	5.39	Farmer 1981
not specified	USA Texas	3.75	5	659 000	554 000	2 ha land	14.64	Johns *et al.* 1981b
P. monodon	South East Asia	4.17	15	260 000(d)	540 000(e)	broodstock ponds, maturation nursery	5.20	IFC 1987
P. vannamei	Ecuador	4.58	7	257 000(d)	766 000(e)	brookstock ponds, maturation	4.68	Ecuador 1985 unpubl. data
P. vannamei	Ecuador	10.00	7?	—	500 000	asking price (built) + land		Al Hajj A. 1989 pers. comm.
P. monodon	Thailand	10.00	20	893 000	1 040 000(e)	broodstock ponds, maturation	7.44	Thailand 1988 unpubl. data
not specified	USA Texas	15.00	5	1 437 000	1 149 000	2 ha land	7.98	Johns *et al.* 1981b
P. vannamei	Ecuador	16.70	7	836 000	1 326 000(e)	broodstock ponds, maturation	4.17	Ecuador 1986 unpubl. data
P. vannamei	USA Texas	16.90	—	—	1 628 000(e)	maturation	—	Lawrence 1985
P. vannamei	Ecuador	20.00	7?	—	2 150 000	asking price (built), 21 ha land, maturation	—	Al Hajj A. 1989 pers. comm.
P. vannamei	Ecuador	30.00	15	—	3 660 000	broodstock ponds, maturation, nursery	—	Ecuador 1987 unpubl. data
not specified	USA Texas	45.00	5	3 311 000	3 562 000	2 ha land	6.13	Johns *et al.* 1981b

(a) Days from metamorphosis. (b) Adjusted for inflation to (1989) US$. (c) Excluding interest, taxes, royalties. (d) Excluding depreciation. (e) Excluding land.

Table 10.7b Components of hatchery investment costs, excluding land %.

Construction	Plant and networks	Tanks	Lab. and other equipment	Technical assistance	Total (b)	Brood-stock ponds (c)	Reference/source
13.6	39.4	30.7	16.3	—	100	—	SEAFDEC 1985
50.2	28.2	9.6	12.0	—	100	—	New *et al.* 1978
41.8	18.4	24.2	15.6	—	100	+26%	AQUACOP 1979
34.4	11.8	17.2	36.6(a)	—	100	+31%	AQUACOP 1982
6.9	14.0	59.4	19.7	—	100	—	Farmer 1981
53.0	23.6	12.0	11.4	—	100	+18%	IFC 1987
49.7	25.2	5.9	12.7	6.5	100	+11%	Ecuador 1985 unpubl. data
34.9	30.4	17.4	13.0	4.3	100	—	Ecuador 1985 unpubl. data

(a) Includes solar heating system. (b) Contingencies usually included at +10%. (c) Percentages represent additional cost above total for hatchery.

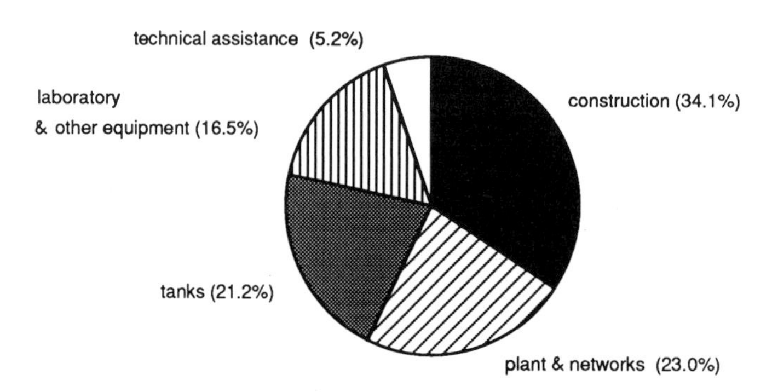

Fig. 10.2a Proportionate investment costs for a shrimp or prawn hatchery (based on a summary of data in Table 10.7b).

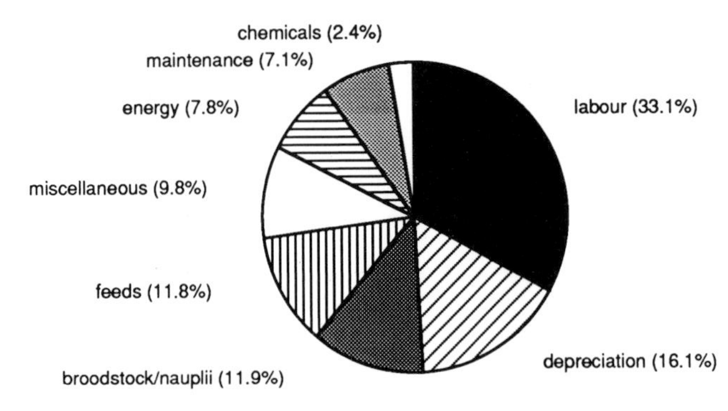

Fig. 10.2b Proportionate operating costs for a shrimp or prawn hatchery (based on a summary of data in Table 10.7c).

Table 10.7c Components of hatchery operating costs* %.

Labour	Feeds	Chemicals	Energy	Misc. (b)	Broodstock/ nauplii	Maintenance	Depreciation	Total	Reference/source
22.8	10.5	5.2	4.1	5.2	35.1	4.5	12.6	100	SEAFDEC 1985
10.9	40.0	1.3	18.6	3.2	10.4	1.3	14.3	100	NACA 1986
54.1	5.4	0.3	20.4	—	—	—	19.8	100	AQUACOP 1979
52.6	10.8	0.7	7.9	—	5.6	—	22.4	100	AQUACOP 1982
31.0	12.2	1.1	—	5.1	4.1	10.2	36.3	100	Farmer 1981
50.8	1.1	2.8	2.5	10.8	22.5	0.5	9.0	100	Johns *et al.* 1981b
48.9	11.8	4.7	6.1	14.9	—	13.6	—	100	IFC 1987
49.9	15.0	2.8	9.2	4.7	0.9	17.5	—	100	Ecuador 1985 (unpubl.)
12.9	11.9	5.6	1.9	33.4(a)	14.4	8.2	11.7	100	Thailand 1988 (unpubl.)

*All costs exclusive of interest payments, taxes and royalties. (a) Includes contingencies. (b) Includes insurance, telephone, administration, packaging, miscellaneous materials and services.

Table 10.8 Listing of works, equipment and services needed to establish a representative medium-sized penaeid hatchery with a monthly capacity of four to eight million post-larvae (PL_{7-10}).

Construction/Buildings
- site preparation, inc. roads
- drainage
- hatchery buildings
- ancillary buildings, e.g.
  - canteen, restroom, accommodation
- fencing

Plant, equipment and networks
- generator and fuel tank
- electric network and installations
- seawater intake
- seawater pumps and distribution network
- freshwater pump and network
- filters and water treatment apparatus
- air compressors and network
- air conditioning units
- effluent treatment

Tanks
- reservoirs
- broodstock holding/maturation
- spawning/hatching
- larvae rearing
- algae culture
- artemia hatching/holding/enrichment
- rotifer culture/holding/enrichment

Laboratory and other equipment
- low power binocular microscope
- stage microscope 40–400X
- salinity refractometer
- pH meter
- haemacytometers
- balance, 0.01 g precision
- balance, 1 g precision
- assorted general laboratory items
- hand nets, mesh screens, sieves, air diffusers
- transport vessel, air/oxygen supply
- autoclave for algae culture materials
- refrigerator
- freezer
- food preparation apparatus e.g. blender, fish cutter
- cleaning equipment e.g. sponges, brushes
- calculators, computer
- telephone/radio
- road vehicles for personnel/delivery
- fire extinguishers
- furniture
- office equipment
- tools for maintenance and repairs

Services
- surveying
- construction
- design/supervision
- consulting/technical assistance

suggests that significant economies of scale exist and are mostly captured at a production of five million or more PLs per month. Economy of scale, however, is not the only factor influencing profitability. Indeed Israel (1987b) studied data from hatcheries in the Philippines and established that small and backyard scale hatcheries gave better financial rates of return than medium and large sized operations, despite lower unit production costs prevalent in large hatcheries. This conclusion may have been influenced by accounting methods in which small family concerns tend to exclude labour costs, but it would account for the proliferation of backyard hatcheries in south-east Asia. The advantage of small hatcheries is their flexibility; they are able to operate in response to favourable conditions such as a seasonal abundance of wild spawners. Larger operations, on the other hand, can only take full advantage of economies of scale when able to maintain year-round production.

10.9.1.2 Penaeid maturation units

A significant expense for most penaeid hatcheries is the purchase of broodstock. Table 10.9 gives examples of broodstock prices in south-east Asia. In order to avoid or reduce dependence on wild spawners, some hatcheries utilize maturation facilities to supplement their supply of nauplii. The cost of setting up a maturation facility to

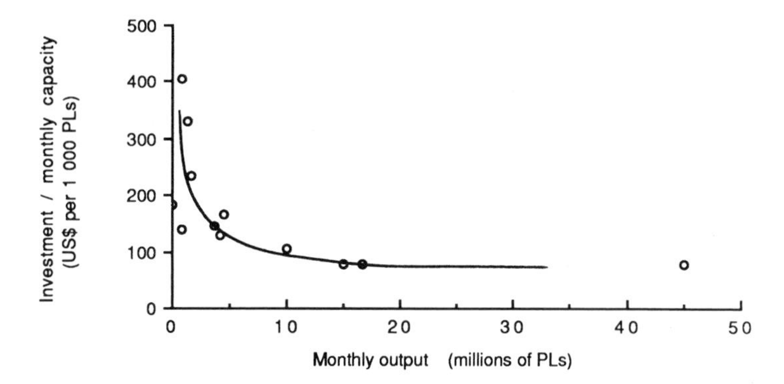

Fig. 10.3a The effect of production capacity on hatchery investment costs (curve fitted by eye to illustrate trend).

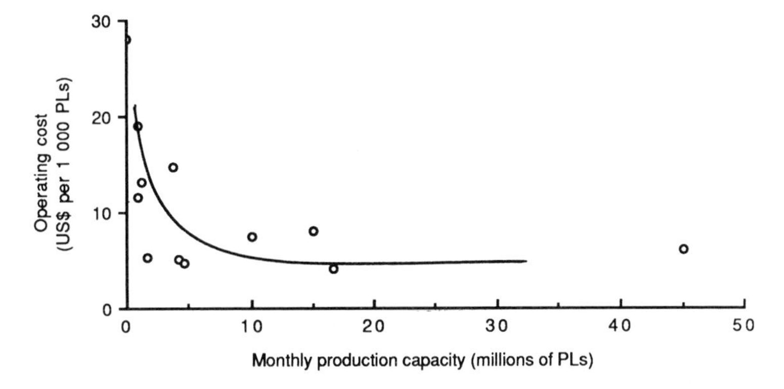

Fig. 10.3b The effect of production capacity on hatchery operating costs (curve fitted by eye to illustrate trend).

Table 10.9 Examples of prices for gravid penaeid broodstock.

Species	Location	Price US$ each	Notes	Year	Reference/source
P. chinensis	China	4.05	—	1988	Rosenberry 1988
P. monodon	Vietnam	5.5–11.0	low demand	1989	Quynh 1990
P. monodon	Vietnam	25–30	high demand	1989	Quynh 1990
P. monodon	Indonesia	8.5–22.5	—	1989	Yap 1990
P. monodon	Thailand	12–50	gravid and non-gravid	1989	Currie, D. 1989 pers. comm.
P. monodon	Thailand	70	large	1989	Currie, D. 1989 pers. comm.
P. monodon	Taiwan	60–150	gravid and non-gravid	1987	Chen 1990
P. monodon	Taiwan	50–500	—	1980	Chen 1990

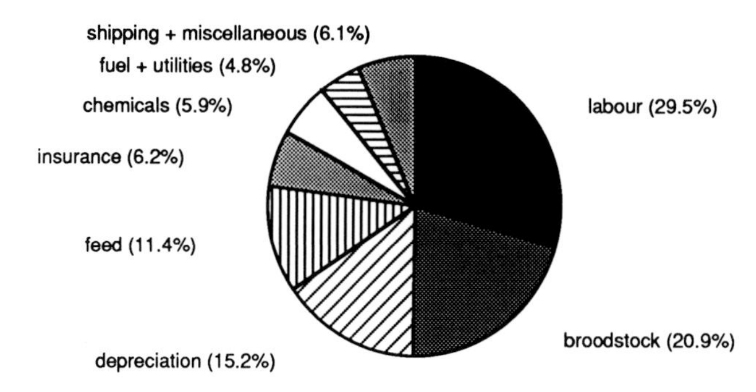

Fig. 10.4 Proportionate operating costs for a shrimp maturation facility (based on Johns *et al.* 1981a, interest and taxes excluded).

produce 1.5 million nauplii daily (*P. vannamei* (assumed)), has been estimated at US$400 000 (Lawrence *et al.* 1985). The 557 m^2 facility envisaged formed part of an integrated project including a larvae rearing unit costing US$700 000.

In another study (Johns *et al.* 1981a), a maturation unit designed to produce on average 3.36 million nauplii (*P. stylirostris*) per day consisted of sixteen circular tanks (3.66 m in diameter, each to be stocked with 70 adults) and was costed at US$331 772. (The greater output of this facility compared to that described by Lawrence *et al.* (1985) can probably be attributed to the fact that the productivity of *P. stylirostris* in an artificial maturation system generally exceeds that of *P. vannamei*. The annual operating costs of this scheme were estimated to total US$243 000 for year-round production. It was assumed that 4% of female shrimp would spawn per night and that a ready market for all production existed. Profitability was ensured if nauplii were sold at US$0.50 per 1000; however, if operation was restricted to only seven months per year a selling price nearer to US$1.00 per 1000 nauplii would have been required to achieve financial viability.

The operating costs of any maturation unit will consist of the major elements illustrated in Figure 10.4. Labour costs include the salaries of suitably trained or experienced managerial staff, and feed costs are high because a high quality expensive diet is usually required (Section 7.2.2). Broodstock are also a major expense (especially if imported or pond reared) and require routine replacement at a rate of anything between 8% and 60% per month.

Despite their capacity to provide a regular supply of nauplii, maturation facilities have not become widely established and have only been included in a few large hatcheries and integrated projects. The principal reason for this is that alternative local or imported supplies of nauplii or wild gravid broodstock can usually be exploited more economically. Faced with strong competition of this type a maturation unit can become a financial liability; even if it is closed down for part of each year, fixed costs must still be covered.

Another obstacle to the establishment of maturation units is doubt about the quality of the nauplii they produce. Lee *et al.* (1988) identified a significant difference between the growth rates of *P. vannamei* larvae from wild females and those from females induced to mature artificially by unilateral eyestalk ablation. In Ecuador a sharp price differential has developed: US$0.5–0.6 per 1000 for nauplii from artificially matured females, compared to US$1.2–1.4 per 1000 for nauplii from wild-caught gravid females (Rosenberry 1989).

Notwithstanding these problems, in the future it is likely that induced maturation will play a more important role in supplying hatcheries, especially if stocks of wild spawners are over-exploited. In China, for example, the government is promoting induced maturation of captive *P. chinensis* in an effort to reduce dependence on the wild fishery, and in some countries seasonal fishing restrictions are already operative (Section 7.2.2). Research continues on improving the reliability of maturation in captive shrimp, and some recent successes have been achieved in France with small (30 g) *P. monodon* females, which may have implications for the closed-cycle breeding of this species (D.A. Jones, 1991, pers. comm.).

10.9.1.3 Nurseries

The stocking of nursery reared juveniles rather than young post-larvae into on-growing ponds improves survival rates, reducing the duration of the on-growing cycle and bringing maximum benefit in areas where the growth season is short (Section 5.2). Thus, whether the nursery phase is performed at the hatchery, at the farm or as an independently managed concern, its economic justification relies on the extra value given to nursed juveniles. In the case of penaeids in the Philippines, the price of juvenile *P. monodon* PL_{40} has been reported as roughly double that of PL_{20} (Primavera 1983), and in Taiwan *P. monodon* PL_{27-29} have fetched around 80% more than PL_{13-15} (Mock 1983).

In these and other countries in south-east Asia many nurseries are operated as independent concerns and their viability is helped by the short culture period (12–25 days) of each batch of juveniles, which permits many production cycles per year. Nursery ponds incorporated into extensive and semi-intensive farms are typically stocked for longer periods of 30–60 days. They are operated to suit the production and harvesting schedule of on-growing, and often serve as juvenile holding ponds. In a semi-intensive shrimp farm, nursery ponds usually represent around 5%–10% of the total surface area and so a similar proportion of the total farm investment costs can be attributed to them (although allowance needs to be made for the extra cost per hectare of building smaller ponds (Section 8.2.2)).

In Texas and other locations with warm temperate or Mediterranean climates, the use of nurseries enclosed within greenhouses has been investigated as a means of producing juveniles in advance of the normal on-growing season. In this way it is possible to boost overall productivity and sometimes obtain two on-growing cycles per year. Smith *et al.* (1983) used enclosed nurseries for the production of *Macrobrachium* juveniles. Although actual commercial data was not available, they calculated that an on-growing operation could reasonably afford to pay US$30 per 1000 for juveniles of 0.3–0.4 g and up to US$45 and US$60 per 1000, respectively, for larger animals of 1 g and 2 g. In another system Juan *et al.* (1988) calculated the cost of producing 1 g penaeid juveniles to be US$15.1–17.21 per 1000. However, they concluded that using these juveniles to obtain two crops of shrimp per year was not as profitable as stocking young postlarvae (PL_5, which cost only US$7.50 per 1000) directly into outdoor on-growing ponds and obtaining a single crop per year.

The greatest expense of running a nursery is the initial cost of the post-larvae, so the survival rate is critical to profitability. At least 50%–70% survival is desirable. If

Table 10.10 Examples of operating and investment costs for shrimp and prawn farms.

Species	Location	Output ($mt\,yr^{-1}$)	Output ($kg\,ha^{-1}\,crop^{-1}$)	Stocking density no. ha^{-1}	Farm size ha ponds	Operating costs US\$ yr^{-1}	Operating cost US\$ kg^{-1}	Total investment US\$ (e)	Reference/source
Mixed penaeids	India	0.4	480		0.78	574(c)	1.53	281	Pai *et al.* 1982
P. monodon	Indonesia	2.0	1 000	25 000	1	1 492(a,b)	0.746	—	Indonesia 1986 unpubl. data
polycult. + fish	India	2.1	525		4	3 094(b,c)	1.47	796	Gopalan *et al.* 1982
P. vannamei	Ecuador	4.5	500	18 000	5	8 665(a,b)	1.93	—	Shang 1983 (1982 data)
M. rosenbergii	Thailand	4.6	1 138	40–70 000	4	13 430(c)	2.95	20 240	Shang 1982 (1980 data)
M. rosenbergii	South Carolina, USA	4.9	1 235	65 000	4	46 220(a,b)	9.36	69 316	Sandifer & Smith 1985
P. indicus	India	7.0	875	200 000	4	9 548(b,c)	1.36	1 562	Gopalan *et al.* 1982
P. monodon	NSW, Australia	7.2	1 200	80 000	6	86 100	11.96	235 400	Hardman *et al.* 1990
M. rosenbergii	Hawaii, USA	9.1	2 273	160–220 000	4	94 430(c)	10.28	91 950	Shang 1982 (1980 data)
P. vannamei	Ecuador	11.0	1 100	40 000	5	19 760(a,b)	1.80	—	Shang 1983 (1982 data)
P. monodon	Thailand	14.0	1 000	150 000	7	44 877(c)	3.21	30 912	Shang 1983 (1982 data)
P. monodon	Taiwan	18.0	6 000	200 000	1.2	126 521(c)	7.03	78 000	Shang 1983 (1982 data)
P. monodon	Qslnd, Aust.	18.5	1 540	80 000	6	132 400	7.16	235 400	Hardman *et al.* 1990
P. vannamei	Ecuador	20.0	2 000	70 000	5	40 875(a,b)	2.04	—	Shang 1983 (1982 data)
P. monodon	Taiwan	20.6	5 648	234 000	1.7	113 305	5.49	37 230	Chiang & Liao 1985
P. monodon	NSW, Australia	22.5	3 750	250 000	6	178 100	7.92	332 700	Hardman *et al.* 1990
P. monodon	Taiwan	24.0	8 000	500 000	1	75 300(a,b)	3.14	60 000	Taiwan 1985? unpubl. data
P. monodon	Taiwan	24.8	7 333	200 000	2.25	155 140(b)	6.25	18 821	Chen 1990 (1987 data)
P. vannamei	Ecuador	25.0	2 500	100 000	5	61 295(a,b)	2.45	—	Shang 1983 (1982 data)
P. monodon	Malaysia	42.0	1 667	67 000	8.4	241 752(c)	3.98	241 752	Shang 1983 (1982 data)
P. monodon	Qslnd, Aust.	45.0	3 750	250 000	6	303 200	6.74	332 700	Hardman *et al.* 1990.
M. rosenbergii	Hawaii, USA	45.5	2 273	160–220 000	20	353 600(c)	7.78	389 720	Sandifer & Smith 1985
Metapenaeus macleayi	NSW, Australia	48.0	800	200 000	20	155 000(a)	3.23	147 800(i)	Maguire & Allan 1985
P. japonicus	Japan	50.0	—	—	2.55	475 000	9.49	660 000(h)	Japan 1975 unpubl. data
P. vannamei	Texas, USA	51.2	1 066	135 580	48	600 826	11.74	503 522	Johns *et al.* 1983
P. stylirostris	Texas, USA	67.6	1 409	200 000	48	521 125	7.71	503 522	Huang *et al.* 1984
P. monodon	NSW, Australia	75.0	3 750	250 000	20	486 900	6.49	769 600	Hardman *et al.* 1990
Penaeid	Hawaii, USA	89.8	1 123	—	40	899 600(c)	10.01	1 445 000	Shang 1983 (1982 data)

P. vannamei and *P. stylirostris*	Texas, USA	92.1	1 918	200 000	48	562 605	6.11	503 522	Huang *et al.* 1984
P. vannamei	Ecuador	113	500	—	90	—	—	1 000 000(d,j)	Al Hajj A. 1989 pers. comm.
P. vannamei	Ecuador	124	500	—	99	—	—	700 000(d,j)	Al Hajj A. 1989 pers. comm.
P. monodon	Qslnd, Aust.	150	3 750	250 000	20	852 100	5.68	769 600	Hardman *et al.* 1990
P. vannamei	Americas	175	700	57 000	100	789 000	4.99	750 000	Lawrence 1985
P. vannamei	Guatemala	189	909	80 000	83	374 000(a,b)	1.98	1 320 000	Mock 1987 (1985 data)
P. vannamei	Americas	250	1 000	57 000	100	911 000	4.12	750 000	Lawrence 1985
P. vannamei	Texas, USA	255	1 571	180 000	162	835 758(c)	3.28	869 940	Shang 1983 (1982 data)
P. vannamei	Texas, USA	273	3 366	140 000	81	1 354 200(c)	4.95	627 280	Parker 1990
P. vannamei	Ecuador	313	500	—	250	—	—	1 500 000(d)	Al Hajj A. 1989 pers. comm.
P. vannamei	Ecuador	319	500	—	255	—	—	2 500 000(d)	Al Hajj A. 1989 pers. comm.
P. vannamei	Ecuador	369	909	60 000	203	1 075 474(c)	2.91	576 114	Shang 1983 (1982 data)
P. vannamei	Ecuador	434	500	—	347	—	—	2 500 000(d,j)	Al Hajj A. 1989 pers. comm.
P. monodon	Malaysia	497	16 000	900 000	10	1 392 000	2.80	1 038 000	Malaysia 1988 unpubl. data
P. vannamei	Ecuador	500	500	—	400	—	—	4 500 000(d)	Al Hajj A. 1989 pers. comm.
P. vannamei	Texas, USA	518	1 850	300–370 000	280	—	—	2 220 400	Jaenike 1989
P. monodon	Thailand	583	3 600	200 000	90	3 776 000(a)	4.04	2 608 000	Thailand 1988 unpubl. data
P. monodon	SE. Asia	612	784	40 000	300	2 445 324(b)	4.00	4 292 296	IFC 1987
P. vannamei	Ecuador	625	500	—	500	—	—	5 000 000(d)	Al Hajj A. 1989 pers. comm.
P. vannamei	Ecuador	775	500	—	620	—	—	4 350 000(d)	Al Hajj A. 1989 pers. comm.
P. monodon	—	1000	1 111	50–150 000	400	—	—	5 295 000	1987 unpubl. data
P. monodon	—	1000	3 000	—	112	—	—	4 672 000	1987 unpubl. data
P. monodon	Indonesia	1152	5 000	375 000	115	6 336 000	5.5	10 972 168(k)	Fuke P. 1990 pers. comm.
P. vannamei	Ecuador	2721	853	60–70 000	1418	9 471 250(a)	3.48	5 672 000	Hirono 1989
P. japonicus	Japan	—	4–5 000	200–500 000	—	—	34.24	—(f)	Yamaha 1989
P. japonicus	Japan	—	6–10 000	400–600 000	—	—	37.67	—(g)	Yamaha 1989
P. japonicus	Japan	—	15–20 000	700–800 000	—	—	44.52	—(h)	Yamaha 1989
P. japonicus	Italy	—	189	14 000	—	—	14.88	—	Lumare *et al.* 1989
P. japonicus	Italy	—	525	30 000	—	—	9.62	—	Lumare *et al.* 1989

(a) Excluding interest. (b) Excluding depreciation. (c) Including land lease/rent. (d) Asking price already built, including land. (e) Excluding land. (f) Partial embankment ponds. (g) Full embankment ponds. (h) Shigueno system. (i) Includes land. (j) Price includes extra land for development. (k) Plastic lined ponds, nucleus/plasma scheme.

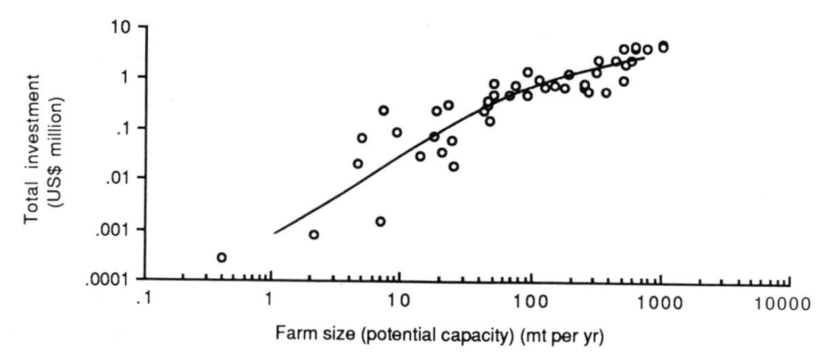

Fig. 10.5 The effect of farm size (potential production capacity) on the total investment required (curve fitted by eye to illustrate trend).

survival falls to 30% the selling price (or value) of the nursed juveniles would need to be more than triple the purchase cost, simply to cover the initial purchase of post-larvae.

10.9.1.4 On-growing investment costs

Table 10.10 shows the operating and investment costs of shrimp and prawn farms and is based on actual case studies as well as projected estimates. The data relates to farms ranging from a small 0.78 ha extensive farm in India producing 0.4 tonnes per year, to large 99–1418 ha semi-intensive operations in Ecuador producing 124–2721 mt yr^{-1}. Although some examples refer to intensive operations from Taiwan and Japan, the bulk of the data relates to semi-intensive farming with stocking densities in the range 50 000–200 000 juveniles ha^{-1} and yields of 500–5000 kg ha^{-1} $crop^{-1}$. The relationship between the total annual output of a farm and the size of the investment is shown in Figure 10.5. When the investment level per tonne of annual output was calculated for each farm, values averaged US$6406 mt^{-1} and nearly half of them fell in the range US$3000–8000 mt^{-1}. The highest value of US$32 694 mt^{-1} corresponded to a hypothetical 6 ha penaeid farm in eastern Australia. This farm was identified as being an unprofitable proposition partly because of its small size and partly because of the limited on-growing season which would permit production of only one crop of shrimp per year (Hardman *et al.* 1990).

Further calculations based on the data in Table 10.10 suggest that the investment levels per hectare are more widely scattered than the levels of investment per tonne of annual output. They range from as little as US$199 ha^{-1}, for a small extensive farm in India, up to US$259 000 ha^{-1} for a super-intensive Shigueno-type farm in Japan. Possibly the highest investment level per hectare of any shrimp farm has been the US$1 000 000 ha^{-1} made in the super-intensive system operated in Hawaii by a company called Marine Culture Enterprises (New 1988) (Section 7.2.6.6).

Normally, by far the greatest expense in establishing a shrimp or prawn farm is construction. Data from a number of the semi-intensive farms listed in Table 10.10 indicated that construction accounted for an average 61% of total investment costs (range 44%–81%). The most important component of this cost is earth moving. In one report, IFC (1987) calculated that 442 720 m^3 of earth would need to be shifted, at an estimated cost of US$2.50 m^{-3}, in order to construct 321 ha of ponds and supply and drain canals. The cost of earth moving can vary greatly with locality and soil type. Clearance of vegetation, roots and rocks can add considerably to the cost. In the USA, earth moving prices quoted to Johns *et al.* (1983) ranged between US$0.58 and US$3.29 m^{-3}. In some large projects it has proved to be worth buying earth-working machinery and employing crews instead of hiring the services of a contractor for the duration of the construction phase.

If ponds are to be made with concrete walls rather than earthen dikes, construction costs per ha will be far higher – around six times higher in the case of Taiwan (Chiang & Liao 1985). The cost of constructing small ponds (0.36 ha) with plastic liners has been put at around US$31 500 ha^{-1} (P. Fuke, 1990, pers. comm.). This type of pond has been used in permeable sandy areas in Oman and Indonesia and is suited to intensive culture (Section 7.2.6.5).

Land purchase can be a major financial outlay. Prices per hectare vary depending on competing land usage, and may increase in areas where shrimp or prawn culture is successful. Values ranging from US$2000 ha^{-1} (IFC 1987) to US$40 800 $^{-1}$ (Chiang & Liao 1985) have been used in various cost analyses, though in many situations land can be leased at economical rates. In case studies from six different countries (Shang

Table 10.11 Listing of works, equipment and services needed to establish a semi-intensive shrimp or prawn farm.

Construction
- earthworks
- inlet/outlet gates
- culverts
- harvest basins
- bank stabilization
- access roads
- dredging

Water supply
- pumps
- piping
- well

Buildings
- feed storage
- office
- workshop
- accommodation
- guard huts

Electrical supply/installations
- generator(s), fuel tank
- network
- lighting

Equipment
- tools
- ice machine
- communications equipment
- office equipment
- excavator
- harvest pump, nets and other equipment
- aerators

Instruments
- oxygen meter
- pH meter
- salinity refractometer
- Secchi disk
- low power binocular microscope
- balance, 500 × 0.1 g
- weighing scales, 200 × 0.5 kg (for harvesting)

Vehicles
- Boats, outboard motors
- tractor, trailer
- pick-up truck
- truck
- cars
- motorcycles

Miscellaneous
- fencing

Services
- design, engineering
- consulting/technical assistance
- surveying
- supervision
- legal

1983), most land costs appeared as annual lease payments representing only around 1% or less of the annual operating budget. The highest figure of 4% came from Hawaii.

A listing of works, equipment and services required to set up a shrimp or prawn farm is provided in Table 10.11. Dredging is sometimes required to enable water from a nearby creek or other body of water to reach the site of a pumping station. Expenditure on the water supply system will include installation and purchase of pumps and piping. Sometimes a well is needed to supply freshwater and in the case of a prawn farm relying on a borehole, well costs can be considerable. For a *Macrobrachium* farm in South Carolina, Bauer *et al.* (1983) estimated a well would represent 36% of the investment costs. Obviously, because of the cost implications, the proximity of the water supply is one of the major considerations in site selection (Section 6.3.1.2).

10.9.1.5 On-growing operating costs

In most cases the major components of the cost of running a shrimp or prawn farm are feed and juveniles, which between them usually account for 50%–70% of expenditure. Figures 10.6a and 10.6b show proportionate costs for operating a large (1418 ha) shrimp farm in Ecuador and a 4 ha prawn farm in Thailand. Manning levels for semi-intensive farms in Ecuador are around 15–25 people per 100 ha (CPC 1989; Hirono 1989). In contrast, a small 2–5 ha shrimp farm in Taiwan may be operated by as few as one skilled technician and one permanent worker, with operations such as harvesting and pond bottom cleaning performed by labourers on temporary contracts (Chien & Liao 1987).

Figure 10.7 shows the relationship between the potential annual output and annual operating costs for farms producing up to 500 mt per year, and suggests that economies of scale are mostly captured in operations producing at least 50–100 mt yr^{-1}. However, little or no relationship and no economy of scale were found when the data were used to compare farm output level and operating cost per kg. This suggests that despite some advantages of operating a large farm, there are in fact profitable production strategies for farms of all sizes, right down to units of <5 ha. Indeed, it appears that some of the very smallest operations (e.g. a 0.78 ha farm in India (Pai *et al.* 1982)) are able to produce at very low costs per kg by adopting extensive culture methods. Although the cost per kg of production is not the only measure of the financial viability of an operation, it is a useful figure because it will largely determine the size of profit margins (Section 10.8). Of the factors affecting operating cost per kg, location has a clear influence (Figure 10.8). High operating costs in some locations can be attributed to higher costs for feed and labour, and in the cases of Texas, Japan, Italy and eastern Australia, also to restrictions on productivity imposed by the limited on-growing season.

Although the data in Table 10.10 mostly refers to semi-intensive culture, it is possible to use it to compare different levels of culture intensity. Figure 10.9 describes the positive correlation between stocking density and the resulting yield per crop. In addition, there is a similar but less clearly defined relationship between the yield per crop and the level of investment per hectare required to achieve it (Figure 10.10). Figure 10.11, which relates operating cost per ha to yield level, also indicates a positive correlation between greater financial input and enhanced productivity levels.

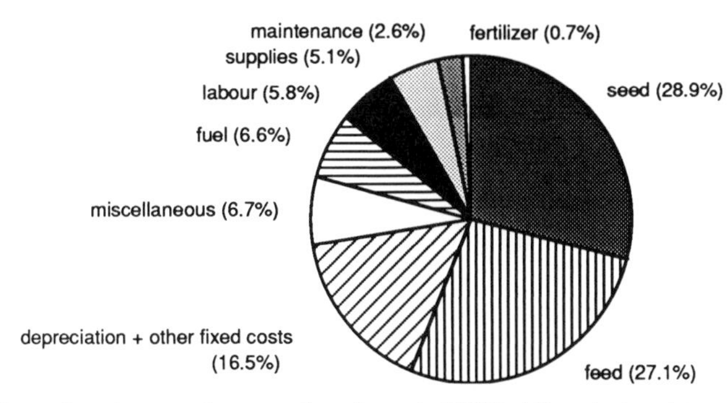

Fig. 10.6a Proportionate operating costs for a large (>1000 ha) Ecuadorian shrimp farm (based on Hirono 1989).

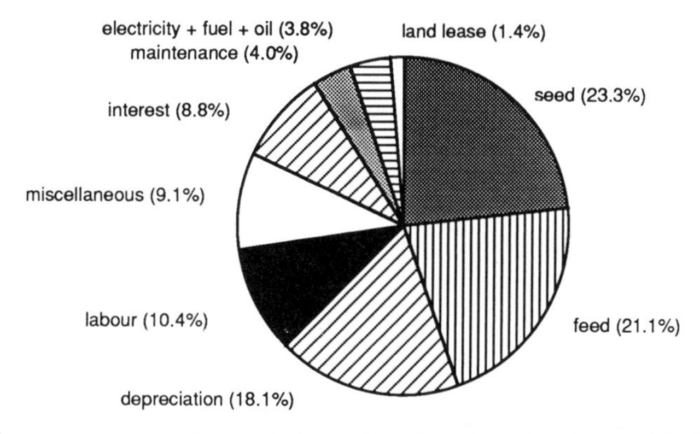

Fig. 10.6b Proportionate operating costs for a 4 ha *Macrobrachium* farm in Thailand (based on Shang 1982).

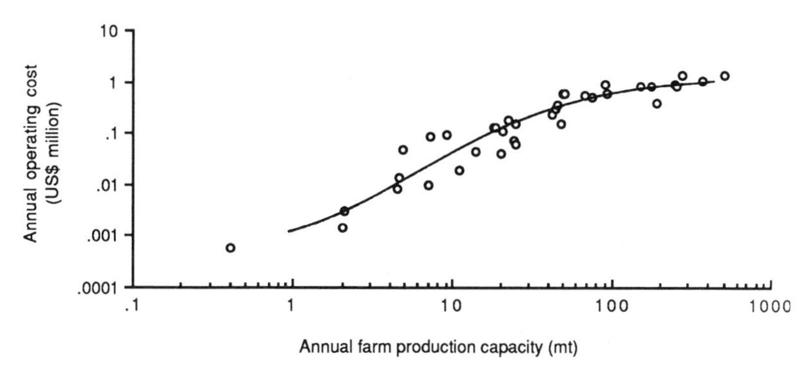

Fig. 10.7 The effect of farm size (potential production capacity) on annual operating costs (curve fitted by eye to illustrate trend).

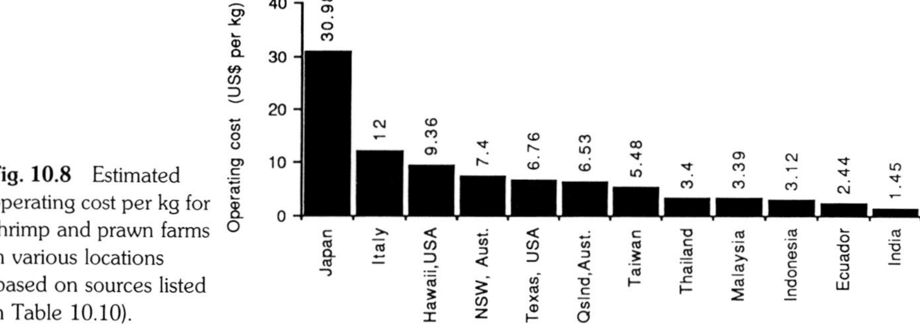

Fig. 10.8 Estimated operating cost per kg for shrimp and prawn farms in various locations (based on sources listed in Table 10.10).

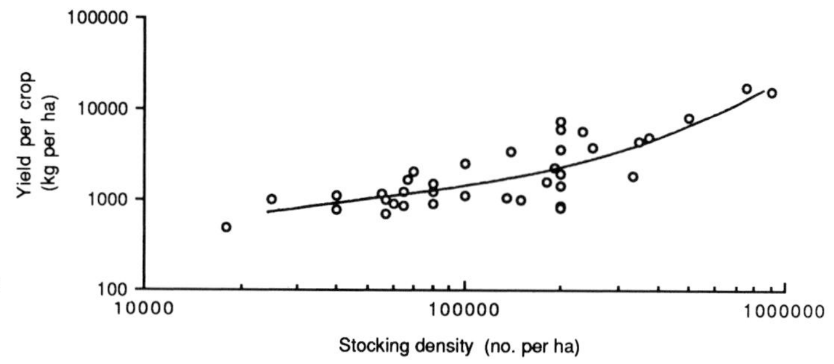

Fig. 10.9 Relationship between stocking density and yield in shrimp and prawn farms (curve fitted by eye to illustrate trend).

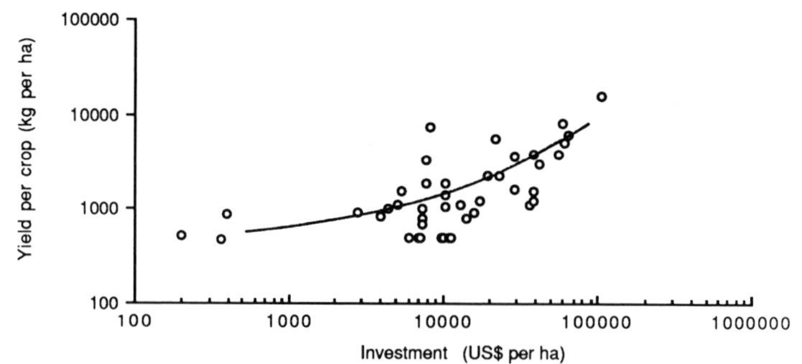

Fig. 10.10 Relationship between investment per ha and yield in shrimp and prawn farms (curve fitted by eye to illustrate trend).

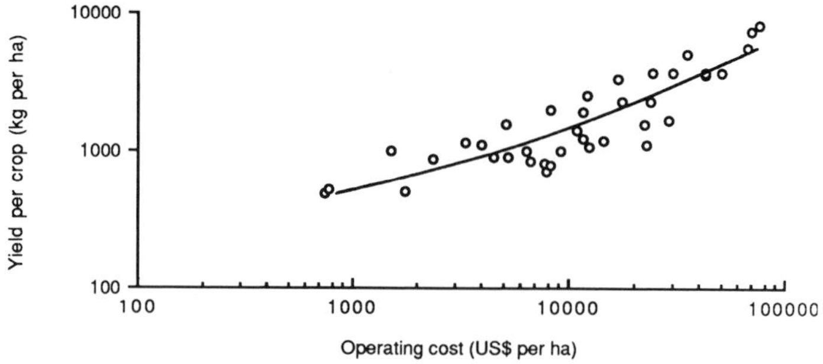

Fig. 10.11 Relationship between operating cost per ha and yield in shrimp and prawn farms (curve fitted by eye to illustrate trend).

However, this figure also suggests that the yield levels per hectare are roughly proportional to the financial inputs and provides evidence that contradicts the arguments laid out in Section 10.8, viz that there are diminishing returns associated with increasing intensity of operation. In view of this, it would seem likely that some of the financial analyses cited in Table 10.10 have failed to adequately account for the full costs and enhanced risks associated with high intensity aquaculture.

10.9.1.6 Stock enhancement

In Japan, inland sea areas and bays are stocked with hatchery reared juvenile *P. japonicus*. The costs of the stocking operations are largely borne by the state or the individual prefectures involved, with the long-term objective of stabilizing or boosting the coastal shrimp fishing industry, rather than making short-term financial gains.

Henocque (1984) reviewed the economics of Japanese stock enhancement programmes. Post-larvae of 10 mm, costing around US$5.95 per 1000 to produce, are reared in government hatcheries and sold to fishermen at an 85% discount. The post-larvae are either released directly to open waters or held in nurseries for around 15 days until they reach 25–30 mm in length. In one investigation a survival rate of 5.5% was estimated following direct release, and a rate of 15.4–22.7% following nursery rearing. Thirty-five per cent of these survivors were estimated to be captured in the fishery, giving overall recapture rates (from early post-larvae) of 1.9% for direct release and 5.4–8% if nursery rearing was used. Obviously each release area has different characteristics, and recapture rates can never be calculated with any precision, but an overall recapture rate of 1.5–2% would justify investment in the long-term objectives, and indeed it does appear that restocking has had beneficial effects on Japanese shrimp fisheries through the stabilization of recruitment levels. Whether the fishing co-operatives would be keen to absorb 100% of the costs is doubtful unless protectable or enclosed areas were established which offered security for the investment.

The concepts of stock ownership and security of investment are also of fundamental importance for commercially-based ranching operations. In one analysis a profitable balance sheet for shrimp ranching was drawn up on the basis of releasing 25 mm nursed juveniles (approximately 0.2 g). The juveniles were estimated to cost US$22.8–38.0 per 1000 and were assumed to be recaptured at 30 g with a fishing cost of US$5.24 kg^{-1} (Oshima 1984). The viability of this hypothetical enterprise rested on the assumed overall recapture rate of 35% (from nursed juveniles, based on 70% survival and 50% capture rate in the fishery); if it were only half this rate, profits would disappear.

10.9.2 Crayfish

10.9.2.1 Hatchery and nursery

Hatcheries do not play any role in the US crayfish industry and so economic data come mainly from Australia and Europe. In Europe the cost of establishing a nursery and simple prefabricated hatchery capable of producing 100 000 juvenile crayfish per year has been estimated at US$21 400 (Arrignon 1981). This unit, which was based on the production of the white-footed crayfish (*Austropotamobius pallipes*) for restocking or on-growing, would require 5000 broodstock per year. Annual running expenses were estimated at a mere US$1780 and the operation was considered viable at a selling price of US$0.18 per juvenile. Other estimates place the cost of producing juvenile crayfish (*Astacus astacus*) at US$0.04 each for hatchlings, and at US$0.25 each for juveniles of one summer old (summerlings) (Huner *et al.* 1987), and it has been noted that in Spain it is possible to produce and sell hatchlings of *P. leniusculus* at a profit of US$0.10 each. However, in one outdoor recirculating tank system in Germany, the cost of producing 3.5 month old summerlings of *A. astacus* was found to be around US$0.63 each (Keller 1988), and there was thought to be little potential for profit in this nursery system since the prices obtained for juveniles in Europe usually fall in the range US$0.5–0.75 each.

In Australia the capital cost of a hatchery designed to produce 2 000 000 yabbie (*Cherax destructor*) juveniles annually, was estimated to be US$21 680 (including the cost of an office) (Staniford & Kuznecovs 1988). This sum represented 5.7% of the total cost of an integrated project which incorporated 10 ha of semi-intensive on-growing ponds. The proposed nursery for the same operation consisted of 475 above-ground pools of 3.66 m in diameter and was costed at US$50 900 (13.7% of the project's total cost).

10.9.2.2 Re-stocking and ranching

It has been estimated that an investment of US$7.10 ha^{-1} would be necessary to establish a viable population of *P. leniusculus* in an extensive fishery in Sweden. Six to seven years later worthwhile catches of 5–10 kg ha^{-1} could be expected and would generate a revenue of around US$142 ha^{-1} (Huner *et al.* 1987). Waters established for such purposes as flood control and irrigation offer potential sites for stocking crayfish; however, the returns from ranching alone would not justify the costs of creating them.

10.9.2.3 On-growing

USA Typically the investment and operating costs of crayfish farms in the USA are kept low. Most farmers make use of marginal land already in their possession and create ponds simply by adapting and raising existing drainage levees (bunds). Pond construction costs (including water control structures) have been estimated at US$499 ha^{-1} in Louisiana (de la Bretonne & Romaire 1989) and US$611 ha^{-1} in Texas (Avault & Huner 1985). These figures are low when compared to an estimated US$3013 ha^{-1} invested in the construction of deeper, more substantial semi-intensive shrimp ponds in Texas (Parker 1990).

Most operators are engaged in farming rice, soybeans or sugar cane as well as crayfish, and the costs of vehicles, farm facilities and equipment can be spread between the different crops. Some new investment in wells, pumps, pipes and traps is, however, usually necessary, and a crayfish combine – a specialised open work-boat propelled by a spiked wheel – may also be required. Figure 10.12 shows the proportionate investment costs for an extensive 16 ha crayfish pond in south-western Louisiana, expected to produce between 800 and 1700 kg ha^{-1} yr^{-1}. Proportionate operating costs for this type of farm are given in Figure 10.13. Major elements, apart from fixed costs, are labour, bait, forage crops (usually millet or rice) and fuel for

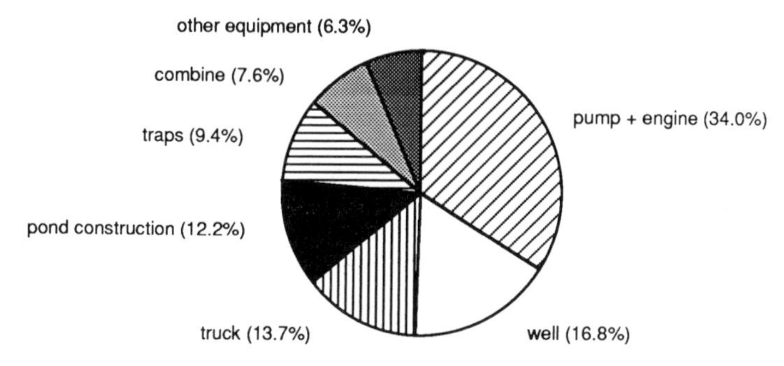

Fig. 10.12 Proportionate investment costs for a 16 ha crayfish farm in south-western Louisiana (based on de la Bretonne & Romaire 1989).

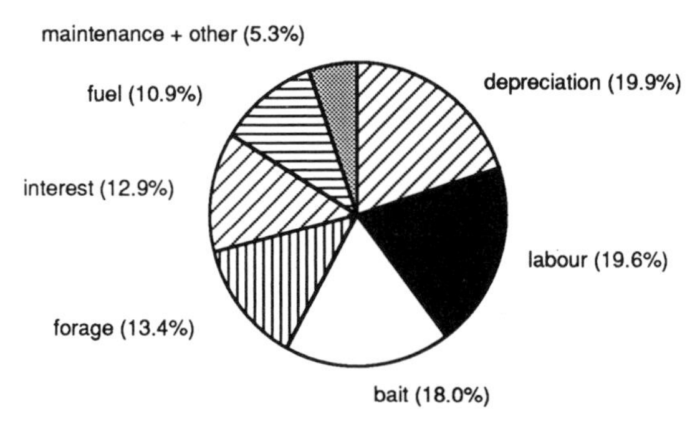

Fig. 10.13 Proportionate operating costs for a crayfish farm in Louisiana (based on Avault & Huner 1985; de la Bretonne & Romaire 1989).

pumps and vehicles. Forage crops are not always provided, though they are an effective way of enhancing productivity and serve as an alternative to the use of feed or fertilizer (Section 7.5.4). Farm owners usually perform most of the labour themselves or use an existing farm workforce. However, outside labour may be required to meet the extra demands of harvesting and this expense can have a major impact on profitability.

Crayfish stocking in the USA is usually only performed in the first year of a pond's operation, after which a self-sustaining population becomes established. Stocking costs are thus only incurred once, in stark contrast to most shrimp and prawn farms where expenditure on juveniles can represent 10–50% of an annual operating budget.

Dellenbarger *et al.* (1988) calculated the break-even prices for crayfish production in Louisiana under a range of different conditions. Table 10.12 gives the results of an analysis based on these data, and indicates that yield levels have the greatest impact on the cost of production. The variation in cost relating to location can be attributed to the differing costs of pumping water; wells in south-west Louisiana are deeper than in the north-east. Double-cropping with rice achieves a drop in costs because some costs can be shared between the two crops.

In South Carolina the average yield of crayfish farms is an estimated 840 kg ha^{-1} yr^{-1} and production costs in upland farms of 2–32 ha have been put at US$1.80–2.00 kg^{-1} (Eversole & Pomeroy 1989). The total capital investment needed for a

Table 10.12 Sensitivity analysis on crayfish production cost (kg^{-1}) in Louisiana, USA.

Factor	Change	Resulting change* in production cost %
Yield	increase by 36%	−26
	decrease by 36%	+57
Farm size	increase by 60%	−20
	decrease by 60%	+20
Location	SW Louisiana	+4
	NE Louisiana	−4
Culture regime	double crop with rice	−12

*Baseline cost US$1.37 kg^{-1}; based on hypothetical 20.2 ha farm in central Louisiana producing 1234 kg ha^{-1} yr^{-1} (based on Dellenbarger *et al.* 1988).

crayfish farm with a single 8 ha pond was calculated at US$24 385 (excluding land). Some farmers in South Carolina improve their financial returns by raising waterfowl as well as crayfish.

Europe In Europe semi-intensive culture in small ponds is carried out along with the stocking of natural or extensive fisheries. Both types of culture rely on a supply of juveniles hatched and reared in captivity. In small semi-intensive canal-type ponds the operating costs comprise labour (30–60%), food (5–30%) and energy (0–15%). They vary depending on the size and intensity of the operation (Clarke 1989). In the UK the cost of excavating a canal-type pond of 0.1 ha has been estimated at US$1550 and the possible gross revenue from ten such ponds has been put at US$9300 per year (Alderman & Wickins 1990).

Australia Crayfish culture in Australia is performed at various levels of intensity. Semi-intensive farms and extensive 'farm dam'-type operations are the most common. The major costs of establishing and running the latter, which are based on self-sustaining crayfish populations after initial stocking, will conform to the patterns for extensive crayfish production in the USA and Europe. Semi-intensive operations, however, make use of purpose-built ponds stocked with nursery-reared juveniles and require greater financial input. For example, investment in semi-intensive marron (*Cherax tenuimanus*) farms in the size range 5–10 ha, averages US$31 500–55 100 ha^{-1} and operating costs average US$3400–11 800 ha^{-1} yr^{-1}. At an output level of 2000–3000 kg ha^{-1} yr^{-1}, it is estimated to cost US$3.9–7.9 to produce a kilogram of marron (O'Sullivan 1988). Huner (1990) has argued that this cost may be too high for Australian crayfish to compete effectively on US markets with alternative products (freshwater prawns and Canadian 'canner' lobsters). While this could well be true if all crustacean products were sold in similar processed forms (e.g. 'chunks'), competition could be effectively reduced if sufficient attention were paid to product differentiation during marketing.

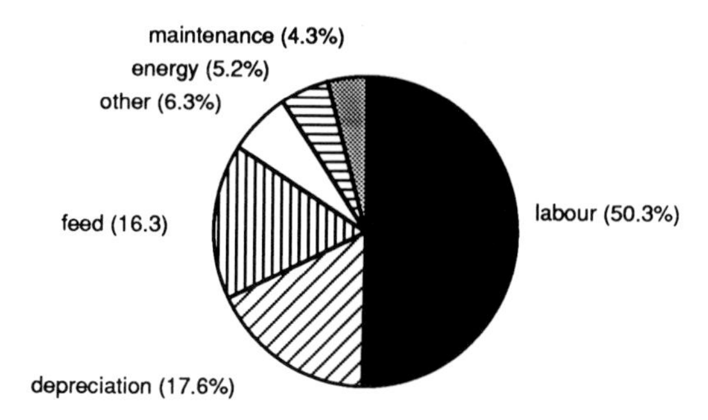

Fig. 10.14 Proportionate operating costs for an integrated crayfish hatchery, nursery and 10 ha farm in Australia (based on Staniford & Kuznecovs 1988, interest not included).

The economics of a hypothetical integrated semi-intensive yabbie (*Cherax destructor*) farm have been investigated by Staniford & Kuznecovs (1988). The investment necessary for a 10 ha (50 × 0.2 ha ponds) farm incorporating a reservoir, hatchery, nursery, processing facility and land was estimated at US\$37 800 ha^{-1}. Pond construction costs were US\$13 600 ha^{-1} and together with the reservoir they represented 44.4% of the total investment. Fences and netting required for predator control were a significant expense, accounting for 11.1% of the total. The baseline yield for the feasibility analysis of this farm was set at an optimistic 3300 kg ha^{-1} yr^{-1}, and resulting production costs were estimated at US\$4.54 kg^{-1} (packaging excluded). The different elements of this cost are shown in Figure 10.14.

Labour costs are the greatest and include the salaries of the owner-operator and a farm manager. They are high when compared to the labour input of an extensive crayfish operation in Louisiana (Figure 10.13), reflecting the additional management input needed in a semi-intensive culture operation. The production cost of US\$4.54 kg^{-1} includes the costs of running the hatchery and nursery. However, if the same operation were to rely on buying juveniles on the open market at US\$0.39–0.43 each (O'Sullivan 1988), the expense would represent US\$2.20 kg^{-1} of production; i.e. a very significant proportion of production costs and in line with the pattern for semi-intensive shrimp and prawn farming (Section 10.9.1.4).

10.9.2.4 Soft shelled crayfish

Soft-shell crayfish production is a lucrative activity increasingly undertaken in the USA as a sideline to crayfish farming (Section 7.5.7). Operations are viable because the prices obtained for soft-shell crayfish are around 15–20 times more than for hard shelled animals (Clarke 1989). The cost of setting up a unit with the capacity to hold 454 kg of animals and produce 1960 kg yr^{-1} of soft crayfish was estimated at US\$15 000, assuming that construction labour was supplied by the owner-operator (Culley & Duobinis-Gray 1989). Production costs, of which 40% represented the purchase of immature crayfish, were estimated to range between US\$9.2 and US\$13.0 per kilogram, depending on how well an operation was managed. Achieving a moulting rate of around 2% or more per day was critical to profitability, and overall productivity was limited by the length of the growing season which could vary between four and seven months depending on the weather.

10.9.3 Lobsters

Only limited data on the economics of clawed lobster culture are available and none were found for spiny lobsters. The clawed lobster data come mainly from Europe and relate to integrated units for producing juveniles for release (see below). All but one of the examples considered relate to real operations and all were based on the purchase of wild egg-bearing females from fishermen.

The cost of establishing and operating a production unit for juvenile lobsters suitable for restocking has been investigated in the UK by the Sea Fish Industry Authority (SFIA, 1990 pers. comm.), the North West and North Wales Sea Fisheries Committee (W. Cook, 1990 pers. comm.) and private organisations (Ingram 1985; P. Franklin, 1990 pers. comm.), and in Norway by the Tiedemanns Group and the Institute of Marine Research (G. van der Meeren, 1990 pers. comm.). Data on investment and operating costs for units producing from nearly 5000 to 100 000 juveniles per year are summarised in Table 10.13. No account was taken of transport and release costs, nor of site surveys. In the absence of significant commercial production, these figures represent some of the best cost estimates available.

10.9.3.1 Broodstock and hatchery

The costings for lobster broodstock and hatchery units were usually incorporated within figures relating to integrated units for producing nursed juveniles (comprising broodstock, larvae culture and nursery facilities). However, it is useful to consider them separately since any future ranching or restocking programmes could use stage four to six juveniles, the cost of which will be crucial.

In one hatchery/nursery unit costing US$163 036 (Table 10.14), investment costs of US$33 830 were directly attributable to the broodstock and larvae rearing phases of operation. By adding to this figure a relevant proportion (40%) of the other costs (excluding the nursery unit), a total of US$65 214 can be derived which approximates to the complete cost of establishing a broodstock and larvae rearing unit alone. The output of the whole unit was rated at 30 000 juveniles per year, produced in two batches to provide lobsters for late and early summer releases. The hatchery and larvae rearing facilities were required to produce a total of 36 000 post-larvae, also in two batches. However, a good deal more than two batches per year could undoubtedly be produced if required.

The major costs of producing three-week-old, stage four to six post-larval lobsters were assessed by Ingram (1985) to be US$293 per 1000, of which labour accounted for 82%. It is interesting to compare this figure with figures for the cost of producing shrimp and prawn larvae of a similar age, which range from US$2.57 to US$18.98 per 1000 animals (Table 10.7a). This much lower cost results from economies of scale achieved in hatcheries which usually number their annual production in millions, compared with tens or hundreds of thousands in lobster hatcheries. Shrimp and prawns are generally more fecund than lobsters and, in addition, larvae culture is feasible at higher densities of 50–200 ℓ^{-1} compared to 30–40 ℓ^{-1} for lobsters (Section 7.8.7).

Table 10.13 Estimated investment and operating costs for lobster hatchery/nursery units.

Output juveniles yr^{-1}	Stage at release	Investment cost(a) (US$)	Operating cost(b) (US$ yr^{-1})	Operating regime (temperature, and no. of batches per year)	Cost per juvenile (US$)	Reference/source
30 000	10–11	247 000	128 250	ambient, 1	4.28	Franklin P. 1990 pers. comm.
30 000	10–11	163 035	56 924	>18°C, 2	1.90	modified from SFIA 1990 pers. comm.
25 200	11	87 780	37 088	>18°C, 2	1.47	"
16 000	11	82 245	37 088	>18°C, 2	2.32	"
8 000	9–12	not given	9 196(c)	22°C, 2–3	1.15	Ingram 1985
4 980	>8	not given	30 400(d)	18–20°C, 2	6.10	Cook W. 1990 pers. comm.(e)
100 000	1 yr old	not given	350 100(f)	20°C, 1	3.50	based on van der Meeren G. 1990 pers. comm.

(a) Exclusive of tax and cost of site. (b) Including depreciation, excluding interest payments. (c) Includes major contributions to production costs only, does not include depreciation. (d) Includes release cost. (e) Refer to text. (f) Heat provided by industrial effluent. (UK£1 = US$1.90).

Table 10.14 Investment costs* for a 30 000 per year lobster hatchery/nursery unit (modified from SFIA (1990) pers. comm.)

	US$
Broodstock unit	19 836
Larval rearing unit	13 994
Nursery unit	50 854
Food preparation area	4 560
Service area	4 290
Sundry	1 862
Installation	19 000
Building (188 m^2)	48 640
Total	163 036

* excluding taxes; (UK£1 = US$1.90).

10.9.3.2 Juvenile production units

We now revert to consideration of integrated juvenile production facilities capable of producing juveniles up to one-year-old. These larger juveniles were actually produced for restocking trials but would be equally suitable for on-growing in battery farms.

Estimates for the cost of building a unit to produce 30 000 stage ten to 11 juveniles per year vary, and have been put at US$247 000 and US$163 035 (Table 10.13). These prices assume that entirely new facilities are constructed, although it seems that considerable savings are possible if existing marine hatchery facilities can be leased and adapted to lobster production. Figure 10.15 shows the proportionate operating costs of a representative unit based on the data from the four British sources. Labour costs are particularly high and result from the individual holding requirements of juvenile lobsters. Attempts to reduce labour input through communal rearing have resulted in much reduced survival rates.

The figures for operating cost per juvenile in Table 10.13 do not include the effect of interest payments. If, for example, interest payments at 10% are included in the figure of US$4.28 per juvenile, the cost increases to US$5.23. Even this figure may be

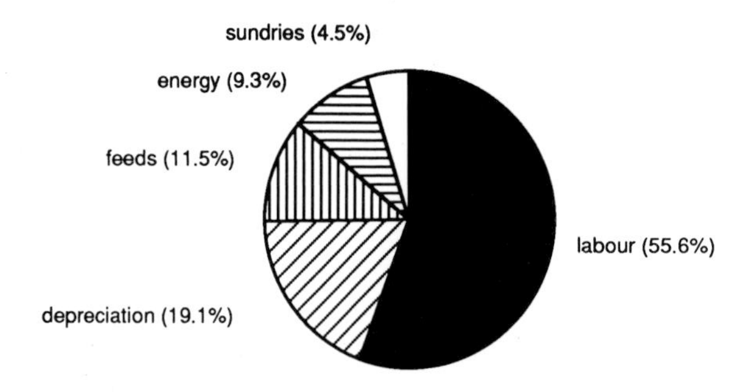

Fig. 10.15 Proportionate operating costs for a lobster hatchery/nursery unit producing 5000–30 000 juveniles per year (based on sources in Table 10.13).

conservative if the full impact of possible disease and equipment failures is taken into account (P. Franklin, 1990, pers comm.).

The cost per juvenile can also be influenced by the overall approach to production. For example, in addition to considering the building of a completely new and independent facility, the option of setting up an operation in a similar style to a 'backyard' shrimp or prawn hatchery has also been investigated (P. Franklin, 1990, pers. comm.). In this case an existing hatchery facility equipped with a seawater supply was occupied for a low rent, and all equipment such as pumps, tanks and piping were purchased second-hand or fabricated on site. The owner/operator performed all installation work, received no salary, and operated the finished unit at ambient temperatures. In this way a unit for producing 3600 stage ten to 11 juveniles per year was set up for a mere US$3800 to provide animals at a production cost of just US$0.49 each. Although this approach is not sustainable (the operator will eventually require an income) it serves to illustrate the potential of this activity as a part-time occupation requiring a relatively low initial investment.

An alternative approach, with potential to cut the cost per juvenile, may be to increase the scale of production. The manager of a Norwegian hatchery/nursery unit has been quoted as saying that interesting profits would be possible at an output level of around one million juveniles per year (Anon. 1987). Table 10.13 does include cost data based on one large Norwegian unit with a capacity for producing 100 000 juveniles annually, but economies of scale are not immediately apparent (US$3.50 per juvenile) since this operation is based on rearing animals for a period of one year rather than the three to four months considered in the other examples. If year-round operation were maintained but lobsters produced at the smaller size, output of three batches per year could in theory reduce the cost per juvenile to around one third of the given value.

A small British unit operated by the North West and North Wales Sea Fisheries Committee, which produced almost 5000 juveniles (of minimum stage eight) in two batches per year, calculated the cost per juvenile to be US$6.10 (including release costs at around 5–10% of this total) (W. Cook, 1990, pers. comm.). However, taking into account the possibility of considerable savings in a larger, more efficient operation, it was estimated that the cost could be reduced to around US$1.90 each (including release). The savings would result from economies of scale, output of three batches per year, increased energy efficiency and reduced labour requirements through less frequent cleaning and the greater use of an artificial diet.

Costings produced by SFIA (1990, pers. comm.) ranged between US$1.60 and US$2.55 per juvenile and were based on establishing a juvenile production unit within an existing aquarium facility where space was rented. When these costings were modified (Table 10.13) by including the cost of an independent building to house the same operations, investment and depreciation costs rose but overall operating costs fell (to US$1.47–2.32 each) as a result of the elimination of rental charges.

10.9.3.3 Restocking and ranching

The economic viability of stock enhancement proposals based on the release of hatchery or nursery reared juvenile lobsters is impossible to assess without knowledge of survival and recapture rates. Even though billions of stage four lobsters have been reared and released in North American waters, the impact of this type of programme remains unquantified. In Britain stage ten to 12 juveniles large enough to be micro-tagged have been released on to lobster grounds and identified four to five years later in the local fishery. The results are encouraging in that three-month-old juveniles survived in the wild and tended to remain at or close to their release site. However, it is still too early to say if genuine enhancement rather than substitution has occurred, or to estimate the scale of releases required to justify the programme on the basis of increased catches. Stage ten to 12 juveniles were chosen for release as the smallest size suitable for micro-tagging and because the nursery rearing could be completed in a single phase without the need for transfer to larger individual rearing compartments. Lobsters of this size are, however, far more expensive to produce than stage four juveniles and it has not been determined whether this extra expense is fully offset by higher survival rates in the sea. The optimum release size is still unknown, though it is critical to the economics of a release programme.

Apart from the cost of producing juvenile lobsters, restocking and ranching programmes would incur significant additional expenditure on site surveys, habitat modification or protection (ranching projects), transportation, and the release operation itself.

10.9.3.4 On-growing

Despite advances in lobster culture technology, significant problems remain with the very high cost of the necessary on-growing installations, and the absence of a cost-effective diet. As a result, lobster culture has yet to become established commercially and real data on economic aspects are unavailable.

10.9.3.5 Holding and fattening

The possibility of holding and fattening under-sized or soft-shell wild-caught lobsters to enhance their value and take advantage of seasonal price fluctuations has stimulated commercial activity and some research. In an economic analysis of a projected operation for fattening 30 000 1–1.5 lb (454–681 g) lobsters, Bishop & Castell (1978) estimated that positive cash flows could be generated based on:

- Post moult weight gain of 70%, boosted by unilateral eyestalk ablation;
- Selling price enhancement of 62.5% per pound above purchase price;
- Mortality rate of 10%;
- Feeding rate of 1% body wt day^{-1}.

They also concluded that two six month crops per year were more profitable than a single holding lasting nine months. However, a holding trial with 3535 lobsters was not a success because it suffered heavy mortalities as a result of interruptions in the water supply.

10.9.3.6 Intensive (battery) culture

The systems developed for intensive lobster culture are land-based and rely on individual containment at least during the final year or so of captivity. Despite the lack of real commercial data, the most important factors affecting the economics of such an approach have been identified. Unfortunately, very high capital and labour costs result from the need for individual confinement, and although automation has the potential to reduce labour costs, any increasing sophistication would incur increased development and capital costs.

The amount of space given to each lobster is important. While excess space increases growth rate, it reduces the number of animals that can be held in a production unit. On the other hand, if conditions are cramped depressed growth will extend the culture period and increase overall costs (Sections 7.8.8 and 7.8.9).

The operating temperature is also critical because it largely determines growth rate and the duration of the culture cycle. Temperatures between 20° and 22°C are considered optimal, and because of the prohibitive expense of heating large volumes of water, the availability of a water supply at or very close to this temperature is considered to be a virtual necessity for any commercial venture. Allen & Johnston (1976) estimated that production costs would be more than doubled if the incoming water temperature were 12°C rather than 20°C.

If a unit does incur significant heating costs, the use of a recirculation system and/or heat recovery apparatus can result in significant savings. Allen & Johnston (1976) estimated that the expense of installing and operating a recirculation system would be offset by heat savings if the difference between the operating water temperature and the water supply temperature were greater than 5–7°C; and Herdman (1988) estimated that using heat exchangers to transfer effluent heat to incoming water was economically justifiable if the temperature difference was greater than 2°C.

The optimum output level for a production unit has been estimated at around 80 000 animals (of 500 g) per month (Allen & Johnston 1976), and although operations of this size capture most economies of scale, they represent very large investments. For example, Coffelt & Wikman-Coffelt (1985) estimated that an initial investment of US$31.2 million would be required for a system with a similar output level. The annual operating costs of this unit, producing one million 450 g lobsters per year, were projected to be US$3.3 million, of which labour represented 38% and feed 30%. However, assumptions underlying these calculations have been questioned, notably the non-existence of a suitable formulated ration (Aiken & Waddy 1989). Other estimates place the cost of producing a 500 g lobster at US$3.38 (Johnston & Botsford 1981) and US$3.80 (Ingram 1985).

It is estimated that it would take on average 27–30 months to produce a lobster of 500 g. Putative investors in lobster culture should expect to wait a minimum of four to five years before a project provides any significant returns on investment. However, the economics of culture could be greatly improved through the use of unilateral eyestalk ablation to reduce the growth period from 27–30 months to about 20 months (Aiken & Waddy, 1989) (Section 12.7). While 450–500 g lobsters represent the established market size, if 200–250 g animals could be successfully marketed production costs for these smaller animals would be significantly lower. This is because growth rate in captivity declines markedly beyond 250 g (Aiken & Waddy 1989), and because the risk of losses for a particular batch due to disease or other problems is reduced by a shorter on-growing period. If success were achieved with 250 g lobsters, further reductions in target harvest size might follow. In all cases, however, culture operations will need to make allowances for large variations in individual growth rates (Table 4.5).

10.9.4 Crabs

Very little information on the economics of crab culture has been published, probably because few farms raise crabs in monoculture and even fewer are of significant size. Some Asian shrimp and fish hatcheries may also produce post-larval crabs when the market is favourable, but again cost data for this particular activity are scarce.

In south-east Asia the mud crab *Scylla serrata* is farmed both in extensive polyculture (with shrimp and milkfish) and alone in monoculture (Section 7.10.4). Extensive systems which rely entirely on the ingress of wild juveniles as tidal ponds are filled, often count crabs only as a by-crop. In other operations, in order to boost the yield of crabs, farmers purchase locally netted wild crab juveniles and thus incur significant stocking costs. Feed costs are usually minimal since only trash fish or sometimes rice bran are added to the pond. Labour costs are typically low since the bulk of the work is performed by owner-operators. Generally, the greater overall productivity possible with polyculture in extensive systems results in greater profits than does crab monoculture (Lapie & Librero 1979). Nevertheless, a recent analysis of *S. serrata* monoculture in the Philippines (Agbayani *et al.* 1990) has indicated that this activity is potentially very lucrative if yields of around 1000 kg ha^{-1} $crop^{-1}$ can be sustained.

Agbayani *et al.* (1990) based their economic projections on results from experimental on-growing trials performed in a shallow (50 cm) earthen pond. Crabs were fed with chopped tilapia and it was established that the best results (survival 88%; growth 2.28 g d^{-1}; feed conversion ratio 1.72:1; yield 1019 kg ha^{-1} $crop^{-1}$; size at harvest 232 g) and financial return (ROI 124%) could be obtained at a stocking density of 5000 juvenile crabs per hectare (mean weight 25.3 g). Higher densities were less profitable because they resulted in significantly lower survival and growth rates and less efficient feed conversion. The cost of setting up a one ha crab farm was put at US$1140, of which 59% represented construction of the pond and another 28% the cost of a perimeter fence. The installations and pond were expected to last for only two to five years before replacement and reconstruction would be necessary. Operating costs were projected at US$1.10 kg^{-1}, with the major components comprising feed, labour, depreciation and seed (Figure 10.16). The attractive financial returns projected, however, relied on the assumption that the yield of 1019 kg ha^{-1} per 90 days, obtained experimentally, could be achieved three times a year in a commercial venture. The authors note that, traditionally, yields per crop from crab monoculture in the Philippines are only around one third of this value.

Crab farmers in Taiwan obtain seed either from fishermen, nursery operators or middlemen, and the prices they pay varying depending on size. Juveniles of 2–3 cm CW may fetch two to five times more than smaller animals of 0.5–1.0 cm (Cowan 1983).

Juvenile crabs (*Portunus trituberculatus*) are produced in Japanese hatcheries at a cost of between US$7.70 and US$21.10 per 1000 (Cowan 1983) and are sold to fishing co-operatives for restocking (Section 7.10.8). In one hatchery, energy and food accounted for 83% of variable expenses and 68% of overall production costs. The exact cost of releasing juveniles in the wild, however, varies depending on whether nursery rearing is used and on the type of release method. Nursery rearing in onshore tanks made of concrete or canvas can more than double the cost per juvenile, and release in open water using artificial habitats of branches or frayed rope suspended from rafts can increase overall costs by a factor of 2.5.

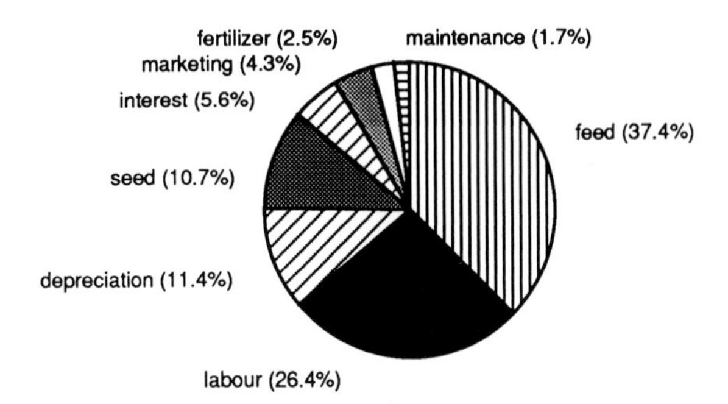

Fig. 10.16 Proportionate operating costs for a 1 ha crab farm in the Philippines (based on Agbayani *et al.* 1990).

In common with all stock enhancement programmes economic viability depends on recapture rates, and although these are notoriously difficult to establish, some estimates have been made. For example, in one small bay in Hiroshima, where no previous fishery existed, a recapture rate of 3.2% has been recorded. In other situations size frequency analysis has been able to identify restocked cohorts, since hatchery seed tend to be larger than their wild counterparts. One such study placed recapture rates at between 4% and 12% (Cowan 1983). Economic viability also hinges on crab market prices. At recapture rates of 1.2–2.9% and stocking costs of between US$14.2 and US$34.5 per 1000 juveniles, economic returns were considered possible at a selling price of US$1.18 per crab.

Crab holding operations in Taiwan are often run by dealers who trade in both farmed and fished crabs. Fertilized females are retained for fattening and are sold at a profit when they reach the highly prized 'red' (full roe) stage (Section 7.10.4). In the USA, shedding systems for blue crabs (*Callinectes sapidus*) can take advantage of the big price differential between soft and hard-shelled animals. Nevertheless, while it is economically viable to take wild crabs on the verge of moulting ('peelers') and hold them for short periods, it is not feasible to take intermoult animals and feed them until they moult (Oesterling & Provenzano 1985). Large production units may have around 100–150 trays and employ about 12 people (Section 7.10.9). The turnover period is about one week.

10.9.5 Processing plant

Although data on the costs of establishing and running crustacean processing plants are generally scarce, some useful information is available regarding the production of frozen shrimp. One detailed study has been prepared by Waits & Dillard (1987), based on processing shrimp or *Macrobrachium* in Mississippi, USA, in the form of raw frozen shell-on tails, packed in 5 lb (2.27 kg) cartons. The total cost of a plant equipped to produce a daily maximum of 6.5 mt was calculated to be US$781 184 (excluding land). The components of this cost are illustrated in Figure 10.17. The cost of the building (1040 m^2) represented nearly 40% of the total, and the blast and storage freezers together accounted for another 27%. An area of land measuring 0.35 ha would be required for the operation, and the purchase of a site in Mississippi with water frontage (to allow boat-caught shrimp to be landed) was estimated to add

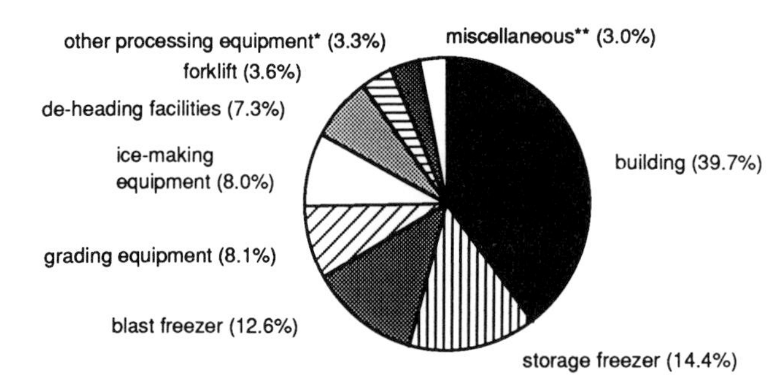

Fig. 10.17 Proportionate investment costs for a processing plant (max. daily output 6.5 mt frozen headless shrimp or prawns) (based on Waits & Dillard 1987, excluding land, including full cost of ice-making equipment). * Weighing, packing, glazing, strapping. ** Office equipment, licences, utility connections.

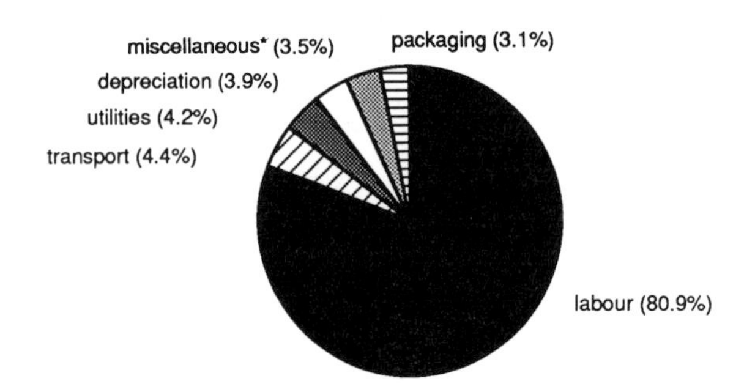

Fig. 10.18 Proportionate operating costs for a processing plant (max. daily output 6.5 mt headless shrimp or prawns) (based on Waits & Dillard 1987, excluding interest and taxes). * Includes maintenance, insurance, waste disposal.

another US$120 000 to the total investment costs. However, a less expensive inland site (US$35 000) could be occupied if incoming product were to be delivered by trucks (usually the case with supplies from aquaculture).

The total operating costs for the processing plant, at an annual output of 1139 mt, were put at US$1 165 018 (excluding taxes and interest), giving a total processing cost of US$1.02 kg^{-1}. Proportionate costs are shown in Figure 10.18 and illustrate the dominance of labour costs (80.9%). The cost of producing whole shrimp rather than tails would be significantly lower because no labour would be required for de-heading, and because the processing yield for whole product is 100% (compared to 40% for *Macrobrachium* tails and 57–68% for shrimp tails). Transporting product by truck from farms to the processing plant was estimated to add another US$0.065 kg^{-1} to the overall costs.

To help minimize the processing cost, the full capacity of a plant must be used. This becomes difficult, however, when output from fisheries or aquaculture is seasonal. The costings given above relate to a plant operating for a total of nine months a year with two months at maximum capacity, and were based on processing wild-caught shrimp as well as product from aquaculture. No doubt cost savings could be made if consistent quantities of product could be processed all year round. On the other hand, if the same plant were operated for only two months a year rather than nine, to produce just 314 mt of tails, total processing costs would increase to US\$1.53 kg^{-1} (excluding taxes and interest).

The bulk of processed crayfish is produced in the USA in the form of fresh and frozen tail meat. Crayfish are peeled after being blanched. The meat yield is a mere 15% and the major processing cost by far is the labour required for peeling by hand. This cost, estimated at around US\$2.75 per kg of meat (Roberts & Dellenbarger 1989), has been identified as a constraint to crayfish marketing since it results in overall processing costs greater than those of competing shrimp products. Peeling machines have been developed to cut labour requirements but have not been successful. An account of US crayfish processing methods is given by Moody (1989).

10.9.6 Feed mill

Since only limited information on the costs of setting up and operating a crustacean feed mill is available, a useful indication of likely expenditure levels can be obtained by considering the situation for an animal (livestock or poultry) feed mill and then taking into account the special requirements of crustacean feeds and their cost implications.

A technical and investment guide for those interested in producing compound feeds on a small-scale in developing countries has been compiled by Parr *et al.* (1988). For a plant producing 4200 mt of pellets for poultry or livestock annually, an investment equivalent to around US\$371 000 was identified, and annual operating costs were put at US\$800 000. Although these totals are likely to underestimate the costs for a crustacean feed mill with an equal capacity, Figures 10.19a and 10.19b, which illustrate the component costs, provide summaries which can be more generally applied.

The special requirements for aquaculture feeds, as compared to animal feeds, are discussed by Barbi (1987). Compounded aquaculture feeds in general, and crustacean feeds in particular, require ingredients that are very finely and uniformly ground and this results in increased investment and operating costs for the necessary milling equipment. In addition, pelleting machines for crustacean diets are fitted with thicker dies with smaller holes. This improves compaction as the feed mixture is extruded to form pellets, and the extra frictional heat helps to gelatinise starches and improve the binding qualities. However, at the same time the output of the machine is significantly reduced. For example, a pelleter equipped to produce 2 mt hr^{-1} of 3.5 mm poultry pellets, would only be able to produce 1.3 mt hr^{-1} of 2.5 mm shrimp pellets (J. Wood, 1990, pers. comm.).

Despite the extra costs of investing in equipment for producing crustacean diets, and the associated increase in maintenance and utility costs, feed ingredients represent some 70% of operating costs (Figure 10.19b) and it is the cost of these ingredients that has the most significant impact on the overall cost of producing a compound crustacean feed. Crustacean diets, especially those with high protein formulations, usually rely heavily on relatively expensive ingredients such as fishmeal, and this is one of the main reasons why they tend to be more costly to produce than animal diets.

There is scope for significant cost savings through alterations in the mixture of ingredients of all compound feeds. Indeed, in the manufacture of poultry and livestock

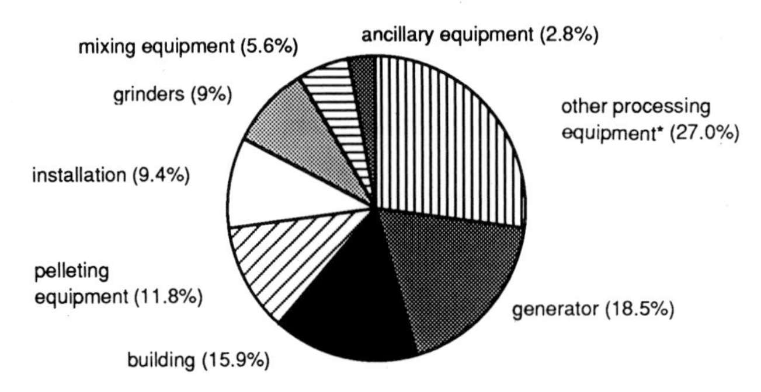

Fig. 10.19a Proportionate investment costs for a feed mill with an annual capacity of 4200 mt (based on Parr *et al.* 1988). *Weighing, elevators, augers, holding bins, grinders, steam production, pellet cooling, bagging, electrical control.

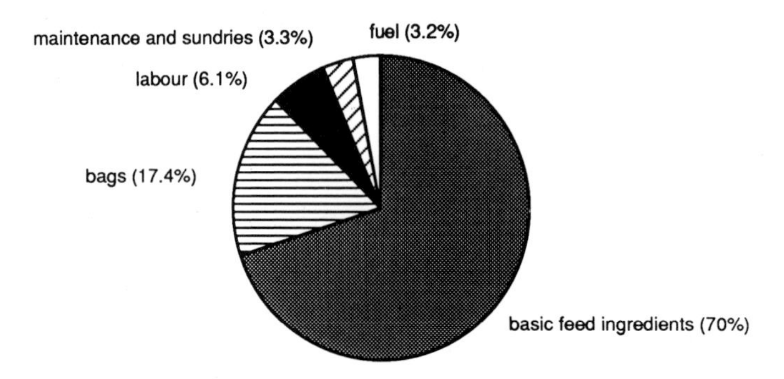

Fig. 10.19b Proportionate operating costs for a feed mill with an annual capacity of 4200 mt (based on Parr *et al.* 1988).

diets, the use of computers to generate the lowest cost formulations has become routine, and enables feed mill operators to take advantage of fluctuations in the price of raw materials to maximize profits, while keeping the overall nutritional profile of their final product largely unchanged. By comparison, the lowest cost formula approach for crustacean feeds is not well developed, principally because of inadequate knowledge of crustacean nutrient requirements and the nutrient availability of different ingredients (Akiyama & Dominy 1989).

In addition to the conventional pelleting process, high quality aquaculture feeds, including crustacean diets, are increasingly being produced by a method known as extrusion cooking. The features and costs of the two processes are compared by Kearns (1989). Extrusion cooking equipment has higher investment and operating costs than conventional pelleting equipment; Barbi (1987) estimated that the investment for an extrusion cooking and drying system would be about 7.5 times greater than for a typical pelleting system, and that the operation and maintenance costs would be about 91% higher. But it does allow more flexible and lower cost feed formulations to be used which can generate significant overall savings.

To produce a stable pellet a conventional pelleting machine requires a formula containing around 25–35% starch ingredients and binders, and this dictates that high cost aquatic meals (largely fish) be incorporated to make up the desired levels of protein and other nutrients. An extrusion cooked mixture, on the other hand, need contain only 10% starch ingredients or binders to form a stable pellet and this allows more flexibility in the diet formulation, and a range of protein sources usually cheaper than fishmeal can be included. Figures provided by Kearns (1989) indicate that an initial investment of US$2 000 000 would be needed for a complete plant equipped with an extrusion cooking system to produce 24 000 mt of feed annually. An equivalent plant with more conventional pelleting equipment would cost US$1 350 000 – significantly less. Although the cost of utilities for the first system was calculated to be greater (US$21.53 mt^{-1} against 11.93 mt^{-1}), for a diet containing 40.7% protein cost savings on the cheaper extrusion formula (largely through the reduction of the fishmeal content from 40% to 10%) amounted to over US$100 mt^{-1}.

10.10 References

Agbayani R.F., Baliao D.D., Samonte G.P.B., Tumaliuan R.E. & Caturao, R.D. (1990) Economic feasibility analysis of the monoculture of the mud crab (*Scylla Serrata*) Forskol. *Aquaculture*, **91** (3/4) 223–32.

Aiken D.E. & Waddy S.L. (1989) Culture of the american lobster, *Homarus americanus*. In *Cold-water aquaculture in atlantic Canada* (Ed. by A.D. Boghen), pp. 79–122. The Canadian Institute for Research on Regional Development, Moncton.

Akiyama D.M. & Dominy W.G. (1989) Penaeid shrimp nutrition for the commercial feed industry. *Proceedings of the People's Republic of China aquaculture and feed workshop*, Sept 17–30, 1989 (Ed. by D.M. Akiyama), pp. 189–236. American Soybean Assoc., Singapore.

Alderman D.J. & Wickins J.F. (1990) *Crayfish Culture*. Lab. Leafl. (62), MAFF Direct. Fish. Res., Lowestoft.

Allen P.G. & Johnston W.E. (1976) Research direction and economic feasibility: An example of systems analysis for lobster aquaculture. *Aquaculture*, **9**, 155–80.

Anon. (1987) Tobacco firm starts smolt production and launches in lobster. *Fish Farming International*, **14** (9) 18–20.

AQUACOP (1979) Intensive larval rearing of *Macrobrachium rosenbergii*. A cost study. *Proc. World Maricult. Soc.*, **10**, 429–34.

AQUACOP (1982) Mass production of juveniles of freshwater prawn *Macrobrachium rosenbergii* in French Polynesia: Predevelopment phase results. In *Proc. Symp. Coastal Aquaculture 1982, Pt. 1*, pp. 71–5. Mar. Biol. Association of India, Cochin, India.

Arrignon J. (1981) *L'écrevisse et son elevage*. Gauthiers-Villars, Paris.

Avault J.W. Jr. & Huner J.V. (1985) Crawfish culture in the United States. In *Crustacean and Mollusk Aquaculture in the United States* (Ed. by J.V. Huner & E. Evan Brown), pp. 1–54. AVI Inc., Westport, Connecticut.

Barbi J.W. (1987) *Shrimp and fish feed processing*. Presented at Aquaculture Europe '87 International Conference, 2–5 June 1987. RAI Amsterdam, (mimeo).

Bauer L.L., Sandifer P.A., Smith T.I.J. & Jenkins W.E. (1983) Economic feasibility of *Macrobrachium* production in South Carolina. *Aquacultural Engineering*, **2** (3) 181–201.

Bishop F.J. & Castell J.D. (1978) 1977 commercial lobster culture feasibility study,

Clark's Harbour, Nova Scotia. *Fisheries and Marine Service Industry Report 102*, Aug. 1978.

Chamberlain G.W. (1989) *Status of shrimp farming in Texas*. Presented at 20th. Meeting World Aquacult. Soc., 12–16 Feb. 1989. Los Angeles.

Chen L.-C. (1990) *Aquaculture in Taiwan*. Fishing News Books, Blackwell Scientific Publications, Oxford.

Chiang P. & Liao I.C. (1985) The practice of grass prawn (*Penaeus monodon*) culture in Taiwan from 1968 to 1984. *J. World Maricult. Soc.*, **16**, 297–315.

Chien Y-H. & Liao I-C. (1987) *Bioeconomic consideration of prawn farming*. Dept. of Aquaculture, National Taiwan College of Marine Science and Technology, Keelung, Taiwan.

Clarke S. (1989) Freshwater crustacean farming: the world scene. S. Australian Dept. Fisheries, *SAFISH*, **13** (4) 10–12.

Coffelt R.J. & Wikman-Coffelt J. (1985) Lobsters: one million one pounders per year. *Aquacultural Engineering*, **4** (1) 51–8.

Cowan L. (1983) *Crab farming in Japan, Taiwan and the Phillipines*. Information Series Q 184009, Queensland Department of Primary Industries, Queensland, Australia.

CPC (1989) *Libro blanco del camarón* (in Spanish). Cámara de Productores de Camarón, Guayaquil, Ecuador.

Culley D.D. & Duobinis-Gray L. (1989) Soft-shell crawfish production technology. *J. Shellfish Res.*, **8** (1) 287–91.

de la Bretonne L.W. Jr. & Romaire R.P. (1989) Commercial crawfish cultivation practices: A review. *J. Shellfish Res.*, **8** (1) 267–85.

Dellenbarger L.E., Schupp A.R. & Avault J.A. (1988) Louisiana's crayfish industry: An economic perspective. In *Freshwater Crayfish* 7, Ed. by P. Goedlin de Tiefenau, pp. 231–7. Musée Zoologique, Cantonal, Lausanne, Switzerland.

Eversole A.G. & Pomeroy R.S. (1989) Crawfish culture in South Carolina: An emerging aquaculture industry. *J. Shellfish Res.*, **8** (1) 309–13.

FAO (1977) *Planning of aquaculture development – an introductory guide*. Fishing News Books, Blackwell Scientific Publications, Oxford.

Farmer A.S.D. (1981) Prospects for penaeid shrimp culture in arid lands. In *Advances in Food Producing Systems for Arid and Semi-Arid Lands*, pp. 859–97. Academic Press, London.

Forster J.R.M. & Beard T.W. (1974) Experiments to assess the suitability of nine species of prawns for intensive cultivation. *Aquaculture*, **3**, 355–68.

Gopalan U.K., Purushan K.S., Santhakumari V. & Meenakshi Kunjamma P.P. (1982) Experimental studies on high density, short term farming of shrimp *Penaeus indicus* in a 'pokkali' field in Vypeen Island, Kerala. In *Proc. Symp. Coastal Aquaculture 1982*, Pt. 1, pp. 151–9. Mar. Biol. Assn. of India, Cochin.

Hanson J.S., Griffin W.L., Richardson J.W. & Nixon C.J. (1985) Economic feasibility of shrimp farming in Texas: An investment analysis for semi-intensive pond grow-out. *J. World Maricult. Soc.*, **16**, 129–50.

Hardman P.J.R., Treadwell R. & Maguire G. (1990) *Economics of prawn farming in Australia*. Presented at International Crustacean Conference, 2–6 July, 1990, Brisbane.

Henocque Y. (1984) Aménagement de la ressource côtière au Japon: effet des repeuplements marins (in French). *Rapports Techniques 11-1984*, Institute Scientifique et Technique des Peches Maritimes.

Herdman A. (1988) Heating of hatchery water supplies. In *Aquaculture engineering technologies for the future*. Institution of Chemical Engineers Symposium series No. 111, pp. 343–56. EFCE Publication series No. 66, Hemisphere, London.

Hirono Y. (1989) Extensive shrimp farm management. In *Proceedings of the south-east Asia shrimp farm management workshop*, 26 Jul.–11 Aug., 1989, (Ed. by D.M. Akiyama), pp. 2–10, Philippines, Indonesia, Thailand. American Soybean Association, Singapore.

Hu Q. (1990) On the culture of *Penaeus penicillatus* and *P. chinensis* in southern China. In *The culture of cold-tolerant shrimp* (Ed. by K.L. Main & W. Fulks), pp. 77–91. The Oceanic Institute, Honolulu, Hawaii.

Huang H.-J., Griffin W.L. & Aldrich D.V. (1984) A preliminary economic feasibility analysis of a proposed commercial penaeid culture operation. *J. World Maricult. Soc.*, **15**, 95–105.

Huner J.V. (1990) New horizons for the crayfish industry. *Aquaculture Magazine*, **16** (5) 65–70.

Huner J., Gydemo R., Haug J., Jarvenpåa T. & Taugbøl T. (1987) Trade, marketing and economics. In *Crayfish Culture in Europe*, pp. 54–62. Report from the workshop on crayfish culture, 16–19 Nov.,1987, Trondheim, Norway.

IFC (1987) *Marine shrimp farming: A guide to feasibility study preparation*. Aquafood Business Associates, International Finance Corp., Charleston, SC 29412, USA.

Ingram M. (1985) *Intensive culture of lobsters and other species*. Clearwater Publishing Limited, Isle of Man.

Israel D.C. (1987a) Economic feasibility of aquaculture projects: a review. *Asian Aquaculture*, **9** (1) 6–9.

Israel D.C. (1987b) Comparative economic analysis of prawn hatcheries. *Asian Aquaculture*, **9** (2) 3–4, 12.

Jaenike F. (1989) Management of a shrimp farm in Texas. In *Proceedings of the south-east Asia shrimp farm management workshop*, 26 Jul.–11 Aug., 1989, (Ed. by D.M. Akiyama), pp. 11–21, Philippines, Indonesia, Thailand. American Soybean Association, Singapore.

Johns M., Griffin W., Lawrence A. & Hutchins D. (1981a) Budget analysis of a shrimp maturation facility. *J. World Maricult. Soc.*, **12** (1) 104–9.

Johns M., Griffin W., Lawrence A. & Fox J. (1981b) Budget analysis of penaeid shrimp hatchery facilities. *J. World Maricult. Soc.*, **12** (2) 305–21.

Johns M.A., Griffin W.L., Pardy C. & Lawrence A.L. (1983) Pond production strategies and budget analysis for penaeid shrimp. In *Proc. 1st Int. Conf. on Warm Water Aquaculture – Crustacea*, 9–11 Feb. 1983, (Ed. by G.L. Rogers, R. Day & A. Lim), pp. 19–33. Brigham Young University, Hawaii.

Johnston W.E. & Botsford L.W. (1981) Systems analysis for lobster aquaculture. In *Aquaculture in heated effluents and recirculation systems* (Ed. by K. Tiews), **2**, 455–64. Heenemann Verlagsgesellschaft, Berlin.

Juan Y., Griffin W.L. & Lawrence A.L. (1988) Production costs of juvenile penaeid shrimp in an intensive greenhouse raceway nursery system. *J. World Aquacult. Soc.*, **19** (3) 149–60.

Kearns J.P. (1989) Advantages of extrusion cooking and comparisons with the pelleting process for aquatic feeds. In *Proceedings of the People's Republic of China aquaculture and feed workshop*, 17–30 Sept 1989, (Ed. by D.M. Akiyama), pp. 245–69. American Soybean Assoc., Singapore.

Keller M. (1988) Finding a profitable population density in rearing summerlings of European crayfish *Astacus astacus* L. In *Freshwater crayfish* 7 (Ed. by P. Goeldlin de Tiefenau), pp. 259–66. Musée Zoologique Cantonal, Lausanne, Switzerland.

Kim J.H. (1990) The culture of *Penaeus chinensis* and *P. japonicus* in Korea. In *The culture of cold-tolerant shrimp* (Ed. by K.L. Main & W. Fulks), pp. 64–9. Oceanic Institute, Honolulu, Hawaii.

Lapie L.P., Librero A.R. (1979) *Crab farming in the Philippines: a socio-economic study*. Research paper series/Socio-economic survey of the Aquaculture Industry (21), SEAFDEC-PCARR Research Programme, Los Banos, Laguna, Philippines.

Lawrence A.L. (1985) Marine shrimp culture in the western hemisphere. In *Second Aust. Nat. Prawn Sem.* (Ed. by P.C. Rothlisberg, B.J. Hill & D.J. Staples), pp. 327–36. NPS2, Cleveland, Australia.

Lawrence A.L., McVey J.P. & Huner J.V. (1985) Penaeid shrimp culture. In *Crustacean and Mollusk Aquaculture in the United States* (Ed. by J.V. Huner & E. Evan Brown), pp. 127–57. AVI Inc., Westport, Connecticut.

Lee D. O'C., Illescas R., Miranda L., Escobar F., Salvador J.A. & Lucien-Brun H. (1988) *An analysis of the influence of dietary and other rearing condition variables on growth and survival rates of larval* P. vannamei *in a commercial hatchery*. Presented at 19th Ann. World Aquacult. Soc. Mtg., Honolulu, Hawaii, 2–9 Jan. 1988 (mimeo).

Leeds R. (1986) Financing aquaculture projects. *Aquacultural Engineering*, **5** (2–4) 109–113.

Liao I.-C. & Chien Y.-H. (1990) Evaluation and comparison of culture practices for *Penaeus japonicus, P. penicillatus* and *P. chinensis* in Taiwan. In *The culture of cold-tolerant shrimp* (Ed. by K.L. Main & W. Fulks), pp. 49–63. Oceanic Institute, Honolulu, Hawaii.

Lumare F., Amerio M., Arata P., Guglielmo L., Casolino G., Marolla V., Schiavone R. & Ziino M. (1989) Semi-intensive culture of the kuruma shrimp *Penaeus japonicus* by fertilizer and feed applications in Italy. In *Aquaculture, a biotechnology in progress* (Ed. by N. de Pauw, E. Jaspers, H. Ackefors & N. Wilkins), pp. 401–7. European Aquaculture Society, Bredene, Belgium.

Maguire G.B. & Allan G. (1985) Development of methods for growing juvenile school prawns, *Metapenaeus macleayi*, in estuarine ponds. In *Second Aust. Nat. Prawn Sem.* (Ed. by P.C. Rothlisberg, B.J. Hill & D.J. Staples), pp. 337–51. NPS2, Cleveland, Australia.

Main K.L. & Deupree R.H. Jr. (1986) Commercial aquaculture in Hawaii, 1986. *Information Text Series 031*, College of Tropical Agriculture & Human Resources, Hawaii.

Main K.L. & Fulks W. (Eds.) (1990) *The culture of cold-tolerant shrimp: Proceedings of an Asian-US workshop on shrimp culture.* Oceanic Institute, Honolulu, Hawaii.

McCoy H.D. II. (1986) Intensive culture systems past, present and future Parts I, II and III. *Aquaculture Magazine*, **12** (6) 32–5; **13** (1) 36–40; **13** (2) 24–9.

McCoy H.D. II. (1988) Aquaculture and the stock market. *Aquaculture*, **14** (1) 20–30.

Mock C.R. (1983) *Penaeid shrimp culture consultation Taiwan*, 1–11 Dec. 1983. Trip report (mimeo).

Mock C.R. (1987) *Penaeid shrimp culture consultation and visit, Guatemala, Central America*, 16–23 Feb. 1987. NOAA, NMFS, US Dept. of Commerce, (mimeo).

Mock C.R. (1990) Shrimp farming in China. *World Shrimp Farming*, **15** (5) 11–12.

Moody M.W. (1989) Processing of freshwater crawfish: a review. *J. Shellfish Res.*, **8** (1) 293–301.

Muir J.F. & Kapetsky J.M. (1988) Site selection decisions and project cost: the case of brackish water pond systems. In *Aquaculture engineering technologies for the future.* Institution of Chemical Engineers Symposium series No. 111, pp. 45–63, EFCE Publication series No. 66, Hemisphere, London.

NACA (1986) A prototype warm water hatchery. *Technology series, selected publication No. 4*, NACA, Bangkok, Thailand.

Nash C.E. (1988) *Investment in aquaculture.* Aquaculture Development and Coordiation Programme, UNDP, FAO, Plenary paper, presented at Aquaculture International Congress, 9 Sept. 1988, Vancouver, Canada.

New M.B. (1988) Shrimp farming in other areas. In *Shrimp '88, Conference proceedings*, 26–28 Jan. 1988, Bangkok, Thailand. Infofish, Kuala Lumpur, Malaysia.

New M.B. (1990) Freshwater prawn culture: a review. *Aquaculture*, **88** (2) 99–143.

New M.B., Sanders S., Brown R.L. & Cole R.C. (1978) The feasibility of farming *Macrobrachium* in Dominica, West Indies. *Proc. 9th Annual Meeting World Maricult. Soc.*, 3–6 Jan. 1978, pp. 67–90. Atlanta, Georgia, USA.

ODA (1977) *A guide to the economic appraisal of projects in developing countries.* Ministry of Overseas Development, HMSO, London.

Oesterling M.J. & Provenzano A.J. (1985) Other crustacean species. In *Crustacean and Mollusk Aquaculture in the United States* (Ed. by J.V. Huner & E. Evan Brown), pp. 203–34. AVI, Westport, USA.

Oshima Y. (1984) Status of fish farming and related technological development in the cultivation of aquatic resources in Japan. *TML Conference Proceedings, No. 1*, pp. 1–11. Tungkang Marine Laboratory, Pingtung, Taiwan.

O'Sullivan D. (1988) *The culture of the marron (*Cherax tenuimanus*) in Australia: a review.* Presented at 19th Ann. World Aquacult. Soc. Mtg., 2–9 Jan., Honolulu, Hawaii, (mimeo).

Pai M.V., Somvanshi V.S. & Telang K.Y. (1982) A case study of the economics of a traditional prawn culture farm in the Northern Kanara district, Karnataka, India. In *Proc. Symp. Coastal Aquaculture 1982, Pt. 1*, pp. 123–8. Mar. Biol. Assn. of India, Cochin.

Parker J. (1990) Shrimp farm investment analysis for outdoor ponds in Texas. In *Texas A&M University shrimp farming short course materials, article 47* (Ed. by G.D. Treece). Texas A&M Sea Grant College Program, College Station, Texas.

Parr W.H., Capper B.S., Cox D.R.S., Jewers K., Marter A.D., Nichols W., Silvey D.R. & Wood J.F. (1988) The small-scale manufacture of compound animal feed. *Bulletin No. 9*, Overseas Development Natural Resources Institute, Chatham.

Primavera J.H. (1983) Prawn hatcheries in the Philippines. *Asian Aquaculture*, **5** (3) pp. 5–7 + 8.

Quynh V.D. (1990) Shrimp larviculture in Viet Nam. *Larviculture and Artemia News-*

letter, (16) 22–6. State University of Ghent, Belgium.

Roberts K.J. & Dellenbarger L. (1989) Louisiana crawfish product markets and marketing. *J. Shellfish Res.*, **8** (1) 303–7.

Rosenberry R. (1988) Shrimp farming in China. *Aquaculture Digest*, **13** (9) 2–4.

Rosenberry R. (1989) Ecuador: No seedstock! No money! Exports slipping. *Aquaculture Digest*, **14** (7) 1–2.

Rosenberry R. (1990) *Shrimp farming in the western hemisphere*. Presented at Aquatech 90, Malaysia, June 1990 (mimeo).

Sandifer P.A. & Smith T.I.J. (1985) Freshwater prawns. In *Crustacean and mollusk aquaculture in the United States* (Ed. by J.V. Huner & E. Evan Brown), pp. 63–125. AVI Inc., Westport, Connecticut.

SEAFDEC (1985) *A guide to prawn hatchery design and operation*. Extension manual No. 9, South East Asia Fisheries Development Centre Tigbauan, Iloilo, Philippines.

Shang Y.C. (1982) Comparison of freshwater prawn farming in Hawaii and in Thailand: Culture practices and economics. *J. World Maricult. Soc.*, **13**, 113–19.

Shang Y.C. (1983) The economics of marine shrimp farming: A survey. In *Proc. 1st Int. Conf. on Warm Water Aquaculture – Crustacea* (Ed. by G.L. Rogers, R. Day & A. Lim), 9–11 Feb. 1983, pp. 7–18. Brigham Young University, Hawaii.

Shigueno K. (1975) *Shrimp culture in Japan*. Association for International Technical Promotion, Tokyo.

Sidarto A.A. (1977) *The financing of small-scale aquaculture projects, the case of brackishwater pond culture in Indonesia*, pp. 175–85. ASEAN77/FA.EgA/Doc.WP1 Directorate General of Fisheries, Ministry of Agriculture, Jakarta.

Skabo H. (1988) Shrimp farming developments in West Africa. In *Shrimp '88, Conference proceedings*, 26–28 Jan. 1988, pp. 95–102, Bangkok, Thailand. Infofish, Kuala Lumpur, Malaysia.

Smith I.R. (1984) *Social feasibility of coastal aquaculture: packaged technology from above or participatory rural development*. ICLARM cont. 225, prepared for Consultation on Social Feasibility of Coastal Aquacult., 26 Nov.–1 Dec. 1984, Madras, India.

Smith T.I.J., Jenkins E.W. & Sandifer P.A. (1983) Enclosed prawn nursery systems and effects of stocking juvenile *Macrobrachium rosenbergii* in ponds. *J. World Maricult. Soc.*, **14**, 111–25.

Staniford A.J. (1988) More on the economics of yabbie farming. S. Australian Dept. Fisheries, *SAFISH*, **13** (2) 10–12.

Staniford A.J. (1989) The effect of yield and price variability on the economic feasibility of freshwater crayfish *Cherax destructor* Clark (Decapoda: Parastacidae) production in Australia. *Aquaculture*, **81** (3–4) 225–36.

Staniford A.J. & Kuznecovs J. (1988) Aquaculture of the yabbie, *Cherax destructor* Clark (Decapoda: Parastacidae): an economic evaluation. *Aquaculture and Fisheries Management*, **19**, 325–40.

Tietze U. (1989) No fish without credit. *Infofish International*, (5) 46–7.

Waits J.W. III & Dillard J.G. (1987) *The costs of processing and hauling freshwater shrimp in Mississippi*. Bulletin 953, Mississippi Agricultural and Forestry Experiment Station, Mississippi.

Yamaha (1989) Prawn culture. *Yamaha Fishery Journal*, (30). Yamaha Motor Co., Shizuoka-ken, Japan.

Yap W.G. (1990) Backyard hatcheries take off in Jepara. *Infofish International*, (2) 42–7.

Zhang N. (1990) On the growth of cultured *Penaeus chinensis* (Osbeck). In *The culture of cold-tolerant shrimp* (Ed. by K.L. Main & W. Fulks), pp. 97–102. Oceanic Institute, Honolulu, Hawaii.

Zhang N., Lin R., Cao D., Zhang W., Gao H. & Liang X. (1983) Preliminary observations on the relationship of body weight, length with the daily food requirements of Chinese shrimp, *Penaeus orientalis* Kishinouye. *Oceanologia et Limnologia Sinica*, **12** (5) 482–7.

Chapter 11
Impact of crustacean aquaculture

11.1 Introduction

The preceeding chapters have dealt with the biological and technical factors that promote or constrain consistent performance and productivity in crustacean aquaculture projects. There are, however, additional considerations equally vital to the long term success of a venture, that are not always readily appreciated by business administrators and their advisers. These are concerned with the social and environmental consequences and the interactions that can arise with institutions as a result of rapid movement into crustacean aquaculture in general, and into tropical shrimp farming in particular. The consequences are widely believed to include a beneficial socio-economic impact with only marginal environmental disruption brought about by implanting a new industry on to otherwise unproductive rural land (NOAA 1988). While this may be true in some instances, a number of popular misconceptions persist, for example that crustacean aquaculture development automatically produces:

- Better standards of living in rural areas;
- More equitable distribution of income, wealth and profits;
- More jobs and job opportunities;
- Easily marketable, high value products;
- Food for rural communities;
- Improved local nutritional standards;
- Significant, sustainable rewards for investors.

Failure to achieve these highly desirable objectives can lead to local disillusionment expressed by increased poaching, vandalism and banditry, perhaps even resulting in project failure. The blame has sometimes been placed on government agencies, developers, scientists or extension workers for not understanding or giving adequate priority to local needs and conditions, or at other times on farmers and local authorities criticized for resisting change or lacking initiative (Weeks 1989).

11.2 Social impact

'The task of government is to liberate technology from its closed class structure and make it accessible to society at large' (Hayashi 1984).

11.2.1 Institutional involvement

National and international policy makers often appear to equate development with increased productivity and economic efficiency when planning aquaculture strategies. It is easier and less costly for a nation to develop aquaculture through large-scale, corporate undertakings financed by development banks at subsidised or below market rates, than through participatory rural development by coastal communities (Smith 1984). Capital intensive technologies are favoured since there is usually a need to increase foreign exchange. Partly as a result of this, governments and international aid agencies have invested heavily in shrimp farming. The Asian Development Bank and World Bank support coastal shrimp projects in at least nine Asian countries, with Japan, Belgium, the UK, the USA and other developed nations providing bilateral support (Bailey & Skladany, in prep.). Much less investment is placed in inland freshwater aquaculture operations generally because of restricted resource availability and because fewer cultivable freshwater crustacean species are known.

Where there is serious concern for the impact that the aquaculture development will have on society, one of the first questions to be asked is whether or not it will reinforce existing socio-cultural and institutional power structures that keep the majority of society members in poverty, or will provide opportunities for a wider spread of benefits (Smith 1984). Many of the areas designated by governments for aquaculture development are economically depressed coastal zones, without local capital and with ground inherently unsuitable or difficult to work. Coastal communities are poor because they have no access to alternative employment, and are likely to remain so because existing community and national structures allow local *élites* to monopolize the benefits of new enterprises. It is also recognised that institutions tend to adapt slowly to changing technologies and may actually inhibit a just distribution of benefits when the technologies are applied. Major social changes thus seem inevitable, particularly when resources are put into the more intensive enterprises which promote rapid economic growth. They will not, however, necessarily lead to the fairer distribution of income.

11.2.2 Land ownership and common resources

In most countries coastal land is state owned but can often be transferred to private ownership if the implementation of a large aquaculture project is judged to be in the national interest. When this happens local communities may be deprived of traditionally common resources vital to their subsistence. Their concept of security changes and they are likely to become increasingly dependent on seasonal jobs in the new industry, that require few skills. Intensive aquaculture operations often employ fewer local people than traditional rice or sugar farms for areas of equivalent size (Bailey 1988). If the project is successful land prices will increase, further enhancing the power of the local landlords, and may encourage alteration of tenancy terms in the landowner's favour. Wages and living standards may also be forced down because workers cannot get employment elsewhere or cannot get enough alternative sustenance from the altered local environment.

Inland rural populations are often far less well nourished than those on the coast which enjoy access to seafish, and those in towns where seafish are marketed. Inland, ownership is clearly defined and freshwater aquacultures tend to be more integrated, family affairs. Because of this, large scale developments can be sustained although opportunities are often more limited because neighbouring land or water supplies are already owned. From the 'cash crop' point of view it is a pity that better crustacean species than *Macrobrachium* (the larvae of the largest species require brackish water) are not available for culture in inland tropical freshwaters (but see Chapter 4). The

Plate 11.1 Adverse environmental impact caused by clearing mangrove forest for shrimp ponds in Kenya but then abandoning the scheme as an uneconomic proposition.

over-zealous conversion of substantial areas of agricultural or aquacultural land, traditionally used for producing staple crops for high value exportable commodities such as coffee or shrimp, is likely to increase the impact of external market forces on local marketing systems and could further exacerbate local nutritional problems (Section 3.4.1).

In reality then, the large scale development of crustacean aquaculture can have considerable social impact. The rural poor frequently depend on a variety of subsistence activities ranging from small scale trading, manufacturing and farming to fishing and gathering, the latter particularly in mangrove areas where building materials, lumber, firewood, charcoal, thatch and a variety of foodstuffs are sought. When large scale shrimp culture moves in, large areas are cleared and local people may be deprived of these traditional resources or access to them. If the soil is later found to be unsuitable, as it often is in cleared mangrove areas, the site is abandoned and the people left with neither employment nor their traditional resources to fall back on. For these reasons alone shrimp farming in the tropics may seem less socially acceptable than farming fish, seaweed, bivalves or crabs.

Often no institution exists to protect a village or community from development. Smith (1984) argued that the allocation of use rights in tropical coastal regions needs to be administered by decentralised institutions, and he claimed that the the lack of decentralisation could be equated directly with lack of effective control over land use, and with environmental deterioration. However, to prevent the rise of local tyrants it would be desirable to maintain public accountability at all levels.

11.2.3 Community relationships

During the initial phases of project implementation some disruption of communities may occur, such as that seen in Ecuador when villagers from highlands migrated to work on the construction of the coastal shrimp ponds. Although much of the initial construction was mechanized, labourers are required almost continuously since 80% repair and reconstruction is necessary every four years (CPC 1989). Similarly, the problems that occurred in the Taiwanese shrimp farming industry during the late 1980s are expected to lead to a greater involvement of Taiwanese technicians in overseas projects in south-east Asia and the Americas.

A large project may employ a significant number of local people at some time or another but it may also decrease opportunities for alternative employment in the region. One result is wage suppression and increased dependence of tenants and smallholders on landlords and project facilities. Beneficial smallholder programmes have been implemented alongside large projects but there is a danger of tying the small holder restrictively into the main company which may, for example, be the only source of seedstock supply or market outlet (Section 9.5.3).

11.2.4 Integration

Prime examples of vertical integration in crustacean aquaculture are to be seen in Taiwan and Thailand, where as a first step in the integration process local fishermen provide 'seed' crustaceans to new farms. The evolution of the 'semillero' industry of Ecuador is perhaps representative. Fishermen in Goa, who traditionally fished post-larval prawns and shrimp from salt works intakes and irrigation sluices for paste manufacture, benefited when the Goan state government developed 500 farms in the area. The fishermen were able to get a better price for their catch by selling them as seed to farmers. Another instance where the advent of crustacean farming brought advantages to local fishermen arose in Louisiana, where unpredictable wild crayfish catches had meant that no solid markets could be established. When large numbers of crayfish were cultivated in ponds the increased reliability of production coupled with the earlier harvesting time allowed the marketing to be improved.

Participation in a new industry, however, may also mean that debt is incurred because of the need to borrow to purchase materials, feeds, fertilizers, equipment or medication. Servicing the loan becomes especially difficult if yields do not come up to expectation, or if there are crop failures while the necessary experience is being gained (Section 10.3).

11.2.5 Customs and conflicts

Quite apart from the impact of aquaculture on societies, local customs can have a strong impact on aquaculture. Severe problems might be created for the unwary investor who did not know about religious prohibitions against killing or eating animals (Shang 1990), castes and status rulings affecting labour and activity, demand fluctuations caused by major festivals, and even anti-aquaculture lobbies (Anon. 1990c). Protection of investment may be difficult in areas where pacifist religions predominate or where threatening someone with a weapon is illegal (Chamberlain 1985). Crustacean aquaculture may also affect, sometimes advantageously, the traditional roles of men and women in the community and in the household. This too needs to be recognised at the planning stage. For example, in Spain the new crayfish industry that developed in the early 1980s changed the lives of many women who were able to add to the family income by making nets and traps, and by selling gear and souvenirs (Lorena 1983).

A new crustacean farming enterprise will stand a greater chance of success if it interacts peacefully with the surrounding community. Fishermen are frequently among the first to feel threatened by aquaculture developments (Anon. 1989). Several examples of real or potential conflict have been reported and they illustrate the diversity of difficulties that can arise. In Laguna de Bay lake, Philippines, bamboo shrimp and fish culture pens grew so numerous that they hampered navigation and altered the natural water circulation. The farming processes polluted the water so much that phytoplankton and water hyacinth flourished to such an extent that fish and human health began to suffer (Lacanilao 1987). In Kenya imported red swamp crayfish (*P.clarkii*) proliferated and destroyed the vegetation on which a local fishery depended. The same species also caused conflict in Spain and Japan when it destroyed rice paddy bunds with its natural burrowing habits.

In the USA shrimp fishermen objected to the taking of shrimp post-larvae for Louisiana State University research projects (Avault 1989), and it is highly likely that east coast lobster fishermen would react similarly if berried lobsters were taken in any quantity for commercial aquaculture (Aiken & Waddy 1985). Local Indonesian and Philippino fishermen sided with shrimp farmers when large foreign trawlers entered coastal waters to trawl for large shrimp, and as a result these countries banned inshore trawling. The Spanish crayfish industry began after the eel fishery declined because of pollution from agricultural pesticides. *P. clarkii* was stocked by one farmer who subsequently made a profit. Indiscriminate stocking of the crayfish in public waters followed and many people made profits, but friction developed when the fishery became overcrowded.

11.2.6 Expatriate influence

Much aquaculture development in the tropics is undertaken by private entrepreneurs motivated largely by financial considerations. To ensure the smooth implementation and commissioning of large and costly enterprises, expatriate managers and technicians are usually employed. Foreign staff living in alien or isolated communities nearly always experience some form of sociological stress which can adversely affect their work or personal relationships after a time. Groups of staff tend to form cliques and these may be may resented locally and may result in reactions that could jeopardize the security of the project. On the other hand, the expatriate community might unwittingly introduce alien moral and materialistic values to the detriment of the indigenous society.

11.2.7 Summary

In general then, large scale shrimp production makes little or no contribution to local food availability, and the opportunities for many coastal communities to survive change in a world that may be entering a period of climatic and ecological uncertainty are being severely eroded.

Superficially it might appear that improvements could be made with a substantial

Plate 11.2 Discussing the value of a fortuitous catch of aquatic insect larvae as food for juvenile shrimp in northern China.

commitment by the authorities to improving extension services and credit availability to local people, but this often results in dealings made primarily or solely with educated and well-financed businesses or landowners in the area. To gain the wider objectives during project implementation – including improvements to local nutrition and income (Lambert 1986) – it is first necessary to identify the social aspirations and expectations of the communities involved. Experience shows that in many tropical regions this should lead to labour intensive, low cost technology that is far more in sympathy with local social structures and environmental resources than large imported turnkey farm businesses enclosing wide areas of land. That is not to say that large intensive or semi-intensive farms do not have their place, but it does highlight the need for public accountability by those responsible for allowing their implementation.

Now that crustacean farming is established, increased accountability could go some way towards ensuring that more rural communities profit equitably from it at the family level and are able to manage their own natural coastal resources. Both the central and local institutions involved would then play a greater role in maintaining national social integrity and stability for the future.

11.3 Ecological and environmental impact

Small-scale, artisanal crustacean culture generally has little impact on its surroundings. In contrast modern developments, particularly in the shrimp and crayfish industries, have had significant impact both in ecological and environmental terms. Conservationists frequently argue that any change brought about by the establishment of a new industry, or the intensification of an existing enterprise, is undesirable. However, it is important to recognize that clear distinctions can be made between the harmful, disruptive consequences of crustacean farming and the other, often marginal changes that arise whenever species interact with their surroundings. Failure to recognize or objectively assess such differences in impact during the planning or feasibility study phases of projects can unduly stifle legitimate and worthwhile developments. The most often reported detrimental impacts caused by crustacean farming are:

Ecological	*Environmental*
Pressure on natural stocks	Site clearance
Transplantations	Water supplies
Diseases	Effluents
Disease treatments	

In addition, climatic changes are expected to have both positive and negative impacts on crustacean aquaculture worldwide.

11.4 Ecological impact

11.4.1 Pressure on natural stocks

Several kinds of activity in coastal regions threaten wild stocks of crustaceans, and in four cases in particular this is exacerbated by the increasing demand for wild broodstock or juveniles to support aquaculture.

11.4.1.1 Broodstocks

Shrimp of larger, and hence more valuable, size than those produced by farms are increasingly sought by fishermen. This places significant pressure on many stocks and it is a sad fact that in several countries the resource is used extravagantly. In the Philippines and Indonesia, large, breeding shrimp occur in inshore waters and are captured for both fishery and aquaculture. Inshore trawling for shrimp once reached such intensity that foreign vessels were banned, to protect the local industries. In Japan, the established fishery for large, live shrimp provides a convenient source of broodstock for hatcheries, but only 10–50% of the selected females spawn (Juario & Benitez 1988). The demand for broodstock *P. monodon* in Taiwan is substantially increased by the wasteful use of nauplii. This occurs partly because low survival rates result from the continued use of *Skeletonema*, which is a poor larvae food (Kurmaly *et al.* 1989), and partly because several broods may be discarded whenever there are not enough nauplii to fill a typical, large Taiwanese rearing tank (Section 7.2.4).

11.4.1.2 Habitat

Destruction of nursery grounds or other habitat through industrial or agricultural pollution, pond construction or other change of use (mineral extraction, harbour schemes), can have catastrophic effects on crustacean populations. In the case of shrimp, which are primarily an annual crop, loss of nursery habitat could cause the sudden collapse of stocks (García & Le Reste 1981; Gulland & Rothschild 1984). The demise of the European crayfish stocks over the past 100 years was due to increased modification and pollution of waterways as well as to plague fungus (Hogger 1988).

11.4.1.3 Incidental fishing

Substantial losses of post-larvae and juveniles are caused when they are taken incidentally into salt evaporation ponds or, in the case of freshwater prawns, sluice controls for irrigation. Of the three examples given above, the pressures on habitat are likely to cause the greatest risks to shrimp, crayfish and freshwater prawn species. At present broodstock overfishing is probably a greater threat to spiny lobsters than many other species.

11.4.2 Transplantations

Much reliance is nowadays placed on the ability to chose the species to be farmed regardless of its natural origin. The advantages such choice brings are that markets may be already established and the product known; the farmer can select the best species for overall commercial gain; and the culture requirements will be known and stock easily obtained. The disadvantages, however, are that escapes are inevitable and could result in the alien species competing with, and possibly displacing, ecological homologues (animals occupying the same habitat in the home environment). The risk of the insidious introduction of alien virus diseases into local fished stocks through such escapes and from hatchery effluents cannot be discounted. Crayfish transplantation and escapes in particular have caused a number of other disruptions to local environments, including the destruction of weed beds, levees and bunds which has sometimes resulted in fishery or crop losses. Interbreeding may occur in the wild but seems unlikely.

Concern has been expressed that repeated breeding from the offspring of a small number of parents originally imported might adversely affect culture performance (Hedgecock 1987). The impact seems to be species specific with greater risks of detrimental affects occurring among penaeid shrimp than with *Macrobrachium* (Section 12.10).

11.4.3 Disease transmission

Several serious diseases of crustacea have been spread through transplantation of species for research and commercial gain. There are about 30 viruses known in crustaceans; six cause serious problems in cultured penaeids and three in blue crabs (*C. sapidus*), but as yet none have been identified as causing serious problems in *Macrobrachium*, lobsters or spiny lobsters (Sindermann & Lightner 1988).

Shrimp virus diseases are known to have been transferred throughout the Americas and between America and the South Pacific (Bell & Lightner 1983; Chamberlain 1988a). The Chinese white shrimp (*P. chinensis*) seems a good candidate for culture in sub-tropical and some warm temperate zones because it grows well and has greater tolerance to lower temperatures than many tropical shrimp. It has already been transplanted to New Zealand, Europe and the USA for culture trials. It is worth noting, however, that it is susceptible to HPV, a virus disease of penaeids in the USA that is not endemic in south-east Asia (Chamberlain 1988a). Similarly, the Australian red claw crayfish (*C. quadricarinatus*), although potentially suitable for culture in tropical regions outside Australia, is susceptible to crayfish plague fungus. This fungus has been widely spread throughout Europe by introductions of North American crayfish. Gaffkaemia, a bacterial infection of lobsters, is not endemic outside North America and strict measures are necessary to prevent European stocks becoming infected.

11.4.4 Disease treatments

Managers of many commercial shrimp hatcheries are often under such pressure to meet production targets that they resort to the routine use of prophylactic doses of antibiotics in larvae culture vessels. One attraction of antibiotics, or indeed any chemotherapeutant, is that they can help overcome bad husbandry. A typical scenario in an Ecuadorian hatchery would be to put a low dose of a broad spectrum antibiotic into the larvae culture water during the protozoeal stages of early season cultures; to increase the concentration as the resistance of the disease organism rises; to change to a new antibiotic, and later to combinations of antibiotics as the season advances, until all the treatments become ineffective. At that stage the whole hatchery is closed down, chlorinated and dried out (Chamberlain 1988a).

It is now widely recognised that this practice may give rise to more disease-susceptible post-larvae because many will have developed from weak larvae that have been artificially protected from disease by the antibiotics. However, there remains a real danger that continued use of antibiotics will encourage the development of lasting resistance in the pathogens. Many hatchery managers do not have the time, facilities or money to identify the causative organism before choosing a course of treatment. It is ironic that, even though broad spectrum antibiotics are used, they may not be effective against the disease, especially if it is caused by a virus.

Many of the antibiotics used prophylactically or curatively are also used in the treatment of human diseases. In several countries hatchery effluents, which may contain antibiotics, resistant bacteria or virulent disease organisms, are freely discharged on to beaches where they present a potential hazard to neighbouring hatcheries and possibly to the public (Brown 1989).

The use of therapeutic and maturation compounds is widely practised in both commercial and research units, but very few chemical agents are approved, for example by the US Food and Drug Administration, for use in animals destined for human consumption. One growth promoting agent – human growth hormone – seems effective in lobsters (Charmantier *et al.* 1989), but care should be taken at the project planning stage that the intended and potential markets will be ready to accept the cultured product should the use of such agents become necessary during production.

11.5 Environmental impact

11.5.1 Site clearance

Often the first sites to be considered for crustacean farming are those popularly regarded as wastelands. Mangrove zones were once viewed in this light and, being

Plate 11.3 A sunken and abandoned Taiwanese house displaying the effects of land subsidence caused by excessive abstraction of ground water for shrimp culture.

near to water supplies, vast areas were destroyed in the construction of aquaculture ponds. It is now widely accepted that destruction of mangroves may destroy not only fish and crustacean nursery grounds but also natural flood and storm protection barriers (Sakthivel 1985). Not all mangrove habitats are important in these respects, however, and proper development may sometimes be prevented by dogmatic adherence to conservationist principles (New & Rabanal 1985).

In some areas the conversion of salt flats or unprofitable salinas, rice or sugar fields to ponds may be preferable as construction is easier, mangrove nursery and common community resources are not destroyed, and the soil and water quality is often better (Singh 1987). Alternatively, ponds may be dug inland and a pumped water supply installed. In the longer term, environmental and sociological changes must be expected when new roads are built to service an isolated project. These can open up an area to change through development of peripheral service businesses and, later on, of permanent communities.

11.5.2 Water supplies

Ground water abstraction may cause saltwater intrusions into domestic fresh water sources. Excessive extraction of ground water to lower salinity in shrimp ponds occurred over a wide area in Taiwan and eventually led to land subsidence. Partly in response to this major impact, an Environmental Protection Agency was established.

The salinization of soils used in marine and brackish water shrimp ponds may prevent them being reconverted for agriculture for several years if the project becomes uneconomic or fails for other reasons (Kayasseh & Schenck 1989).

11.5.3 Effluents

Effluents can take the form of discharges from ponds and hatcheries or water passing through culture pens and cages. When this occurs in confined areas or where mixing and dispersion is reduced, there are increased risks of eutrophication, toxic algal blooms and disease transfer between farms and hatcheries. In addition, specific disease resistant organisms or antibiotic-resistant bacteria may accumulate (Brown 1989). Discharges of saline water from inland *Macrobrachium* hatcheries in Thailand into public water canals (klongs) may also have undesirable effects.

11.5.4 Climate

Unusual climatic variations are apparent in many areas of the world today. They pose threats ranging from increased flooding risks in Bangladesh to failure of rains causing higher salinities in ponds, for example in Burma, Sri Lanka and Kenya (Chamberlain 1988b), or droughts affecting crayfish production in Spain (Lorena 1986). Some of the effects may be beneficial, especially when warmer pond temperatures allow longer growing seasons. Unpredictability, however, is likely to be the most significant factor in terms of project appraisal. It will be impossible, for example, to predict changes in salinity patterns that may occur in estuaries and mangrove areas and their effect on the distribution and availability of seed and broodstocks. Increased engineering (and pumping) costs may also be incurred if correct salinities and exchange rates are to be maintained. A wise precaution, already being taken in Taiwan, is to investigate the temperature and salinity responses of a range of cultivable species other than those currently farmed, and to pay particular attention to the existence of different physiological strains of valuable species that might perform well under different salinity and temperature regimes (Section 4.7).

11.6 Institutional interactions

Planners and assessors of crustacean farming investments or development strategies will quickly become aware of the necessity to interact productively with numerous institutions, national and international, commercial and private, if their objectives are to be achieved. Aspects of some of these interactions are discussed here in the light of recent events, to give the reader a feel for the variety and scope of problems and benefits that may arise. For convenience the aspects are grouped under financial, managerial and legislative considerations.

11.6.1 Financial considerations

11.6.1.1 Land/water costs

It was once thought that waste land unsuitable for agriculture would be suitable for modern aquaculture developments, but this is not necessarily so. Fertile clay or loam soils are needed for semi-intensive and intensive cultures and usually have a high value. Water (especially freshwater) supplies are also competitively sought. Several developing nations have positive commitments to crustacean aquaculture and have made land available for leasing to local fishermen and farmers, frequently in conjunction with supportive extension and training schemes. In the tropics this is often marginal land formerly of low value and may only be suitable for the less intensive farming strategies. Attempts to intensify production from such areas may yield unreliable harvests and will not make best use of the land.

In contrast, most coastal land in developed countries is difficult to obtain and expensive because it is competitively used and is usually considered to be a public amenity. Sites near unpolluted water supplies and those with access to geothermal or industrial waste heat can be particularly valuable. The use of heated supplies to extend the growing season or make possible the culture of warm water species in more temperate climates has been discussed in Section 5.3. The point to be considered during the planning of such projects is the financial undesirability of having capital equipment for heat transfer, heat exchange or back-up lying idle for large parts of the year. The same consideration applies to other installations, such as processing plant, which may be under-utilized during the growing season (Section 10.9.5).

In the context of using industrial 'waste' heat for aquaculture, it cannot be too strongly emphasized that, firstly, the temperature, flow and chemical composition of the heated medium will be manipulated to suit the primary industry's purposes, not those of the culture enterprise; and secondly, many industrialists may tolerate the presence of an aquafarm simply to alleviate public fears of pollution or to project a 'green' image. Thirdly 'waste' heat is not free; it has to be piped, monitored and transferred directly or indirectly to the farm. Finally, and perhaps most revealingly, there is at least one report in the trade press of charges (royalties) suddenly being demanded for 'waste' heat after several years of aquaculture operation (Anon. 1988).

11.6.1.2 Credit/loans

Whether it is the commencement of a new business, the conversion to crustacea from a different crop, or the up-grading of an existing crustacean farm, adequate finance will be vital (Section 10.3). In developing countries long-term loans at reduced interest rates, together with free technical and management training, are helping many to participate in small-scale shrimp culture (Sakthivel 1985). Where there is a lack of local capital, co-operatives eligible for government subsidies may be formed. However, in many countries low interest loans and grants may be difficult to obtain, especially if land or some other form of security is not owned. Too often credit is seen to be the monopoly of big business, and in the west, paradoxically, the minimum size of grant or loan on offer may be too large for the individual or family business (Rosenberry 1983). Some support may be available in a number of developed countries for small to medium sized enterprises too far from the market place to survive unaided. Publicity and advice about the schemes varies from good to poor, and gaining acceptance for high risk projects like crustacean farming will be difficult.

11.6.1.3 Investment and insurance

The high risks inherent in all crustacean farming operations have not deterred entrepreneurs in the past, and crustacean aquaculture has long enjoyed a high degree of 'investor appeal'. Significant or repeated failures, however, will give any industry a bad name and will discourage support, both financial and occasionally from publicly funded research programmes. The problems are not always technical and commonly include over or under-capitalization, over-optimistic expectations of scale and timing of returns, and no allowance made for an adequate learning period or for crop failures. The need to attract investors in developing countries has led some governments to grant worthwhile tax holidays during the early years of a project. In Brazil, however, the high level subsidies made available for shrimp farms were reported to have been made at the expense of the funding of research and development so vital to the future success of the industry (Chamberlain 1988b).

Protection of investment by appropriate insurance is common practice in many established industrial activities. In crustacean farming and aquaculture, however, it is rare. Most firms and many companies either do not insure or cannot afford insurance. By nature all crustacean farmers are optimists, but reputable insurance companies exist and will offer reasonable terms provided they are satisfied that adequate working practices and precautions are taken to prevent losses. Usually insurers will undertake a risk management exercise (Hatch *et al.* 1987), in which they visit farms to see where and how severe the weaknesses are (Secretan 1986; 1988). In view of the current climatic instabilities, perhaps all should be encouraged to insure! However, insurance against loss of stock in the fish farming industry is reported to be reaching the point of non-viability following increased storm and disease outbreaks (Anon. 1990a). Shrimp crop insurance is available (Anon. 1990b) but it is likely that in future only the best husbandry practices and best risk management programmes will be insurable at acceptable rates.

11.6.1.4 Markets and production costs

Cultured shrimp probably contribute little to the diets of the poor since the majority are more profitably sold to Japan, USA and Europe. Freshwater prawns, crayfish and crabs, together with some shrimp, may be sold for good prices in city markets in the country of origin, particularly where there is a tourist trade. Most shrimp farmers, however, whether they be smallholders or large corporate enterprises, are at the mercy of world commodity markets and fluctuations in international exchange rates. They are vulnerable regardless of how efficient they are in production. The value of the crop may also be affected by temporary over-supply of a specific size group, as was the case with medium sized shrimp in the late 1980s, or of a specific market. Examples where over-production caused dramatic price decreases during the 1980s include freshwater prawns in Thailand and hatchery-reared shrimp post-larvae in Taiwan.

When profit margins are eroded by a decline in market value of the crop or an increase in interest rates or running costs, the intensive farm with its narrow profit margins will lose profitability first. The extensive shrimp farm run at a low productivity rate may also suffer, and it may be only the semi-intensive farms that combine moderate productivity (yields >1 t ha^{-1} yr^{-1}) with adequate profit margins that can survive (Section 10.8).

The future costs of crustacean feeds may rise in the face of increasing competition for world fishmeal supplies (Pike 1989). It has been estimated that aquaculture will use 15–17% of world supply by the year 2000 (New & Wijkstrom 1990). The increases in feed costs will mainly affect intensive and semi-intensive producers and Wijkstrom & New (1989) recommend that increased attention should be given to the prospects for ranching. However, improvements in fish flesh stripping are making fish-meal production more efficient while new low-temperature processing promises to provide diets that are more effectively assimilated and cause less nitrogen to be excreted. Such diets may considerably reduce pollution levels in the ponds and farm effluent. Investments in value-added crustacean products are also likely to be an advantage as feed costs rise, but their success requires the creation of an individual market image (Section 3.2.3).

11.6.2 Managerial considerations

The rapidity with which crustacean culture has evolved throughout the world is unprecedented. Entrepreneural enthusiasm has combined traditional skills with modern research results to produce viable culture technologies that, when applied in developing countries, can satisfy national goals of generating export revenue and increasing employment. The new technologies are being extended to traditional farmers and fishermen with the aid of demonstration shrimp hatcheries and production units that have been established both commercially and with national or overseas government aid. In a number of areas, however, events appear to have moved too quickly and have revealed what Smith (1984) described as 'a lack of institutional preparedness'.

For example, the uncontrolled encroachment of Taiwanese shrimp farmers on to public land, and their profligate use of ground water as the development of the industry peaked, are well documented instances. In Laguna de Bay in the Philippines, no agencies stopped the extensive construction of fish and shrimp pens that eventually led to severe eutrophication and restricted navigation in the lake. In Great Britain uncontrolled importation of signal crayfish was, predictably, followed by outbreaks of crayfish plague fungus in stocks of the native white-footed crayfish. Throughout Asia there remain areas where training in both technical and management skills and extension services are insufficient to ensure that maximum benefits will be obtained from crustacean culture. Great reliance is still placed on technological packages from overseas, which may not always be sustainable in the long term.

11.6.2.1 Extension services

Many governments and some of the larger projects and feed companies recognise the value of providing rural extension services to interface between the policy makers, researchers or company sales agents and the farmer. To be successful, systematic transfer of information must occur in both directions so that the farmer is aware of new developments and the donors gain feedback concerning the impact and results of their ideas or products. Many extension services, however, do not pay sufficient attention to acquiring feedback information and the exchange becomes unbalanced and unsatisfactory for all concerned (Blakely & Hrusa 1989). Not all societies or communities react in the same way when confronted with new ideas. Industrialized societies seem able to accept and assimilate new ideas more rapidly than traditional societies, where new ideas may be perceived as threatening stability.

The approach adopted by the extension agent will differ according to whether the task is to increase public awareness of a farming opportunity or to encourage the adoption of improved management or a new technique in an existing practice. The approach must be in sympathy with the state of aquaculture development within the locality and properly targeted so that some sections of the community are not ignored in favour of others.

11.6.2.2 Consultants/researchers

The services of skilled specialists may be engaged at any or all stages of project implementation from the initial surveys to the completed application of the culture technology (Section 9.5.2). Few specialists deal solely with crustacean ventures and will typically return to more conventional activities after each appointment. Although the specialists will have considerable expertise in their own field, they may not have much expertise, or experience of applying it, in crustacean farming projects. Since their work is likely to be subject to commercial confidences, they are prevented from sharing valuable experiences. This reduces the opportunities for others to learn from past mistakes (Huguenin & Colt 1986).

Progress is further impeded because there is insufficient feedback from pilot or commercial operations to those in research and development (R&D). The vast majority of farms cannot support or pay for R&D programmes and would not, in any case, wish to reveal their results and difficulties to others for fear competitors might take advantage of the knowledge. The result is 'research by crisis', and is hardly a satisfactory way of supporting a farming industry. Few scientists see time spent on farms trying to identify the real problems as the kind of productive research that will improve their career prospects. So improving communication between extension, technical and aid workers acting as intermediaries between the farmers and the researchers, is of paramount importance.

Unethical practices exist in the crustacean culture business as in many other businesses. There are reports that some international consultancy groups have actively recruited skilled and even semi-skilled hatchery operators and hired them back to the developing industry at exorbitant rates. The potential farmer or investor should of course be wary of organisations and individuals making false claims and unrealistic proposals (Rosenberry 1984; 1985). It is always wise to check the credentials and track record of advisers before they are employed, and to note to what extent they are covered by professional indemnity insurance (Secretan 1988).

11.6.2.3 Managers

There are numerous reasons why projects fail to meet targets or come up to expectations, but many are simply due to poor or inexperienced management. Common problems include:

(1) Ineptitude at anticipating requirements during project implementation or during operation (Huguenin & Colt 1986);
(2) Lack of appreciation of start-up problems;
(3) Failure to look after electrodes and properly standardize chemicals used in making water quality measurements;

(4) Over-confidence in computerized pond management and alarm systems (EIFAC 1986; Hadley 1989);
(5) Ignoring the advice of hired specialists;
(6) Paying insufficient attention to staff needs and expectations;
(7) Failure to modify jobs in the light of experience (Section 8.7.2)
(8) Failure to motivate staff and provide suitable incentives (Haughton 1990);
(9) Inadequate liaison with surrounding businesses, especially those involving pesticide spray programmes against crop pests and for the eradication of bilharzia and malaria;
(10) Lack of awareness of market forces and the desirability to diversify to new species.

11.6.3 Legislative considerations

Many laws governing aquaculture activities were not designed for aquaculture and those responsible for project feasibility studies and assessments should investigate fully the relevant national and local legislative implications at an early stage. Often the laws applied relate to agriculture and are not suitable for aquaculture, let alone crustacean farming. Effluent laws, for example, may not distinguish between biodegradable fish farm discharges and non or slowly degradable chemical discharges. An unnecessary burden of compliance may result (Howarth 1990).

In some countries no laws may be available to protect investment in a crustacean farm or hatchery. Spanish law states that the first two metres of land above the high water mark are public land, so if the Guadalquivir river floods, the entire delta becomes public land and no private aquaculture ventures can be sustained (Lorena 1986).

11.6.3.1 Ownership

The clear delimitation and subsequent protection of investment is not always as straightforward in developing as in developed countries. Moslem inheritance laws for example decree that land is divided among descendents, but since a pond cannot be divided multiple ownership frequently becomes a constraint to development (Shang 1990). In general terms, the ownership of inland freshwater ponds is readily defined and, although limited, the prospects for large-scale development are sustainable. On the coast, however, there is often no security of tenure and one frequently needs political contacts or bribes to get the necessary permits to build ponds. Family ties and land ownership may seem excessively complex to the foreign investor, who probably will not be able to own land directly in any case. Legislation for the protection of traditional rights and custom lands may or may not exist, but the existence of traditional rights will doubtless be perceived as a reality by local inhabitants who will justifiably expect them to be honoured.

Crustacean ranching involves the release of cultured stock into natural bodies of water and presents particular problems in terms of legal ownership and policing of the site to protect the investment. Attempts to ranch crustaceans are only likely to be of commercial interest in areas where legislation permits leasing or ownership of the submerged land, and where the majority of the stock can be maintained within a defined boundary or can be indisputably distinguished from wild stocks. As with pond cultures, marine or brackish water projects are less easily upheld from a legal viewpoint than inland waters.

The production of juveniles for national, non-commercial restocking and ranching programmes may of course be conducted commercially. Indeed, the maintenance of a hatchery and a programme of releases may be imposed as conditions when granting permission for certain kinds of waste disposal, the justification being to compensate for losses due to spoiled habitat or, if the waste is non-toxic and suitably configured, to create new or extend existing fishable resources for the public good.

11.6.3.2 Protection or constraint?

The lack of a national policy in the US and elsewhere during the early days of aquaculture development resulted in difficulty in getting permits and dedicated land. Many conflicts arose with environmental and recreational interests since, while all parties wanted protection, none wanted constraint. The duty of government is to reconcile the expectations of the aquaculturists with those of other commercial, amenity and national interests using natural resources. Sometimes the resulting laws become so complex or numerous that not only is compliance difficult but it may even be difficult to get an enterprise going (Idyll 1986).

In developed nations like the UK (Howarth 1990), New Zealand and the USA there is a plethora of departments and organisations that become involved with aquaculture applications and project implementation. Together they present formidable difficulties for the newcomer who has often to make separate applications to each bureaucracy dealing with land use, water abstraction, disease and quarantine measures, effluent discharges, building plans and health and safety, to name but a few. Considerable improvements in co-ordination between all those involved are needed. In addition there will be objections from private interests and conservation and amenity groups, many of whom regard themselves as guardians of the national heritage and who see it as their duty to object to any change whatsoever in the environment. In Britain, attempts have even been made to turn such confusion to personal advantage by using a crayfish farm proposal as a means of obtaining planning permission for a house in the countryside!

Yet another area where laws create conflicting perceptions of protection or constraint between consumer and farmer is that of chemical treatments. The long history of unchallenged use of a chemical does not indicate that the substance is safe or has been properly registered by the US Food and Drug Administration (or other national authority) for use on crustaceans intended for human consumption. Indeed, compounds widely used in fish culture may even be toxic or harmful to crustaceans, regardless of whether they have been approved. It is not easy to obtain approval to use a chemical in the culture of an animal destined for the table. It may take six years and US$4 million to get FDA approval for a therapeutant or other additive. If the compound has already been approved for fish, approval for crustaceans may still cost US$1 million and take three to four years to process (Idyll 1986). The current status of FDA registrations affecting crustacean farmers is described by Schnick (1988).

11.6.3.3 Positive attitudes and legislation

In some countries, notably Taiwan (Lee 1988), Thailand (Akrasanee 1988), Spain (Santaella 1989) and the Philippines (Idyll 1986), aquaculture is given preferential status in a number of areas such as investment, grants and loans and tax holiday incentives. In the Philippines valuable incentives were granted to agriculture and fishery industries (which includes aquaculture) because of the greater risks to investment due to the vagaries of the weather and the high incidence of spoilage and pest damage. Good publicity and publications were distributed to promote awareness of documentary requirements to start aquaculture operations, and much of the income from the legislation goes back to the industry as funding for extension and research support.

New laws or improvements to existing legislation may be implemented in the light of aquaculture developments or may affect aquaculturists even though they may be implemented primarily for other reasons. For example, in many Indian coastal states laws exist to extend and protect agricultural land by preventing the ingress of seawater. To this end, the construction of bunds not only effectively prevents shrimp culture but in one case led to the destruction of a fishery for *Macrobrachium*. This impounded land eventually proved to be inadequate for profitable rice production and the law had to be amended to allow brackish water aquaculture development (Sakthivel 1985). Laws to curb the importation of disease organisms with live shrimp, and the over-exploitation of natural stocks of large broodstock *P. monodon*, led to smuggling between countries like Malaysia, Thailand, Taiwan and the Philippines (Wickins 1986), and prices of up to US$1800 for one live female have been reported (Chiang & Liao 1985).

The Ecuadorian government attempted to reduce the country's dependence on US markets by introducing exchange control and import restriction policies. While the broad objective had some merit, the effect was to cause hardship to many shrimp farmers (NOAA 1988). In other countries, such as Mexico, shrimp culture permits were reserved, at least up to 1981, for fishermen's co-operatives, so inhibiting applications from industrialists. Successful shrimp farms are almost invariably operated by private investors, as are most hatcheries, and as a consequence Mexico has lagged behind in shrimp culture. Recently, however, the Mexican government initiated a national plan for fisheries development and passed legislation that permitted private sector shrimp farming and 49% foreign ownership of shrimp farms (Rosenberry 1990). As a result, in 1989 104 farms containing 6513 ha of ponds produced an average yield of 582 kg ha^{-1} (Garmendia & Nuñez 1990).

11.7 References

Aiken D.E. & Waddy S.L., (1985) Production of seed stock for lobster culture. *Aquaculture*, **44** (2) 103–14.

Akrasanee N. (1988) Investment outlook for shrimp farming. In *Shrimp '88, Conference proceedings*, 26–28 Jan. 1988, pp. 186–90. Bangkok, Thailand. Infofish, Kuala Lumpur, Malaysia.

Anon. (1988) UK power plant farm closes: pioneer project hit by royalties demand. *Fish Farming International*, **15** (6) 1, 5.

Anon. (1989) Shrimp pen farms fail in India. *Fish Farming International*, **16** (11) 12.

Anon. (1990a) Risks and claims on the upward path. *Fish Farmer*, **13** (2) 9.

Anon. (1990b) Shrimp crop insurance. *Fish Farmer*, **13** (2) 31.

Anon. (1990c) Monks want ban. *Fish Farming International*, **17** (4) 44.

Avault J.W. Jr. (1989) Social/political aspects of aquaculture. *Aquaculture Magazine*, **15** (4) 70–3.

Bailey C. (1988) The social consequences of tropical shrimp mariculture development. *Ocean and Shoreline Management*, **11**, 31–44.

Bailey C. & Skladany M. (in prep.). Aquaculture development in tropical Asia: the path untaken. (Seen in draft).

Bell T.A. & Lightner D.V. (1983) The penaeid shrimp species affected and known geographic distribution of IHHN virus. In *Proc. 1st Int. Conf. on Warm Water* Aquaculture – Crustacea, 9–11 Feb. 1983 (Ed. by G.L. Rogers, R. Day & A. Lim), pp. 280–90. Brigham Young University, Hawaii.

Blakely D.R. & Hrusa C.T. (1989) *Inland aquaculture development handbook*. Fishing News Books, Blackwell Scientific Publications, Oxford.

Brown J.H. (1989) Antibiotics: their use and abuse in aquaculture. *World Aquaculture*, **20** (2) 34–5, 38–9, 42–3.

Chamberlain G.W. (Ed.) (1985) Sociological factors limit yields from Asian farms. *Coastal Aquaculture*, **2** (4) 11–12.

Chamberlain G.W. (Ed.) (1988a) Disease control. *Coastal Aquaculture*, **5** (1) 6–9.

Chamberlain G.W. (Ed.) (1988b) Shrimp culture news from around the world. *Coastal Aquaculture*, **5** (1) 13–17.

Charmantier G., Charmantier-Daures M. & Aiken D.E. (1989) Accelerating lobster growth with human growth hormone. *World Aquaculture*, **20** (2) 52–3.

Chiang P. & Liao I.C. (1985) The practice of Grass prawn (*Penaeus monodon*) culture in Taiwan from 1968 to 1984. *J. World Maricult. Soc.*, **16**, 297–315.

CPC (1989) Libro blanco del camarón. *Cámara de productores de camarón*, May 1989, Guayaquil, Ecuador.

EIFAC (1986) Flow-through and recirculation systems. Report of the working group on terminology, format and units of measurement. *EIFAC Tech. Pap. (49)*.

García S. & Le Reste L. (1981) Life cycles, dynamics, exploitation and management of coastal penaeid shrimp stocks. *FAO Fish. Tech. Pap.* (203).

Garmendia E.A. & Nuñez A. L. G-S. (1990) Shrimp culture status in Mexico 1989. *Abstracts from World Aquaculture 90*, 10–14 June 1990, Halifax, Nova Scotia.

Gulland J.A. & Rothschild B.J. (1984) *Penaeid shrimps – their biology and management*. Fishing News Books, Blackwell Scientific Publications, Oxford.

Hadley P.R. (1989) Instrumentation and microcomputors. *World Aquaculture*, **20** (2) 9.

Hatch U., Sindelar S., Rouse D. & Perez H. (1987) Demonstrating the use of risk programming for aquacultural farm management: the case of penaeid shrimp in Panama. *J. World Aquacult. Soc.*, **18** (4) 260–9.

Haughton P. (1990) Spread a little happiness among the 'wage slaves'. *Fish Farmer*, **13** (2) 63–4.

Hayashi T. (1984) Manpower in the diffusion of technology. *United Nations University Newsletter*, **8** (1) 8.

Hedgecock D. (1987) Population genetic bases for improving cultured crustaceans. *Proc. world symp. on selection, hybridization, and genetic engineering in aquaculture*, 27–30 May 1986, **1**, 37–58. Bordeaux.

Hogger J.B. (1988) Ecology, population biology and behaviour. In *Freshwater crayfish, biology, management and exploitation* (Ed. by D.M. Holdich & R.S. Lowery), pp. 114–44. Croom Helm, London.

Howarth W. (1990) *The law of aquaculture*. Fishing News Books, Blackwell Scientific Publications, Oxford.

Huguenin J.E. & Colt J. (1986) Application of aquacultural technology. In *Realism in Aquaculture: Achievements, Constraints, Perspectives* (Ed. by M. Bilio, H. Rosenthal & C.J. Sindermann), pp. 495–516. European Aquaculture Society, Bredene, Belgium.

Idyll C.P. (1986) Aquaculture legislation and regulation in selected countries. In *Realism in Aquaculture: Achievements, Constraints, Perspectives* (Ed. by M. Bilio, H. Rosenthal & C.J. Sindermann), pp. 545–63. European Aquaculture Society, Bredene, Belgium.

Juario J.V. & Benitez L.V. (1988) Perspectives in aquaculture development in south-east Asia and Japan. *Proc. seminar on aquaculture development in south-east Asia*, 8–12 Sept. 1987, Iloilo City, Philippines. S.E. Asian Fisheries Development Center, Iloilo, Philippines.

Kayasseh M. & Schenck C. (1989) Reclaimation of saline soils using calcium sulphate residues from the titanium industry. *Ambio*, **18** (2) 124–7.

Kurmaly K., Jones D.A., Yule A.B. & East J. (1989) Comparative analysis of the growth and survival of *Penaeus monodon* (Fabricius) larvae, from protozoea 1 to postlarva 1, on live feeds, artificial diets and on combinations of both. *Aquaculture*, **81** (1) 27–45.

Lacanilao F. (1987) Managing Laguna lake for the small fishermen. *Asian Aquaculture*, **9** (3) 3–4, 8, 12.

Lambert G. (1986) Social indicators of the aquaculture-based social laboratory project. *Asian Aquaculture*, **8** (3) 6–8.

Lee J-C. (1988) Government policies and support for shrimp-farming. In *Shrimp '88, Conference proceedings*, 26–28 Jan. 1988, pp. 175–81. Bangkok, Thailand. Infofish, Kuala Lumpur, Malaysia.

Lorena A.S.H. (1983) Socio-economic aspects of the crawfish industry in Spain. In *Freshwater crayfish 5* (Ed. by C.R. Goldman), pp. 552–4. AVI Publishing Co., Westport, Connecticut.

Lorena A.S.H. (1986) The status of the *Procambarus clarkii* population in Spain. In *Freshwater crayfish 6* (Ed. by P. Brinck), pp. 131–3. Int. Assoc. Astacology, Lund, Sweden.

New M.B. & Rabanal H.R. (1985) A review of the status of penaeid aquaculture in south-east Asia. In *Second Aust. Nat. Prawn Sem.* (Ed. by P.C. Rothlisberg, B.J. Hill & D.J. Staples), NPS2, pp. 307–26. Cleveland, Australia.

New M.B. & Wijkstrom U.N. (1990) Feed for thought. *World Aquaculture*, **21** (1) 17–19, 22–3.

NOAA. (1988) *Latin American shrimp culture industry, 1986–90*. IFR/88-38, F/IA23:DW, Foreign Fisheries Analysis Branch, US National Marine Fisheries Service.

Pike I. (1989) Fish meal industry should meet farm demands. *Fish Farming International*, **16** (7) 86–8.

Rosenberry R. (1983) Prawn farming in Hawaii. *Aquaculture Digest*, **8** (8) 2–4.

Rosenberry R. (1984) Cornelius Mock. *Aquaculture Digest*, **9** (11) 23–4.

Rosenberry R. (1985) Corny Mock at large. *Aquaculture Digest*, **10** (11) 5–6.

Rosenberry R. (1990) Understanding shrimp farming in Mexico yesterday, today and tomorrow. *Aquaculture Digest*, **15** (7) 5–7.

Sakthivel M. (1985) Shrimp farming – a boon or bane to India. *ICLARM Newsletter*, **8** (3) 9–10.

Santaella E. (1989) Implementation of national aquaculture programmes. In *OECD, Aquaculture: A review of recent experience*, pp. 187–92. OECD Paris.
Secretan P.D. (1986) Risk insurance in aquaculture. In *Realism in Aquaculture: Achievements, Constraints, Perspectives* (Ed. by M. Bilio, H. Rosenthal & C.J. Sindermann), pp. 536–41. European Aquaculture Society, Bredene, Belgium.
Secretan P.D. (1988) Aquaculture insurance. In *Shrimp '88, Conference proceedings*, 26–28 Jan. 1988, pp. 181–6 Bangkok, Thailand. Infofish, Kuala Lumpur, Malaysia.
Schnick R.A. (1988) The impetus to register new therapeutants for aquaculture. *Progv. Fish. Cult.*, **50** (4) 190–96.
Shang Y.C. (1990) Socio-economic constraints of aquaculture in Asia. *World Aquaculture*, **21** (1) 34–5, 42–3.
Sindermann C.J. & Lightner D.V. (Eds.). (1988) Disease diagnosis and control in North American marine aquaculture. *Dev. Aquacult. Fish. Sci.*, **17**, 1–431.
Singh T. (1987) Mangroves and aquaculture – striking a balance. *Infofish International*, (5) 20–22.
Smith I.R. (1984) Social feasibility of coastal aquaculture: packaged technology from above or participatory rural development? Discussion document prepared for a consultation on social feasibility of coastal aquaculture, 26 November–1 December 1984, Madras, India. *ICLARM contribution No. 225* (mimeo).
Weeks P. (1989) *Aquaculture development: an anthropological critique.* Paper presented at the World Aquaculture Society meeting, Aquaculture 89, 12–16 Feb 1989, Los Angeles (mimeo).
Wickins J.F. (1986) Prawn farming today: opportunities, techniques and developments. *Outlook on Agriculture*, **15** (2) 52–60.
Wijkstrom U.N. & New M.B. (1989) Fish for feed: a help or a hinderance to aquaculture in 2000? *Infofish International*, (6) 48–52.

Chapter 12
Technical constraints, research advances and priorities

12.1 Introduction

Scientific research led to the establishment of present day crustacean hatcheries and much of the on-growing technology. It is, however, axiomatic that the extraordinary rate of growth of the crustacean farming industry owes so much to the versatility and innovative skills of technicians, and to managers and technicians alike being driven by commercial pressures to take a number of intuitive technical short-cuts. The lack of scientific knowledge underlying many of the practices that have evolved in this way, only becomes of concern when animal survival, growth or reproductive performance is not as expected, or when production economics become critical in the face of changes brought about by market forces.

Table 12.1 shows the main areas in which technical constraints to progress exist in each of the groups considered in this book, and these are discussed below.

12.2 Penaeid shrimp

Compared to other groups, the technical constraints to the advancement of shrimp culture were, until about 1988, few and relatively minor. To those presently in the industry, however, broodstock availability and the diseases now capable of afflicting all life cycle stages must pose the most serious threats to future expansion.

Wild broodstock are under increasing pressure since they are perceived by hatchery operators as producing a higher and more consistent quality of larvae than pond-reared or artificially matured stock, even though this has not always been scientifically established. Proven techniques are available for broodstock production as well as for the control of maturation, artificial impregnation and spawning of most, if not all, cultured species. Yet until scarcity or cost of wild females forces hatchery owners to produce their own broodstocks, profligate use of the natural resource will continue. Transportation of live females from country to country, together with any diseases that they may be carrying, will thus remain and will further delay efforts to develop commercially satisfactory techniques for large-scale, reliable broodstock production.

Recent research aimed at improving control over broodstock quality and production is centred on:

(1) Induction of spawning in freshly-captured wild ovigerous females (most spawning results from the shock of capture or a change in environmental conditions at the hatchery);
(2) Induction of maturation in pond-reared adults and wild non-ovigerous females (see below);

Table 12.1 Areas in which technical constraints seem likely to affect culture progress (**** = major constraint).

Species/group.	Broodstock	Larvae	Diet	Disease	On-growing	Harvesting
Penaeids	*			**		
Macrobrachium				*	**	**
Crayfish						
European				***	**	***
Australian			**	**	**	*
USA					*	***
Clawed lobsters	**		**		****	
Spiny lobsters	?	****	**		?	
Crabs		**	**		**	**

(3) Rematuration and conditioning of previously spawning wild adults (mainly by ablation) and pond stock (mainly through dietary influences, since they will already have been ablated).

Maturation is widely induced in shrimp by removing glands that secrete gonad inhibiting hormones (Section 2.4). It is possible that the process may also affect hormones involved in the mobilization of food reserves and may lead to reduced egg quality. Investigations are therefore being conducted to find other means of inducing maturation that are less likely to have unwanted side effects.

Many studies involve a search for natural triggers through control of light intensity, wavelength and photoperiod, usually in association with increases in temperature. Positive effects seem to occur more readily in *P. chinensis* which migrates to spawn in response to changing daylength and temperature, and in burrowing species like *P. japonicus*. In some other species (including *P. monodon*) the effects of these triggers are minimal but are slightly enhanced when a penetrable substrate is present to reinforce cyclic behaviour patterns.

Water quality is important and inhibition of maturation by dissolved organic matter, low pH and calcium levels have been indicated (Oka 1967; Ogle 1988). Placing *P. japonicus* in seawater previously exposed to ultra violet irradiation may also induce maturation (Yano & Tanaka 1984). There is no evidence that males stimulate spawning in closed thelycum species. Other research approaches include the injection of hormones and their precursors (Charniaux-Cotton & Payen 1988) and the implantation of ganglia and/or glands from mature lobsters (Yano *et al.* 1988).

Shrimp diseases will become a greater constraint to progress as semi-intensive and intensive farms strive to become more competitive. Outbreaks are increased by the stresses induced when culture densities are increased and when the quality of farm inputs are reduced to save production costs. In the hatchery the first lines of defence are quarantine of imported stocks and hygiene. A code of practice has been prepared (Turner 1988) governing the introduction of marine species but it is considered by some to be somewhat idealistic (Sindermann 1988). Even so, the repeated introduction of disease-carrying shrimp into an area could ultimately pose a serious threat to shrimp fisheries, the livelihood of those dependent on them and perhaps foreign exchange earnings (Section 11.4.2). Destruction of all susceptible stock once infected animals are found is common practice among farmers of mammalian and avian domestic stock, but no financial compensation is available following the destruction of crustacean stock. One big Ecuadorian hatchery decided it was better to live with the viruses endemic in wild broodstock and minimise the risks by improved husbandry, than to keep killing expensive broodstock.

Recently stocks of shrimp (*P. vannamei*) claimed to be free of all known diseases were established on Hawaii (Anon. 1990a) and will be used initially to provide research material. Certification of 'disease free' shrimp, and indeed juvenile crayfish, for shipment abroad is not uncommon but doubts exist as to the validity of some of the claims. Vaccines are now being developed for shrimp (Itami *et al.* 1989) and at least two companies offer vaccines for use in shrimp larvae cultures. One of these vaccines is given in the diet (Jaffa 1990). Their effectiveness under commercial conditions remains to be judged in the light of experience. The lack of approved drugs and chemicals needed to manage shrimp diseases and pests is seen as a constraint to the expansion of the industry in the USA (Williams & Lightner 1988). The registration and legitimate, safe use of antibiotics and hormones in the hatchery are the subject of much concern (Section 11.4.4).

Perhaps the most topical concerns are for the increasing incidence of the reversible conditions known as soft-shell or crinkle-shell and blue-shell that arise during on-growing. The current indications are that soft shelled shrimp result from an inability to store and mobilize calcium and phosphorus properly, while carotenoid metabolism seems suspect in blue shrimp (Baticados 1988) (Section 3.3.1.2). In some cases, notably in Taiwan and the Philippines, occurrence of both conditions has been linked to the increased use of variable quality feed ingredients, the incorporation of substitute feed materials, and to feed formulation changes hastily made in an attempt to survive in an increasingly competitive industry (Sheeks 1989). In other situations, in India, Malaysia and Taiwan for example, the chemical composition of both pond bottom and water is also suspected and may be particularly important when the ponds are used very intensively (Kwei Lin 1989).

The research conducted into pond bottom eco-chemistry in Hawaii (Chamberlain 1988) and Israel (Krom *et al.* 1985) has considerably improved understanding, but results are not always directly transferrable due to differences between ponds even at the same site (Section 8.3.5). Increased intensification of pond use and prolongation of the on-growing period to produce larger shrimp have brought about a corresponding reduction in the time for pond bottom rejuvenation between crops. There is growing evidence that crop and farm failures are arising as a direct result of such practices. This, coupled with pollution from industry, agriculture and shrimp farm effluents, was the main cause of the collapse of Taiwanese shrimp production in 1988. At least two other major producing nations, Ecuador (Guayas estuary) (Aiken 1990) and Thailand (Bight of Bangkok) seem likely to experience significant pollution problems in the near future, which could affect productivity.

In a wider sense the continued study of interactive effects between fertilization, pond water exchange rates and production is important because of the greater potential profitability of extensive and semi-intensive ventures in the face of increasing climatic and market instabilities (Moriarty & Pullin 1990). Additional studies in western, warm temperate regions would also seem worthwhile in the light of increasing interest in the culture of cool water penaeids in the Mediterranean (New 1988), Atlantic France (Clement 1989) and the USA (Main & Fulks 1990). The costs of treating effluent may be worth considering in new projects as environmentally oriented legisla-

Plate 12.1 Deep plastic-lined pond used for intensive shrimp culture in Oman. The crop is being harvested with a seine net and then packed in ice to be loaded into the truck waiting on the embankment. (Photo courtesy of P. Fuke.)

tion increases. The use of effluent lagoons containing filter feeding bivalves (e.g. mussels) seems promising.

Recent advances with potential for additional support of shrimp farming and allied research include the cryopreservation of nauplii (Taylor 1990); the production of inert diets for larvae – spray-dried algae (Biedenbach *et al.* 1990), microcapsules, flakes, flocks and microparticulate formulations; and novel feed attractants (Anon. 1990b). Under some circumstances (e.g. sandy or acid sulphate soil locations) use of plastic pond liners may be worthwhile but their use will probably mean abandoning traditional concepts of pond management. Specific differences will include deeper water (1.5–2 m) and increased responsiveness to water management needs in relation to phytoplankton control. At present some 700 ha of plastic-lined ponds are reported to be in production in Oman and Indonesia (P. Fuke, 1990 pers. comm.) (Plate 12.1) (Section 7.2.6.5).

12.3 *Macrobrachium*

It is widely recognized that the two major constraints to *Macrobrachium* farming centre on its heterogeneous growth and the fact that it yields some 20% less tail meat than penaeid shrimp. The latter presents a marketing and economic constraint while the former limits yield and increases the cost of both on-growing and harvesting techniques (Section 7.3.7). While the extended larval life is only a minor handicap, the requirement for brackish water throughout larval development can be restrictive in terms of site, water transport and storage requirements or the cost of artificial seawater salts. Where salinas remain in operation, sea salt or concentrated brine may be purchased and diluted for hatchery use. Recirculation systems are used to conserve water and increase control over culture conditions in a number of French owned or designed hatcheries (New 1990).

Considerable effort has gone into the study of male heterogeneous growth and has resulted in real prospects for the production of single-sex populations Malecha *et al.* (in prep). Suggest that all-female or mainly all-female broods can be obtained by implanting an androgenic gland from a mature male into a very young female and mating these 'neomales' with normal females. Israeli studies, however, indicate that all-male, pond reared populations provide higher yields, but it is recognized that hand-sexing would not be commercially feasible. Early size grading also affects yield characteristics and the selection of initially fast growers resulted in an increased net income over slower growing prawns during polyculture with tilapia and carp (Karplus *et al.* 1987). The discarding of the slow growers would waste the investment in their nursery phase (New 1990), but even so this seems a more practical approach than the routine removal of the claws which are used to such effect during the establishment of hierarchies (Karplus *et al.* 1989). The production of mono-sex populations through surgical and genetic manipulation would seem the most likely way forward, but the production of all male progeny has yet to be achieved.

Faster growth and increased broodstock productivity following eyestalk ablation has been demonstrated in India with *M. malcomsonii*. Ablated males increased growth by 36%, females by 41%, while moult frequency increased by 24% and the number of eggs per brood rose 31%. A higher percentage hatch was also claimed (Murugadass & Marian 1989). Similar results were obtained with *M. nobilli* (Kumari & Pandian 1987) and extrapolation from these results indicates that scope for the enhancement of reproductive performance through ablation may also exist in *M. rosenbergii*, although the technique is not used commercially (New 1990).

There are interesting prospects for substituting Australian red claw crayfish (*C. quadricarinatus*) for *Macrobrachium* in inland regions where seawater is scarce or costly. Red claw crayfish grow well over a wider temperature range than *Macrobrachium* (Jones 1990) but are susceptible to crayfish fungus plague.

12.4 Crayfish: USA

One of the most important problems in swamp crayfish ponds is the depletion of the forage substrate necessary for good growth during winter and early spring (Brunson 1989). The depletion is exacerbated by under-fishing and often results in stunting of the remaining stock (Romaire 1989). The supply of forage cannot be guaranteed when too much reliance is placed on volunteer plants. Selected varieties of rice and other crops produce better results but there is still a need to identify other semi-aquatic plants which will provide forage throughout the harvesting season, and to extend studies on food webs within the pond ecosystem. Many growers would be interested in compounded feeds but the dietary requirements of crayfish are, as yet, inadequately known and the likely costs of artificial feeds could not be justified without intensifying production methods (de la Bretonne & Romaire 1989); but the crayfish burrowing habit and the need to harvest by trapping precludes marked intensification of production.

The recent development of an effective, pelleted, artificial bait has saved the industry significant time and cost in the buying, preparation and storage of highly perishable fish-based baits. Trapping costs contribute 60–80% of production costs (Romaire 1989) (approximately 35–40% of total operating costs), even though significant improvements in harvesting boats and machinery have occurred since 1980. There are prospects for using water flows to attract crayfish to selected areas of suitably designed ponds where they can be netted (Romaire 1989), but these seem more likely to be effective with *Cherax* spp. than with *Procambarus* (Section 7.7.7).

Soft-shell crayfish do not enter traps as they do not feed. Electrified trawl nets have been developed to aid the capture of these valuable animals which now make up the most rapidly growing sector of the market. Research that would contribute significantly to the production of soft-shell crayfish (and indeed all other crustaceans for which such a market exists or can be created) is centred on the prediction of when moulting will occur, how moulting can be induced and how hardening rates can be controlled. The prospects for double-cropping crayfish with prawns (*M. rosenbergii*) in Louisiana is being investigated. Of particular interest is the opportunity to produce 15–17 g prawns for the soft-shell market during mid-May to mid-October when crayfish ponds are not in use (Avault 1990).

12.5 Crayfish: Europe

The major constraint to the farming of all native species is the plague fungus disease. Imported North American species are farmed instead but at least one of these, the signal crayfish, may become susceptible to the plague fungus if severely stressed. Drought and increasing agricultural pollution has constrained the Spanish production of red swamp crayfish. On-growing technology beyond extensive cultures is unproven in many countries although the industry is expanding in Sweden (Persson 1989). Research into the diets that will be needed for more intensive culture is low key and has only just begun. Since the farmed species burrow, trapping is the main method of harvesting, although good results can be achieved with fyke nets (D. Holdich, 1990 pers. comm.). The burrowing habit would seem to limit the prospects for significantly increasing stocking densities. Trials to optimize on-growing conditions and verify the yields claimed to be possible for signal crayfish are urgently needed (Alderman & Wickins 1990). Given the increasing interest in crayfish culture in Europe, it is surprising that several important aspects of basic crayfish biology seem to be absent from the literature. For example, we were unable to find data on the meat yield obtainable from *P. leniusculus* or a comparison of the growth rates of *P. leniusculus* and *A. leptodactylus*.

12.6 Crayfish: Australia

There would seem to be considerable potential for the culture of at least three species: red claw, marron and yabbie. More may await 'discovery' and there are some indications that different physiological strains exist. So little is known about the culture requirements, tolerance and yields of these animals (other than the marron and yabbie) that lack of knowledge may itself constitute a constraint. Outside Australia the major concern to an investor would be the threat of plague fungus disease. The rapid rate of reproduction and the possible existence of different strains may hold prospects for genetic improvement in the future. At present the need is to quantify nutritional and water quality requirements and assess the degree of reliability with which semi-intensive cultures of red claw can operate. As with the culture of signal crayfish in Europe, commercial scale trials to optimize on-growing conditions and verify the yields claimed to be possible for red claw crayfish, are urgently needed.

12.7 Clawed lobsters

The most important technical constraint to lobster farming is the need to rear lobsters in individual isolation to prevent fighting and cannibalism. The addition of pacifying agents like lithium, the induction of synchronous moulting, routine periodic claw ablation and genetic selection have all been considered, but no commercially viable advances have yet been achieved. Redesign of containers or other novel habitat is needed but research into system design appears to have ceased. Better prospects for profitability might arise if the lobsters could be sold at a small, 250 g size (Waddy 1988) or as a soft shelled product, but these options would need new market development as well as new legislation. Despite considerable knowledge of lobster nutritional requirements, no really satisfactory compounded diet capable of supporting maximum growth to market size is available. Fortunately, nutrition does seem to be one area in which research is still financed (Section 8.8). Bilateral eyestalk ablation is known to increase growth rate in lobsters but survival is poor unless a near perfect diet is fed (Koshio *et al.* 1989). Unilateral ablation has also been found to reduce on-growing time by as much as seven months (Peutz *et al.* 1987). Growth increases of 10–20% have also been achieved in intact lobsters injected with human growth hormone (Charmantier *et al.* 1989) – a useful research procedure but not one likely to be acceptable on a commercial farm.

Interest has returned to ranching in several European countries (Sections 7.8.11 and 11.6.3) with between 4% (Aberystwyth, UK) and 50% (Tiedemanns, Norway) of commercial catches in specific fisheries being of experimentally released lobsters. The main constraint to commercial uptake is legislative, in that arrangements for the control or protection of investment must exist after the juveniles have been released on to the seabed or artificial reef. The most pressing biological or research need is to know the detailed spacial (stocking frequency and density) and habitat needs (food availability, replenishment rate and thus carrying capacity) for lobsters of different sizes to survive and grow within a defined area (Bannister & Wickins 1989; Bannister *et al.* 1989).

While much is known about the reproductive cycles of male and female lobsters, present practical experience is probably insufficient to establish adequate control or accurate cost estimates. Manipulating temperature and photoperiod to control maturation, egg extrusion and incubation period in captive broodstock, in order to produce large and predictable numbers of eggs each month, will be a complex process, and for continuous production under battery conditions will require computer support for implementation. The computers will be needed to manage several independent controlled-environment stock rooms as well as to provide farm managers with stock movement, feeding and mating schedules.

Control over mating is useful in breeding research programmes and it is now possible to artificially inseminate lobsters by electro-ejaculation of the spermatophore and manual implantation into the female. This technique has been successful in the laboratory (Section 7.8.3) with *H. americanus* (eggs have reached the eyed stage of development), *H. gammarus*, and their hybrids (though the sperm were infertile) (Talbot *et al.* 1983), and with some prawn species. Work with *H. gammarus* (Beard & Wickins, unpublished data) indicated that shelters designed to be in sympathy with the behavioural requirements of wild females during incubation improved the proportion of eggs carried to full term and hence the numbers of larvae produced per female. The shelter configuration permitted natural grooming and egg care behaviour to occur with minimal disturbance from neighbouring lobsters.

12.8 Spiny lobsters

Considerable advances in spiny lobster larvae culture have been made in the past five years. It is now possible to rear tens, perhaps hundreds, of juveniles from the egg in laboratory upwelling recirculation or static water systems, and several hybrid species have been reared experimentally (Section 7.9.3). No significant reduction in the eight to nine month duration of the larval life has been reported and success has only been possible through unstinting and meticulous attention to maintaining good water quality, albeit within conventional limits (Sections 8.4.3, 8.4.4 and 8.5) and feeding husbandry practices. The larvae feed on newly hatched and partially grown *Artemia* as well as mussel flesh, and do not seem to need any unusual micronutrients.

Despite the creditable progress made from 1988 to 1990 it seems likely to be several years yet before any commercial culture is undertaken on a significant scale. Renewed effort is being spent on elucidating the dietary needs of the puerulus and juvenile stages, but again it may be some time before a complete diet for slow growing species like spiny lobsters is available. The key research requirements are now:

(1) To determine the temperature and salinity tolerance of larvae from a wide range of tropical and temperate species in order to determine which of the fastest growers have the shortest and most hardy larval phase;
(2) To compare the optimum and maximum stocking densities that may be employed during nursery and on-growing phases;
(3) To begin investigations to determine the prospects for breeding hybrids that have greater suitability for culture;
(4) To test to see if shelters 'casitas Cubanas' can be arranged to transplant juveniles to areas conducive to good growth.

12.9 Crabs

Investors have not paid the same attention to crabs as they have to shrimp, crayfish and lobsters, possibly because crabs lack the solid tail meat so attractive to consumers, or because of their cannibalistic tendencies, or perhaps because of the protracted larval life of some species (Sections 4.4 and 7.10). Some species do have a short larval life (e.g. *Portunus pelagicus*) and may be more suited to cultivation than many of those presently fished, or farmed from wild caught juveniles. The full range of cultivable crab species has yet to be determined but prospects for extensive culture and soft-shell production of some large, valuable species are now being evaluated (*Mithrax spinosissimus* in the Caribbean and *Portunus pelagicus* in Australia). However, luxury markets for the products will need development.

Collection of wild Dungeness crab megalopae (*Cancer magister*) may be feasible in some areas (Jamieson & Phillips 1990). Experiments are required to evaluate the prospects for their transplantation to areas of modified habitat for on-growing. The most promising outlet seems to be the market for soft-shell crabs. Where this sector of the industry is established the most pressing research needs are to develop methods to predict and induce moulting and gain knowledge of shell hardening rates at different temperatures and salinities so that the hardening can be controlled. Moult inducing hormones have been tested on *Callinectes*, but it is unlikely that products so treated could be marketed, at least in the USA or on a large scale in the EC.

12.10 Domestication

Surveys of electrophoretic variation, useful for distinguishing experimental broodstocks and their progeny, are well advanced in a range of genera and several reports have described non-destructive, external, morphometric measurements that could be used in selective breeding programmes. Distinct natural genetic variations have been found among geographically separate populations of *Macrobrachium rosenbergii* which might be exploited for breed improvements in the long term (Malecha 1983). Studies on penaeids in Australia and the USA found only low levels of genetic variation within species, which would possibly be enough for selective breeding purposes but progress would be slow (Sandifer 1986).

Attempts have been made to assess the potential for the development of selective breeding programmes to improve tail weight in penaeids (Lester 1983) and crayfish (Lutz & Wolters 1989) and to gain faster growth (Fairfull & Haley 1981) and improved tolerance of lobsters to crowding (Finley & Haley 1983). Most of these concluded that although selection would be possible, improvments would only be moderate and may take a long time to achieve. In lobsters interspecific hybridization might offer the most immediate prospects for introducing variability into broodstock, but crosses have not regularly yielded families showing markedly improved characteristics.

Interspecific crosses between *Macrobrachium* species, as well as between several penaeid species, have been successfully achieved but no significant improvements to culture seem to have resulted.

Repeated inbreeding from one stock may result in a reduction of genetic variation (Sbordoni *et al.* 1986) or loss of culture performance (Malecha *et al.* 1980). Indeed, husbandry practices that seem likely to select unintentionally for adverse traits have been reported. Most prevalent are the selection of broodstock from among the first pond raised prawns or shrimp to mature or spawn regardless of size (Doyle *et al.* 1983; Sbordoni *et al.* 1986), and the practice of leaving slow growing crayfish in the ponds to become next season's broodstock (Lutz & Wolters 1989).

Most crustacea, unlike most domesticated animals, come from natural populations with large stores of genetic variation, have high fecundity, and offer the prospect of labile sexual differentiation, all advantageous features in the search for genetic improvement. On the other hand, moulting, combined with territorial instincts and plasticity of growth rate, makes them one of the least suitable groups for cultivation and eventual domestication (Wickins 1984). The existence of a short-lived, water-borne substance capable of inhibiting growth in lobsters (Nelson *et al.* 1980), and the behaviourally induced, morphological changes occurring in dominant male *Macrobrachium,* are just two examples of crustacean features constraining the establishment of domestication programmes. At present, growth rate, meat yield and reproductive potential will be more readily improved by control of the culture environment, surgical manipulations, sex reversal (Malecha *et al.* in press), and possibly hybridization, rather than by true domestication.

12.11 References

Aiken D. (1990) Shrimp farming in Ecuador. An aquaculture success story. *World Aquaculture*, **21** (1) 7–16.

Alderman D.J. & Wickins J.F. (1990) *Crayfish Culture*. Lab. Leafl. (62), MAFF Direct. Fish. Res., Lowestoft.

Anon. (1990a) Disease-free shrimp population established on Hawaii. *Oceanic Institute Newsletter*, **3** (1) 1–2.

Anon. (1990b) Welsh worms whet shrimp appetites. *Fish Farming International*, **17** (4) 14–15.

Avault J.W. Jr. (1990) Some recent advances in crawfish farming research. *Aquaculture Magazine*, **16** (5) 77–81.

Bannister R.C.A. & Wickins J.F. (1989) A new perspective on lobster stock enhancement. *Proc. 20th Ann. Shellfish conf.*, 16–17 May 1989, pp. 38–47. Shellfish Association of Great Britain.

Bannister R.C.A., Howard A.E., Wickins J.F., Beard T.W., Burton C.A. & Cook W. (1989) A brief progress report on experiments to evaluate the potential of enhancing stocks of the lobster (*Homarus gammarus* L.) in the United Kingdom. *ICES Shellfish Committee CM 1989/K:30* (mimeo).

Baticados C.L. (1988) Diseases of prawns in the Philippines. *Asian Aquaculture*, **10** (1) 1–8.

Biedenbach J.M., Smith L.L. & Lawrence A.L. (1990) Use of a new spray-dried algal product in penaeid larviculture. *Aquaculture*, **86** (2/3) 249–57.

Brunson M.W. (1989) Forage and feeding systems for commercial crawfish culture. *J. Shellfish Res.*, **8** (1) 277–80.

Chamberlain G.W. (Ed.) 1988. Rethinking shrimp pond bottom management. *Coastal Aquaculture*, **5** (2) 1–19.

Charmantier G., Charmantier-Daures M. & Aiken D.E. (1989) La somatotropine humane stimule la croissance de jeunes homards americains, *Homarus americanus* (Crustacea, Decapoda). *C.R. Acad. Sci. Paris*, **308** (3) 21–6.

Charniaux-Cotton H. & Payen G. (1988) Crustacean reproduction. In *Endocrinology of selected invertebrate types* (Ed. by H. Laufer & R.G.H. Downer), pp. 279–303. Alan. R. Liss. Inc.

Clement O. (1989) Aquaculture: a bright future for the wetlands of western France? In *Aquaculture, a review of recent experience*, pp. 122–7. Organisation for Economic Co-operation and Development, Paris, France.

de la Bretonne L.W. Jr. & Romaire R.P. (1989) Commercial crayfish cultivation practices: a review. *J. Shellfish Res.*, **8** (1) 267–75.

Doyle R.W., Singholka S. & New M.B. (1983) 'Indirect selection' for genetic change:

a quantitative analysis illustrated with *Macrobrachium rosenbergii. Aquaculture*, **30** (1–4) 237–47.

Fairfull R.W. & Haley L.E. (1981) The early growth of artificially reared American lobsters. *Theor. Appl. Genet.*, **60**, 269–73.

Finley L.M. & Haley L.E. (1983) The genetics of aggression in the juvenile American lobster, *Homarus americanus. Aquaculture*, **33** (1–4) 135–9.

Itami T., Takahashi Y. & Nakamura Y. (1989) Efficacy of vaccination against vibriosis in cultured kuruma prawns *Penaeus japonicus. J. Aquat. Animal Health*, **1** (3) 238–42.

Jaffa M. (1990) Health care for shrimp in hatcheries. *Fish Farming International*, **17** (6) 42–4.

Jamieson G.S. & Phillips A.C. (1990) A natural source of megalopae for the culture of Dungeness crab, *Cancer magister* Dana. *Aquaculture*, **86** (1) 7–18.

Jones C.M. (1990) *The biology and aquaculture potential of the tropical freshwater crayfish* Cherax quadricarinatus. Information Series QI90028, Queensland Department of Primary Industries.

Karplus I., Hulata G., Wohlfarth G.W. & Halevy A. (1987) The effect of size grading juvenile *Macrobrachium rosenbergii* prior to stocking on their population structure and production in polyculture. II. Dividing the population into three fractions. *Aquaculture*, **62** (2) 85–95.

Karplus I., Samsonov E., Hulata G. & Milstein A. (1989) Social control of growth in *Macrobrachium rosenbergii.* 1. The effect of claw ablation on survival and growth of communally raised prawns. *Aquaculture*, **80** (3/4) 325–35.

Koshio S., Haley L.E. & Castell J.D. (1989) The effect of two temperatures and salinities on growth and survival of bilaterally eyestalk ablated and intact juvenile American lobsters, *Homarus americanus*, fed brine shrimp. *Aquaculture*, **76** (3/4) 373–82.

Krom M.D., Porter C. & Gordin H. (1985) Causes of fish mortality in semi-intensively operated ponds in Elat, Israel. *Aquaculture*, **49** (2) 159–77.

Kumari S.S. & Pandian T.J. (1987) Effects of unilateral eyestalk ablation on moulting, growth, reproduction and energy budget of *Macrobrachium nobilli. Asian Fish. Sci.*, **1**, 1–17.

Kwei Lin C. (1989) Prawn culture in Taiwan: What went wrong? *World Aquaculture*, **20** (2) 19–20.

Lester L.J. (1983) Developing a selective breeding program for penaeid shrimp mariculture. *Aquaculture*, **33** (1–4) 41–50.

Lutz C.G. & Wolters W.R. (1989) Estimation of heritabilities for growth, body size, and processing traits in red swamp crawfish *Procambarus clarkii* (Girard). *Aquaculture*, **78** (1) 21–33.

Main K.L. & Fulks W. (Eds.) (1990) *The culture of cold-tolerant shrimp. Proceedings of an Asian-US workshop on shrimp culture.* Oceanic Institute Honolulu, Hawaii.

Malecha S.R. (1983) Crustacean genetics and breeding: an overview. *Aquaculture*, **33** (1–4) 395–413.

Malecha S., Sarver D. & Onizuka D. (1980) Approaches to the study of domestication in the freshwater prawn, *Macrobrachium rosenbergii* with special emphasis on the Anuenue and Malaysian stocks. *Proc. World Maricult. Soc.*, **11**, 500–28.

Malecha S.R., Nevin P.A., Ha-Tamaru P., Barck L.E., Lamadrid-Rose Y., Masuno S. & Hedgecock D. (in prep.). Production of progeny from crosses of surgically sex-

reversed freshwater prawns, *Macrobrachium rosenbergii*: implications for commercial culture (draft).
Moriarty D.J.W. & Pullin R.S.V. (1990) Detritus and microbial ecology in aquaculture. *Proceedings of the conference on detrital systems for aquaculture in 1985.* Bellagio, Como, Italy.
Murugadass S. & Marian M.P. (1989) Maximization of seed production by eyestalk ablation technique in *Macrobrachium malcolmsonii*. Aquaculture 89 Abstracts, p. 103, from World Aquaculture Society Meeting 1989, Los Angeles.
Nelson K., Hedgecock D., Borgeson W., Johnson E., Dagget R. & Aronstein D. (1980) Density dependent growth inhibition in lobsters, *Homarus* (Decapoda, Nephropidae). *Biol. Bull.*, **159**, 162–76.
New M.B. (1988) Shrimp farming in other areas. In *Shrimp 88, Conference proceedings*, 26–28 Jan. 1988, pp. 102–22. Bangkok, Thailand. Infofish, Kuala Lumpur, Malaysia.
New M.B. (1990) Freshwater prawn culture: a review. *Aquaculture*, **88** (2) 99–143.
Ogle J. (1988) Maturation tips. *Coastal Aquaculture*, **5** (1) 2.
Oka M. (1967) Studies on *Penaeus orientalis* Kishinouye – 4, Physiological mechanism of ovulation. *Bull. Fac. Fish. Nagasaki Univ.*, **23**, 57–67.
Persson R. (1989) Crayfish farming in Sweden. In *Aquaculture, a review of recent experience*, pp. 82–91. Organisation for Economic Co-operation and Development, Paris, France.
Peutz A.V.H.A., Waddy S.L., Aiken D.E. & Young-Lai W.W. (1987) Accelerated growth of juvenile American lobsters induced by unilateral eyestalk ablation. *Bull. Aquacult. Assn. Can.*, **87** (2) 28–9.
Romaire R.P. (1989) Overview of harvest technology used in commercial crawfish aquaculture. *J. Shellfish Res.*, **8** (1) 281–6.
Sandifer P.A. (1986) Some recent advances in the culture of crustaceans. In *Realism in Aquaculture: Achievements, Constraints, Perspectives* (Ed. by M. Bilio, H. Rosenthal & C.J. Sindermann), pp. 143–71. European Aquaculture Society, Bredene, Belgium.
Sbordoni V., De Matthaeis E., Sbordoni M.C., La Rosa G. & Mattoccia M. (1986) Bottleneck effects and the depression of genetic variability in hatchery stocks of *Penaeus japonicus* (Crustacea, Decapoda). *Aquaculture*, **57** (1–4) 239–51.
Sheeks R.B. (1989) Taiwan's aquaculture – at the crossroads. *Infofish International*, (6) 38–43.
Sindermann C.J. (1988) Disease problems created by introduced species. In Disease diagnosis and control in North American marine aquaculture (Ed. by C.J. Sindermann & D.V. Lightner). *Dev. Aquacult. Fish. Sci.*, **17**, 394–8.
Talbot P., Hedgecock D., Borgeson W., Wilson P. & Thaler C. (1983) Examination of spermatophore production by laboratory-maintained lobsters (*Homarus*). *J. World Maricult. Soc.*, **14**, 271–8.
Taylor A. (1990) A double bonus for shrimp farmers. *Fish Farmer*, **13** (2) 39–40.
Turner G.E. (1988) Code of practice and manual of procedures for consideration of introductions and transfers of marine and freshwater organisms. *ICES Cooperative Res. Rept. (159).*
Waddy S.L. (1988) Farming the Homarid lobsters: state of the art. *World Aquaculture*, **19** (4) 63–71.

Wickins J.F. (1984) Crustacea. In *Evolution of domesticated animals* (Ed. by I.L. Mason), pp. 424–8. Longman, London.

Williams R.R. & Lightner D.V. (1988) Regulatory status of therapeutants for penaeid shrimp culture in the United States. *J. World Aquacult. Soc.*, **19** (4) 188–96.

Yano I. & Tanaka H. (1984) Effects of ultraviolet irradiated seawater on induction of spawning of kuruma prawn *Penaeus japonicus*. *Bull. Jap. Soc. Sci. Fish.*, **50** (9) 1621–3.

Yano I., Tsukimura B., Sweeney J.N. & Wyban J.A. (1988) Induced ovarian maturation of *Penaeus vannamei* by implantation of lobster ganglion. *J. World Aquacult. Soc.*, **19** (4) 204–9.

Chapter 13
Conclusions

Crustacean farming is a high risk industry with good prospects for worthwhile profit but also potential for serious loss. It is characterized by high investor appeal and over-optimistic predictions and aspirations. The industry has significantly increased the opportunities for a diversity of employment and trading activities, created high value exportable products and, in doing so, provided the justification for improvements to national and regional infrastructures. On the other hand, farmed crustaceans seldom contribute to the diets of the poor and their large-scale culture may cause significant changes in the surrounding communities and environment. The potential for detrimental impact in tropical regions has now been recognized, and plans to minimize additional adverse effects have been made in some affected countries.

Major constraints to continued growth in the industry include unreliable supplies of good quality seed and broodstock, the patchy quality of processing, and increasing pollution which is reducing the number of prime sites. Extension workers and technicians are in short supply and are often inadequately trained. There is, however, an abundance of 'consultants' with no real experience of crustaceans or of crustacean farming. Feed costs seem likely to rise globally and this will have greatest effect on the profitability of the more intensive farms.

Serious setbacks occurred in some shrimp and crayfish sectors of the industry during the late 1980s. With hindsight these were not altogether surprising considering the rates at which expansion and intensification were proceeding. The technical problems that emerged highlighted specific areas where practice had exceeded the boundaries of supportive scientific knowledge. Disease, stunting, shell malformation and associated mortalities were the signs most frequently observed. On most occasions their incidence was linked individually or severally to pond bottom constitution, water chemistry and diet composition. Clearly the interaction of these factors on crustacean shell mineralization and internal ionic regulation is inadequately understood and demands further research at the cellular and molecular level. Disease identification is well advanced but controlling their spread while maintaining a flexible and vigorous industry will be difficult and will require special vigilance from hatchery and farm managers.

A number of exciting new developments are currently in prospect with considerable advances already made in several important areas. Examples include the identification of alternative shrimp and crayfish species suitable for catch crops, or for temperate, freshwater or high salinity environments, which will be useful insurance against climatic and market instabilities; the control of sex ratio and the consequent reduction in growth variability in farmed *Macrobrachium* populations; the culture of a range of spiny lobster larvae species and their hybrids which could open up a new culture industry; the encouraging results now emerging from ranching trials; and, more importantly, the potential of some related schemes to ameliorate negative impacts of inshore waste disposal.

The demand for crustaceans seems set to expand into the next century, for as long as the economies of consumer nations (Japan, USA and Western Europe) remain buoyant, and with increasing wealth and tourism in the tropical producer countries. Policy makers should urgently consider the provision of material and legislative resources for an expansion of low cost, extensive farming and ranching operations as a precaution against the impending rises in farm input costs and climatic changes. For the immediate present, the industry needs to expand the European markets while also diversifying into new species and a range of good quality value-added crustacean products.

Appendix 1
Summary of biological data and examples of typical culture performance

Marine shrimp

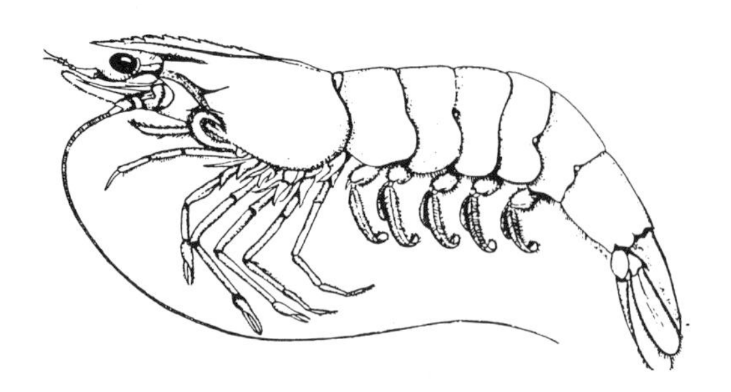

Species	***Penaeus chinensis* (Osbeck 1765)**; formerly *P. orientalis* (Kishinouye 1918)	***Penaeus indicus* (H. Milne-Edwards 1837)**
Common names	Chinese white shrimp, fleshy prawn, pok, taishou-ebi	Indian white prawn
Home range	Yellow Sea, Gulf of Bohai, Korean Bight	India, S.E. Asia
Culture temperature range	16–28°C	22–33°C
Culture salinity range	11–38 ppt.	15–25 ppt.
Fished tonnage	255 077 (1988) (FAO 1990a)	120 000–140 000 (Anon. 1988)
Culture tonnage	199 520 (1988) (FAO 1990b); 139 000 (1990) (Rosenberry 1990)	682 (1988) (FAO 1990b)
Culture methods and location	Semi-intensive ponds, North China; Korea	Extensive and semi-intensive often in polyculture, India, Indonesia, Phillipines
Source of broodstock or seed	80% wild; 20% captive females	Wild or pond-raised seed or broodstock
Special culture features	One operation per year due to climate; only 10% use nursery phase	Non burrowing
Growth rate (on-growing)	25 g in <5 months	4–11 g in 70–120 days
Survival	25–55%	32–91%
Yield (kg ha^{-1} $crop^{-1}$)	314–2308	231–1000

Species	***Penaeus japonicus* (Bate 1888)**	***Penaeus merguiensis* (de Man 1887)**
Common names	Kuruma prawn, Japanese tiger shrimp	Banana prawn
Home range	Indo-west Pacific from Red Sea, east and south-east Africa to Japan and Malay archipelago, east Mediterranean	South-east Asia, Thailand, Indonesia

Culture temperature range	18–28°C	25–30°C
Culture salinity range	35–45 ppt.	15–33 ppt.
Fished tonnage	10 822 (1988) (FAO 1990a)	79 420 (1988) (FAO 1990a)
Culture tonnage	Taiwan 4000; Japan 3020; Korea Rep. 79; Spain 55; France 14; other 3747 (1988) FAO (1990b)	30 464 (1988) (FAO 1990b)
Culture methods and location	Semi-intensive ponds, Japan and elsewhere; super-intensive Shigueno tanks, Japan	Extensive and semi-intensive ponds also in polyculture, Indonesia; Thailand
Source of broodstock or seed	Wild-caught females, hatchery seed	Wild seed, some pond-raised broodstock
Special culture features	Live sales; one operation per year in Japan, more in the tropics; ranching programmes	Often a catch crop
Growth rate (on-growing)	25 g in six months	7–12.5 g in 76–112 days
Survival	40–70%	47–73%
Yield (kg ha^{-1} crop^{-1})	300–30 000	200–5850

Species	***Penaeus monodon* (Fabricius 1798)**	***Penaeus penicillatus* (Alcock 1905)**
Common names	Jumbo tiger shrimp, black tiger prawn, grass prawn, sugpo, udang windu (Indonesia)	Red prawn, red-tailed prawn
Home range	Indo-west Pacific, east and south-east Africa, Pakistan to Japan, Malay archipelago and North Australia	Indo-west Pacific, Pakistan to Taiwan and Indonesia
Culture temperature range	24–34°C	15–32°C
Culture salinity range	5–25 ppt.	15–32 ppt.
Fished tonnage	59 410 (1988) (FAO 1990a)	Not available
Culture tonnage	118 114 (1988) FAO (1990b); 305 000 (1990) Rosenberry (1990)	3500 (1988) (Liao & Chien 1990)
Culture methods and location	Semi-intensive and intensive ponds, Thailand; 60% extensive, 25% semi-intensive, 15% intensive, Philippines; intensive and super-intensive ponds, Taiwan	Semi-intensive ponds, Taiwan
Source of broodstock or seed	Wild-caught and captive (ablated) females	Wild broodstock, hatchery seed
Special culture features	Ablation essential, fastest growing penaeid species	Seasonal operation in Taiwan
Growth rate (on-growing)	21–33 g in 80–225 days	9–21 g in three to five months
Survival	30–80%	45–90%
Yield (kg ha^{-1} crop^{-1})	250 Philippines; 1125 Thailand; 5000–14 500 Taiwan	3400–12 300

Species	***Penaeus stylirostris* (Stimpson 1874)**	***Penaeus vannamei* (Boone 1931)**
Common names	Blue shrimp	Western white shrimp, whiteleg shrimp
Home range	Eastern Pacific, Mexico to Peru	Eastern Pacific, Mexico to Peru
Culture temperature range	22–30°C	26–33°C
Culture salinity range	25–30 ppt. (estimated)	5–35 ppt.

Fished tonnage	186 (1988) (FAO 1990a)	233 (1988) (FAO 1990a)
Culture tonnage	1710 (1988) (FAO 1990b); Panama 1164 (1987) (FAO 1989); Ecuador 3650, Honduras 2475 (1990) (Rosenberry 1990)	82 086 (1988) (FAO 1990b); Ecuador 69 000 (1990), Americas (excluding Ecuador) 24 000 (1990) (Rosenberry 1990)
Culture methods and location	Extensive and semi-intensive ponds, Ecuador; Peru; semi-intensive ponds, New Caledonia	Extensive and semi-intensive, Ecuador; semi-intensive elsewhere
Source of broodstock or seed	Wild-caught and captive females	Wild and hatchery-reared post-larvae and broodstock
Special culture features	Performs less well than *P. vannamei*	Acclimatisation (hardening-off) nurseries
Growth rate (on-growing)	28 g in eight months	7–23 g in two to five months
Survival	5–70%	40–90%
Yield (kg ha^{-1} crop^{-1})	300–2500	500–1500 Latin America; 3000 USA

Freshwater prawns

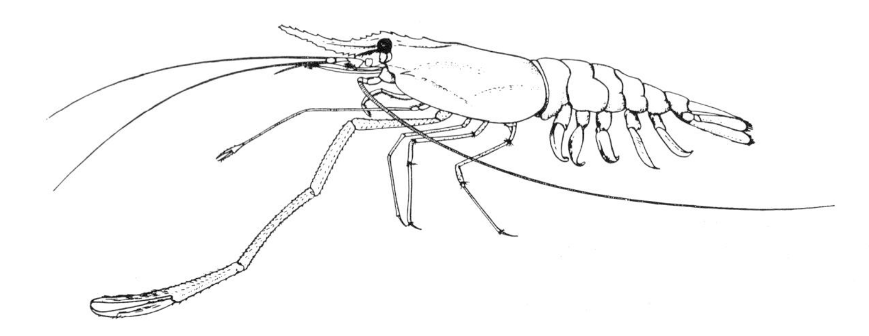

Species	***Macrobrachium malcolmsonii* (H. Milne Edwards 1844)**	***Macrobrachium rosenbergii* (de Man 1879)**
Common names	Indian river prawn	Giant freshwater prawn, udang galah (Malaysia), koong yai (Thailand)
Home range	Pakistan, India, Bangladesh	Indo-west Pacific, North-west India to Vietnam, Philippines, North Australia, Papua New Guinea
Culture temperature range	26–32°C	26–32°C
Culture salinity range	0–5 ppt. adults, 12 ppt. for larvae	0–2 ppt., 12 ppt. for larvae
Fished tonnage	Not available	9875 (1988) (FAO 1990a)
Culture tonnage	Limited	19 307 (1988) FAO (1990b); 26 765 (New 1990)
Culture methods and location	Experimental, India	Semi-intensive ponds, Thailand, Taiwan, Hawaii, Mauritius
Source of broodstock or seed	Wild seed	Pond-raised broodstock, hatchery seed
Special culture features	Will hybridise with *M. rosenbergii*	Larvae require brackish water, wide variety of stocking/ harvesting strategies used
Growth rate (on-growing)	30–40 g in four months with repeated culls	25–45 g in three to five months; batch or repeated harvests
Survival	44–57%	40–60%
Yield (kg ha^{-1} crop^{-1})	475–605	1000–2500

Crayfish: USA and Europe

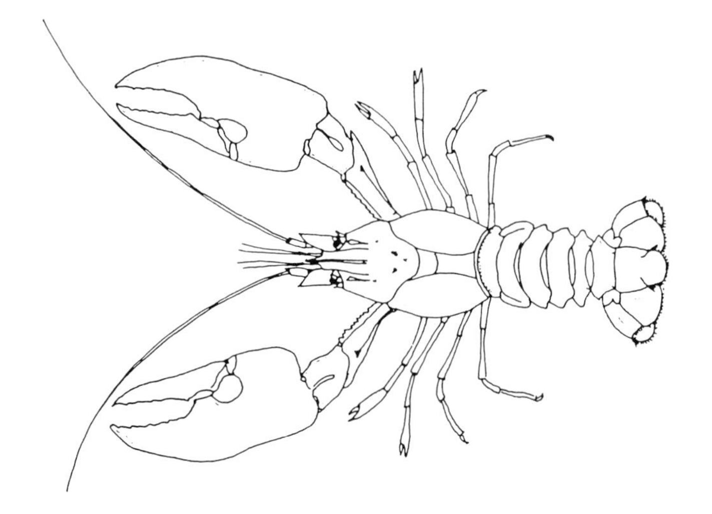

Species	***Astacus leptodactylus* (Eschscholz 1823)**	***Pacifastacus leniusculus* (Dana 1852)**
Common names	Slender clawed crayfish, Turkish crayfish	Signal crayfish
Home range	Eastern Europe, USSR, Turkey	USA
Culture temperature range	10–18°C	16–22°C
Culture salinity range	Freshwater	Freshwater
Fished tonnage	<1000 mt, previously 8000 mt from Turkey	Approx. 5000 mt
Culture tonnage	Not available	<10 UK mt
Culture methods and location	Extensive, mainly hatchery supported fisheries	Extensive and semi-intensive ponds, canals; juveniles raised for restocking
Source of broodstock or seed	Wild or pond-raised	Pond-reared broodstock, hatchery juveniles
Special culture features	Fast growing, susceptible to plague fungus	three to four months nursery, 100 m^{-2}
Growth rate (on-growing)	60–100 g in one to two years	60–100 g in one to two years
Survival	60%	30–40%
Yield (kg ha^{-1} $crop^{-1}$)	500–1000	500–1000

Species	***Procambarus clarkii* (Girard 1852)**
Common names	Red swamp crayfish, crawdad, red swamp crawfish
Home range	Southern USA
Culture temperature range	18–25°C
Culture salinity range	Freshwater (0–5 ppt.)
Fished tonnage	14 000–25 000 world estimate (Huner 1989); 14 600 (Roberts & Dellenbarger 1989); 43 182 (1988) (FAO 1990a)
Culture tonnage	35 000–50 000 world estimate (Huner 1989); 30 400 (Roberts & Dellenbarger 1989); 32 234 (1988) (FAO 1990b)

Culture methods and location	Extensive ponds, South USA, Spain
Source of broodstock or seed	Wild or ponds
Special culture features	Self-sustaining populations, destructive burrower; also soft-shell production
Growth rate (on-growing)	30–80 g in one year
Survival	47–87%
Yield (kg ha^{-1} yr^{-1})	250–3000 (mean 600) USA (Eversole & Pomeroy 1989; de la Bretonne & Romaire 1989)

Crayfish: Australia

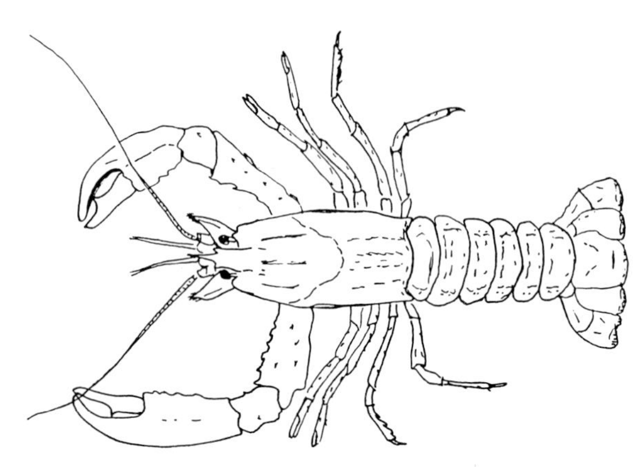

Species	***Cherax destructor* (Clarke 1936)**	***Cherax tenuimanus* (Smith, 1912)**
Common names	Yabbie, yabby	Marron
Home range	South-east Australia	Western Australia
Culture temperature range	15–30°C	13–23°C
Culture salinity range	Freshwater (0–5 ppt.)	Freshwater
Fished tonnage	<500 (Huner 1989)	400 (Morrissy 1978)
Culture tonnage	Australia 9 (1988) (FAO 1990b); <50 (Huner 1989)	Australia 3 (1988) (FAO 1990b)
Culture methods and location	Semi-intensive ponds, South-east Australia	Farm dams, extensive, semi-intensive ponds, Western Australia
Source of broodstock or seed	Wild or ponds	Pond-reared broodstock, hatchery juveniles.
Special culture features	Destructive burrower	Nursery advisable
Growth rate (on-growing)	50–100 g in four to 12 months	40–120 g in 12–24 months
Survival	18–36%	60%
Yield (kg ha^{-1} crop^{-1})	1500 (Villarreal 1988)	1000–2500 Morrissy (1988)

Species	***Cherax quadricarinatus* (Von Martens 1868)**
Common names	Red claw, Queensland red claw, tropical blue crayfish (the name Queensland marron is not correct)
Home range	Queensland, North Australia
Culture temperature range	24–32°C

Culture salinity range	Freshwater
Fished tonnage	<5
Culture tonnage	Australia 7, (1988) (FAO 1990b)
Culture methods and location	Semi-intensive pond culture, Australia, Caribbean
Source of broodstock or seed	Wild and captive
Special culture features	Two weeks attached to mother; 20–50 days nursery culture to 0.1–0.3 g
Growth rate (on-growing)	40–100 g in six to 24 months
Survival	49–94%
Yield (kg ha^{-1} $crop^{-1}$)	1000–4000

Clawed Lobsters

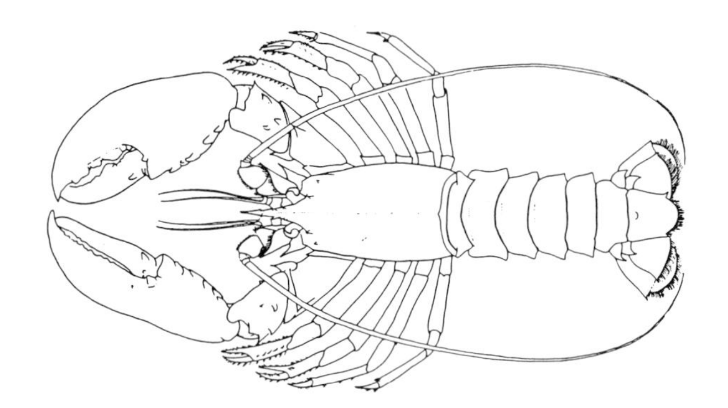

Species	***Homarus americanus* (H. Milne Edwards 1837)**	***Homarus gammarus* (Linnaeus 1758)**
Common names	American lobster, Canadian lobster	European lobster, hummer (Denmark)
Home range	Atlantic Canada, USA	Mediterranean, France, British Isles, Norway, North Africa
Culture temperature range	18–21°C	18–23°C
Culture salinity range	30–35 ppt.	30–35 ppt.
Fished tonnage	62 457 (1988) (FAO 1990a)	2052 (1988) (FAO 1990a)
Culture tonnage	0	0
Culture methods and location	Experimental and pilot battery and fattening systems	Experimental and pilot hatchery and nursery systems for ranching studies
Source of broodstock or seed	Wild ovigerous females	Wild-caught ovigerous females
Special culture features	Individual confinement from metamorphosis	Individual confinement of juveniles
Growth rate (on-growing)	400 g in two years	350 g in two years
Survival	80% (post-nursery)	Unknown after release
Yield (kg ha^{-1} $crop^{-1}$)	Up to 11 400 (experimental battery trials)	Unknown (ranching)

Spiny lobsters

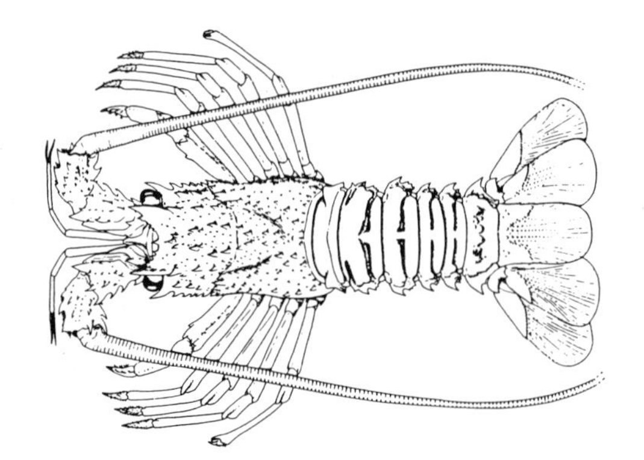

Species	***Panulirus spp.; Jasus* spp.**
Common names	Spiny and rock lobsters, crawfish
Home range	Spiny lobsters, Pacific north America, Caribbean, Indian Ocean; Rock lobsters, Australasia; crayfish, Mediterranean, Spain, Atlantic France
Culture temperature range	18–28°C depending on species
Culture salinity range	30–35 ppt.
Fished tonnage	78 633 (1988) (FAO 1990a)
Culture tonnage	Taiwan 13.2 (1987) (Chen 1990); World 49 (1988) (FAO 1990b); Singapore 24 (1988) (Lovatelli 1990)
Culture methods and location	Cage fattening, Singapore; pond fattening, Taiwan
Source of broodstock or seed	Wild-caught juveniles
Special culture features	Larvae difficult to culture; prospects of ranching
Growth rate (on-growing)	350 g in two to three years.
Survival	80% Taiwan
Yield	45 kg m^{-3} in Singapore cages; 10 000 individuals ha^{-1} in Taiwanese ponds

Crabs

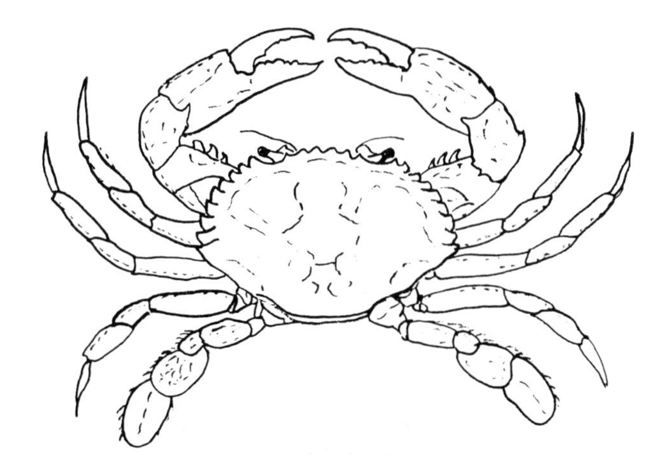

Species	*Scylla serrata* (Forskol 1775)	*Portunus triturbiculatus* (Miers 1876)
Common names	Mud crab, mangrove crab	Blue swimming crab
Home range	South-east Asia, Mauritius	Western Pacific, Japan, Korea, China
Culture temperature range	26–30°C	20–28°C
Culture salinity range	25–35 ppt.	30–33 ppt.
Fished tonnage	11 532 (1988) (FAO 1990a)	35 380 (1988) (FAO 1990a)
Culture tonnage	World 3249 (1988) (FAO 1990b)	Indonesia 20 (1988) (FAO 1990b); Japan, 33.7 million post-larvae released annually
Culture methods and location	Extensive polyculture, Philippines, Taiwan	Hatchery juveniles for release to sea, Japan
Source of broodstock or seed	Wild-caught juveniles, Philippines; hatchery and wild, Taiwan	Wild-caught females
Special culture features	Fattening/ripening of maturing females for gourmet Singapore market	Fish or shrimp hatcheries used; pre-release crabs, stage two to four, held in sea enclosures for one to three weeks, 20–40% survival
Growth rate (on-growing)	8–9 cm CW in three to six months	Stock supplementation (Japan)
Survival	30–70% Taiwan	3–12% recapture
Yield (kg ha^{-1} $crop^{-1}$)	340 (Cowan 1983)	Unknown (ranching)

References

Anon. (1988) The world shrimp industry. In *Shrimp 88, Conference proceedings*, 26–28 Jan. 1988, pp. 1–6. Bangkok, Thailand. Infofish, Kuala Lumpur, Malaysia.

de la Bretonne L.W. & Romaire R.P. (1989) Commercial crawfish cultivation practices: a review. *J. Shellfish Res.*, **8** (1) 267–75.

Chen L.C. (1990) *Aquaculture in Taiwan*. Fishing News Books, Blackwell Scientific Publications, Oxford.

Cowan L. (1983) *Crab farming in Japan, Taiwan and the Philippines*. Information series Q184009, Queensland Dept. of Primary Industries.

Eversole A.G. & Pomeroy R.S. (1989) Crawfish culture in South Carolina: an emerging aquaculture industry. *J. Shellfish Res.*, **8** (1) 309–13.

FAO (1989) Aquaculture production (1984–1987). *FAO Fisheries Circular No. 815, Rev. 1*, Rome.

FAO (1990a) Aquaculture production (1985–1988). *Statistical tables FAO Fisheries circular No. 815*, Rev. 2, FIDI/C815 rev. 2.

FAO (1990b) *Fishery Statistics 1988, catches and landings*, **66**.

Huner J.V. (1989) Overview of international and domestic freshwater crawfish production. *J. Shellfish Res.*, **8** (1) 259–65.

Liao I.-C. & Chien Y.-H. (1990) Evaluation and comparison of culture practices for *Penaeus japonicus, P. penicillatus*, and *P. chinensis* in Taiwan. In *The culture of cold-tolerant shrimp* (Ed. by K.L. Main & W. Fulks), pp. 49–63. Oceanic Institute, Honolulu, Hawaii.

Lovatelli A. (1990) *Regional seafarming resources atlas*. FAO/UNDP Regional seafarming development and demonstration project, RAS/86/024 Jan. 1990.

Morrissy N.M. (1978) The amateur marron fishery in Western Australia. *Fisheries Research Bulletin (21)*. W. Australian Marine Research Laboratories, Dept. of Fisheries and Wildlife, Perth.

Morrissy N.M. (1988) Marron farming – current industry and research developments in Western Australia. *Proc. 1st Australian Shellfish Aquacult. Conf.*, Perth 1988 pp. 59–72. Curtin University of Technology.

New M.B. (1990) Freshwater prawn culture: a review. *Aquaculture*, **88** (2) 99–143.

Roberts K.J. & Dellenbarger L. (1989) Louisiana crawfish product markets and marketing. *J. Shellfish Res.*, **8** (1) 303–7.

Rosenberry R. (1990) *World shrimp farming 1990*. Rosenberry, San Diego.

Villarreal H. (1988) Culture of the Australian freshwater crayfish *Cherax tenuimanus* (marron) in Eastern Australia. In *Freshwater Crayfish* 7 (Ed. by Goeldlin de Tiefenau) pp. 401–8. Musee Zoologique Cantonal, Lausanne, Switzerland.

Appendix 2
Shrimp counts

Marine shrimp are customarily graded by size counts. Raw, head-off shrimp or tails are conventionally, but not universally, counted in pieces to the pound (454 g). Raw, whole head-on shrimp are counted in pieces to the kilogram (1000 g) (ITC 1983).

The count groupings and their tolerances may vary slightly between countries. Equivalent relationships between tail weight and counts of caridean prawns or shrimps will differ from marine shrimp because carideans have a larger head relative to the tail (Figures 2.2a,b).

No. of tails per pound	Approximate weight (g) of	
	Tail	Whole shrimp
over 70	under 6	under 10
61–70	6–7	10–11
51–60	7–9	12–13
41–50	9–11	14–16
36–40	11–12	17–18
31–35	13–14	19–21
26–30	15–17	22–26
21–25	18–21	27–32
16–20	22–27	33–42
11–15	28–40	43–65
under 10	over 40	over 65

No of whole shrimp per kg	Approximate weight of whole shrimp (g)
71–90	11–13
61–70	14–16
51–60	17–20
41–50	21–24
31–40	25–32
21–30	33–49
16–20	50–62
11–15	63–91
under 15	over 91

Reference

ITC (1983) *Shrimps: a survey of the world market.* ITC Publications, International Trade Centre, UNCTAD, GATT, Geneva.

Appendix 3
Glossary of terms

The following descriptions have been freely adapted to explain some of the terms used in this book, and so are not all-embracing, text-book type definitions. Further guidance on terminology may be obtained from Holmes (1979) and Rosenthal *et al.* (1990).

Ablation Surgical removal of glands (eyestalk) to stimulate maturation or growth. (Also called extirpation and enucleation Section 7.2.2.5.)

Acclimation, acclimatization Gradual exposure to new environmental conditions in order to minimize shock or stress.

Acid sulphate soils Acidic soils typically found where areas of mangrove have been cleared. Generally unsuitable for pond construction (Section 6.3.3.5).

Activated charcoal, carbon Finely divided carbon material capable of adsorbing organic molecules.

Ad libitum Feeding until individuals or populations seem satisfied.

Aeration The mechanical mixing of air and water, generally to increase oxygen content but also to remove excess carbon dioxide.

Aggression Hostile act or display to protect territory, family or establish dominance.

Algal bloom Rapid increase in unicellular alga population(s), manifested visually as a change in colour or turbidity of the water.

Algicide Chemical that kills unicellular and macroalgae, e.g. copper sulphate (Section 7.2.6.5).

Alkalinity The concentration of basic minerals (e.g. carbonates) in the water, capable of neutralizing excess hydrogen ions (acidity).

Ammonia The main nitrogenous excretion product of crustaceans. High toxicity which increases with pH increase.

Anaerobic Chemical processes occurring in the absence of oxygen. In ponds, an indication of undesirable substrate conditions.

Antenna The second and longer whip-like head appendage of crustaceans used for sensing the environment.

Antibiotic Natural or synthetic compound capable of inhibiting or killing (susceptible) micro-organisms.

Antioxidant Substance added to crustacean feeds to prevent or delay breakdown of fats (lipids). Maintains food value of diet and increases shelf life.

Artemia Small brine shrimp which lays drought-resistant cysts. The cysts can be stored for several years and will hatch when placed in seawater to produce a nutritious nauplius larva – an ideal food for many crustacean larvae (Section 8.8.1).

Artificial impregnation Manual transfer of spermatophore from male to female, placing near or inserting spermatophore into genital structures of the female so that sperm contact with spawned eggs is inevitable (Section 7.2.2.5).

Artificial seawater A solution of salts, resembling to a greater or lesser extent that of natural seawater. Often used in inland *Macrobrachium* hatcheries.

Artificial tidelands Intertidal ponds used as shrimp or crab nurseries prior to release (Section 7.2.9).

Atterberg limit The moisture content at which a soil sample changes from one consistency to another. Liquid limit: the percentage moisture content at which a soil changes (with decreasing wetness) from liquid to plastic consistency. Plastic limit: the percentage moisture content at which a soil changes (with decreasing wetness) from plastic to semi-solid consistency.

Automatic feeder A device that dispenses feed pellets at pre-selected times, usually electrically operated.

Autotroph An organism that manufactures its own food from inorganic constituents using energy from light or chemical reactions (Section 8.4.5).

Backyard culture Trade press term for Australian 'hobbyist' crayfish growers.

Backyard hatchery Small family owned and run hatcheries.

Batch culture Method of culture in which organisms are stocked and grown without grading or culling until harvesting (Section 7.3.7), *q.v.* Continuous culture.

Battery culture Culture of crustaceans in individual compartments or multiple tanks in a controlled indoor environment.

Benthic organisms, benthos Organisms living on or in the bottom sediments.

Bentonite A very fine grained clay with a high shrink/swell potential, often used to seal ponds.

Berried Female crustacean carrying eggs under her abdomen during a period of incubation.

Billion (US) One thousand million, 1×10^9.

Binder Natural or artificial substance added to bind and hold finely ground dietary ingredients together when in water.

Biofilter, biological filter Part of a water treatment system in which there is a large surface area occupied by micro-organisms that oxidize dissolved organic matter, ammonia and nitrite to less harmful products.

Biomass The total quantity of living organisms or specific organisms in a defined body of water.

Bivalve Molluscs with an openable shell – oysters, clams, mussels.

Black mud A foul-smelling marine sediment, rich in hydrogen sulphide and organic content, typically occurring in poorly managed ponds and uncleaned tanks and pipework.

Blanched Processing term meaning precooked or parboiled, e.g. at 65°C for 15–20 seconds (Section 3.3.2).

Bloom See Algal bloom.

Blue-clawed male Large, dominant male *Macrobrachium rosenbergii* that has developed blue chelae (Section 7.3.7).

BP *Baculovirus Penaei*, a virus disease of shrimp.

Brackish water Seawater diluted with freshwater, for example in an estuary.

Breeding cycle A period between hatching and the first spawning of a given generation.

Broodstock Populations of maturing or mature and breeding animals.

Buffer (see also Alkalinity) A substance or substances which resist or counteract changes in the acid or alkali concentrations in water.

Bund Raised embankment separating two bodies of water.

Buster Colloquial term for a crab that has just started to moult (Section 7.10.9).
Butterfly shrimp A form of prepared, value-added shrimp (Section 3.2.3.1).

Cage culture Growing crustaceans in mesh cages either floating or staked to the bottom.
Canner Small-sized north American clawed lobster suitable for canning, also exported to Europe.
Cannibalism Consumption of one crustacean by others of the same species.
Carapace The one-piece shell structure covering head and thorax of crustaceans.
Caridea Taxonomic group (infraorder) of shrimp and prawns.
Carrying capacity The population of a given species that an area or volume will support without undergoing deterioration.
Cash crop Crustaceans grown for high sale value rather than for use as food locally.
Casitas, casas Cubanas Latin American term for artificial shelters used to attract young (casitas) or adult (casas) spiny lobsters.
Cast net Fine circular throwing net weighted at its circumference, attached at its centre to a thin rope which is held by the fisherman.
Catch crop A subsidiary population, often of a different species, grown between crops of the main cultured species to maximize revenue.
Chela(ae) The pincer claws of a crustacean.
Chloramphenicol One of several broad spectrum antibiotics used, often unwisely, in shrimp hatcheries.
Chlorine, chlorine solution Chemical disinfection agent available in form of powder (calcium hypochlorite) or liquid bleach (sodium hypochlorite solution).
Cholesterol Parent compound for manufacture of many steroids; an essential component of crustacean diets.
CL Carapace length, usually measured from the eye notch to the posterior mid dorsal margin of the carapace.
Clay Fine grained portion of a soil that can exhibit plasticity within a range of water contents and which exhibits considerable strength when air-dry.
Cobble Rock fragment usually rounded or smooth with average dimensions of eight to 30 cm. Also boulder – similar but larger rocks.
Colourmorphs Genetically determined, colour variations of a species, useful in identifying individuals.
Compaction Increasing the density and lowering the porosity of a soil by mechanical manipulation, essential in the construction of ponds.
Complete diet A diet capable of supporting good growth and survival throughout a specified culture phase, e.g. larval life or on-growing.
Continental climate Any climate in which the difference between summer and winter temperatures is greater than the average range for that latitude because of the influence of a large land mass.
Continuous culture, stock, cull On-growing method used for farming *Macrobrachium rosenbergii* in which the fastest growing animals are selectively harvested from the main population at intervals and replaced with new juveniles (Section 7.3.7), *q.v.* Batch culture.
Count A measure used to describe a size-graded crustacean product. For example the number of tails per pound or whole shrimp per kilogram (Section 3.3.1.2 and Appendix II).

Crackers, shrimp or *prawn* Shrimp/prawn-flavoured starch-based snack food.
Crawfish Alternative name for marine Palinuridae, also spiny and rock lobsters. Commonly but incorrectly applied to freshwater crayfish in the USA.
Crayfish Freshwater Astacidae. Occasionally but incorrectly applied to marine rock lobsters, e.g. in New Zealand.
Croquettes, shrimp or *prawn* Peeled and de-veined tails of shrimp/prawns prepared with a seasoned coating.
Cryo-preservation Specialized process for freezing microscopic organisms or gametes for long term storage in a dormant condition. Normal life functions are resumed on subsequent thawing.
Cull Partial harvest of (usually) largest individuals in a population.
Culture To grow an organism or population; a thriving population of micro-organisms.
CW Carapace width, a standard measure of crab size; other farmed crustaceans (shrimps, lobsters) are measured by carapace length.
Cyst Drought-resistant egg-like stage in the life of the brine shrimp *Artemia*. Produced instead of normal eggs in response to drying out of the shrimp's environment.

Dead spot Area or volume of pond bottom or water where circulation is minimal and where sedimentation and anaerobic conditions develop.
Decapsulation Removal of the outer shell of *Artemia* cysts by dissolution in chlorine solution (Section 8.8.1).
Denitrification The chemical reduction of nitrate to nitrogen by certain micro-organisms (see also nitrification).
Detritus Fragments of organic matter or other disintegrated material. Often forms a food resource for juvenile crustaceans, particularly crayfish.
Diatoms Single-celled planktonic plants (see also phytoplankton) covered with two overlapping porous shells of silica.
Dip A treatment or disinfection bath into which one or more animals are placed for a short time (1–60 mins).
Dip net Small, hand-held net on a wooden or wire frame, used for sampling.
DO Dissolved oxygen.
Domestication The adaptation of an organism for life in intimate association with man. Purposeful selection away from the wild type is implied (Section 12.10).
Double cropping Production of two crops from the same pond, simultaneously (polyculture) or alternately (crop rotation). One of the crops need not be an aquaculture product, e.g. rice or salt.
Drop net See Lift net.

Ecdysis The act of casting the external skeleton or shell of crustaceans (Section 2.4).
Ecosystem The interactions of communities of organisms and their physical environment.
EDTA A chemical chelation agent added to seawater during larval cultures to favourably adjust the availability of mineral ions.
Effluent Water which is discharged from a hatchery, farm or other industrial unit.
EIRR Economic internal rate of return (Section 10.5)
Electro-fishing, harvesting Application of an electric pulse to a specially adapted push net which makes shrimp jump out of the substrate and into the water column above the bottom of the pond.

El Niño, La Niña Unseasonable oceanic currents which strongly influence the occurrence of penaeid post-larvae off the coast of Ecuador. El Niño: warm current setting south along the coast which favours shrimp productivity in the wild and in farms (abundant wild post-larvae; rapid growth due to higher temperatures). La Niña: a cool northbound current with the reverse effects.

Embayment A shoreline indentation that forms an open bay which has been fenced or screened for aquaculture purposes (Sections 5.6.2 and 7.2.9).

Endemic Specific or indigenous to an area; applies both to farmed species and to diseases.

Enhancement See Stock enhancement.

Epibiotic Living organisms infesting the outer covering of an animal or plant, e.g. severe infestations of stalked protozoans which can smother crustacean larvae. Also epifauna, epiphyte.

Epipelagic Inhabiting oceanic water at depths not exceeding ca 200 m.

Estuary The lower reaches of a river influenced by ocean tides and mixing with seawater.

Etang French coastal lagoon.

Etiology Assignment or study of the causes of a disease.

Euryhaline Adaptable to a wide range of salinity.

Eutrophication Natural or artificial enrichment (fertilization) of water, usually characterized by excessive blooms of phytoplankton.

Exoskeleton The external shell or covering of a crustacean (Section 2.4).

Expatriate Person from another country; usually an employee, manager or consultant contributing special skills.

Extension service, worker Organisation or person forming the vital two-way link between the farm and the aid or research organisation.

Exuvium(a) The cast shell(s) of a crustacean.

FAO Food and Agriculture Organization.

Farm dam Man-made reservoir providing water for cattle or sheep in Australia, sometimes stocked with crayfish.

Fat Processing term sometimes used for mid-gut gland (see also hepatopancreas).

Fatty acids A group of straight-chain carbon compounds which form the building blocks of fats, oils, waxes and, in conjunction with other components, cell membranes.

FCR Food conversion ratio, usually measured as the weight of dry food fed to the weight of live animal produced.

Feasibility study Comprehensive evaluation of a proposal to farm crustaceans prior to making an investment decision (Section 9.5).

Fecundity The number of eggs produced by a female, commonly but incorrectly used to denote the number of viable larvae produced.

Feeding rate The amount of feed offered to crustaceans in a specified time. The amount may be given in several discrete doses.

Fertilizer A natural (e.g. manure) or chemical material added to water or soil to increase natural productivity.

Filter, biological Part of a water treatment system in which there is a large surface area occupied by micro-organisms that oxidise dissolved organic matter, ammonia and nitrite to less harmful products.

Filter, mechanical Part of a water treatment system which mechanically strains or collects suspended particulate material from the water.

Fish meal A dehydrated and often defatted ground/processed fish material, used in animal feed manufacture.

Flagellate alga Single-celled planktonic plant which swims by means of a whip-like flagellum.

Flake A soft form of ice used in processing delicate crustacean flesh.

Flock Fragments of bacterial or other biological growths sloughed off from the surfaces of biological filters and aquaculture tanks.

Foam fractionation, separation Water treatment methods for the removal of dissolved and colloidal organic material and bacteria from water, usually accomplished by inducing a counter current of water (downwards) and fine air bubbles (upwards) in a vertical cylinder. The resulting foam (foamate) is discarded (Section 8.4.3).

Food conversion ratio see FCR.

Forage To search for food; the plant material actually consumed by a grazing or detritus feeding crustacean (e.g. red swamp crayfish).

Fouling The assemblage of organisms that attach to and grow on underwater objects; also the deleterious accumulation of dissolved and particulate material in a body of water or pond bottom.

Freeboard The distance between the water surface and the top of the surrounding vessel or bund (pond embankment).

Frequency distribution An arrangement of data (often size measurements) grouped into classes, which shows the number of observations falling within each category.

Fuller's earth A variety of clay or marl containing 50% silica, used in treating batches of water for use in hatcheries.

Fyke nets Traps made of netting held under tension by a series of hoops.

Gastrolith 'Stomach stone': mass of calcium carbonate found each side of the cardiac stomach, especially in crayfish. Forms reservoir of calcium for shell remineralization after moulting.

Gearing ratio The relative proportion of loan capital to risk capital (Section 10.2.2).

Geothermal water Naturally warm water from below ground.

Gill net see Tangle net.

Glazing A thin layer of ice covering frozen crustaceans which gives them an attractive shiny appearance.

Gravid Female with eggs.

Ground water Water that has percolated through the soil into porous bed-rock.

Growout North American term describing the period for which crustaceans are grown from the post-nursery phase to market size. English equivalent is on-growing.

Habitat The locality, site and particular type of local environment occupied by an organism (Sections 4.2 and 5.6.2).

Haemacytometer Graduated glass microscope slide and cover slip used for counting blood cells, ideal for counting microalgae (unicellular phytoplankton).

Hapas Net cages suspended in the water between poles, not usually in contact with the bottom.

Hardness In practical terms, a measure of the amount of calcium and magnesium ions in water. Frequently expressed as a calcium carbonate equivalent.

Hatchery Building or tanks used for the maintenance and conditioning of broodstock and for the culture of their larvae. A nursery facility may be included.

Heat exchanger Shell and tube or parallel plate type devices for the transfer of heat from one fluid to another, used to recover waste heat or to boost the temperature of incoming water.

Heat pump Electrically driven device used to transfer heat from one area or body of water to another; acts like a domestic refrigerator transferring heat from inside to outside.

Hectare Metric unit of area, 10 000 square metres or 2.46 acres.

Hepatopancreas The major digestive gland in crustacea, also referred to as the mid-gut gland.

Hermaphrodite A species capable of producing male and female sex cells either synchronously or by changing from one sex to the other.

Heterogeneous growth Different rates of growth occurring in the same population of animals leading to a wide range of sizes at the time of harvesting (Section 7.3.7).

Heterotrophic organism Organisms that are dependent on organic matter for food (Section 8.4.5).

Hierachy Behaviourally maintained system of dominance among crustaceans.

HPV Hepatopancreatic parvo-like virus disease of shrimp.

HUFA Highly unsaturated fatty acid; more than four double bonds.

Husbandry The art of keeping organisms alive and fit (Section 8.7.1).

Hydraulic load The daily rate of flow of water through a given volume or over a given surface area of a filter, usually expressed as m^3 of water per m^3 or m^2 of filter respectively (e.g. $m^3\ m^{-3}\ d^{-1}$).

Hydrodynamic survey Gathering of information about tides, currents, direction and volume of flow of a body of water.

Hyperbolic bottomed tank Deep, approximately oval larviculture tank whose bottom has a hyperbolic shape over long and short axes to aid even dispersal of larvae and suspended food.

IHHN Infectious hypodermal and haematopoietic necrosis, a viral disease of shrimp.

Impregnation In crustaceans the deposition of a spermatophore on or in the female. Alternatively the treatment of porous soils with other material to prevent leaks.

Inbreeding Mating or crossing of individuals more closely related than average pairs in the population.

Incubation The holding of eggs between spawning and hatching.

Infestation The presence of organisms growing on or in a host species.

Inorganic Chemical compounds not containing carbon as a principle element (except carbonates).

Insolation The amount or duration of sunshine (Section 6.2.1.5).

Integrated, vertically integrated Applied to a crustacean farm which maintains or has control over its own support facilities, broodstock and feed supplies, hatchery, nursery, processing, marketing.

Inter-moult The period between each ecdysis during which the crustacean is hard shelled.

IQF Individually quick frozen, a processing step in the preparation of high quality crustaceans.

IRR Internal rate of return (Section 10.5).

Kill-chill To dip in ice water, then precook at 65°C for 15–20 seconds. See Blanched (Section 3.3.2).

Kosher Food fulfilling the requirements of Jewish law.

Kreisel A cylindrical vessel with a concave bottom and an upwardly spiralling water flow developed to keep clawed lobster larvae in suspension during culture (Section 7.8.7).

Lab-lab Name used in the Philippines to describe a dense mat of aquatic plant and micro-animal communities that develops on the bottom of ponds.

Larva(ae) Usually the planktonic, free-swimming stage of cultured crustaceans, although some larval stages only exist in incubated eggs.

Levee See Bund.

Lift net Net fixed to a circular, square or cross-shaped frame which is positioned beneath the water and lifted swiftly to catch shrimp and fish which settle on or above it. Sometimes attached to a wooden pole or fixed frame to assist in lifting, and sometimes baited.

Lime Calcium oxide, used as a disinfectant in ponds. Commonly but incorrectly applied to forms of calcium carbonate, such as powdered limestone, used to increase pH levels in ponds (Sections 7.2.6.5 and 8.3.5.1).

Liner Clay or plastic sheet applied to a vessel or pond to stop leaking or diffusion of acidic minerals into the water.

Lipid Name given to dietary fats.

Macrobenthos The larger organisms (from 1 mm upwards) living in or on the pond bottom.

Malachite green Analine dye effective in the control of external fungal and protozoan infections.

Management Planning and supervision of hatchery and farm operations; assessment and manipulation of water flow to maintain good water quality; assessment and manipulation of fertilization, feeding and water exchange to control phytoplankton density within desirable limits.

Mangrove A tidal salt marsh community dominated by trees and shrubs, mainly *Rhizophora* spp. If cleared for pond construction, the underlying soil is usually found to be strongly acidic (Section 6.3.3.5).

Marl A general term for calcareous clay or calcareous loam.

Maturation Ripening, the cell divisions by which gametes are produced, to enter a phase of reproductive competence.

Maturation facility, unit Part of a farm or hatchery for the production or conditioning of breeding crustaceans.

Megalopa Name given to a late stage crab larva between the zoeal and first crab stage.

Metabolite The product of any chemical change occurring in living organisms; waste metabolites from one species may be used as food by another.

Metamorphosis The marked change in form that occurs between life cycle phases during the development of crustaceans.

Micro-encapsulated diets Very small particles containing compounded ingredients and surrounded by a digestible coat.

Micro-particulate diets Very small particles of food manufactured for larvae.

Micro-wire tag, Microtagged A small length (1 mm × 0.25 mm dia.) of magnetized steel wire, etched with a binary code and injected into crustacean muscle issue to identify hatchery reared animals among wild stock during restocking and ranching trials. Detection involves passing fished animals through a sensitive metal detector.

Milk fish *Chanos chanos*; an important food fish grown extensively in ponds in south-east Asia.

Moist pellet A compounded diet with a moisture content of around 30%. May also contain fresh ingredients.

Monk A water control structure governing pond depth at a pond exit, and water flow rate at a pond entrance. Screens prevent escape of stock and entry of predators and competitors (Section 8.2.2.2). Same function as Sluice gate.

Monoclonal antibody A specific antibody derived from a culture of a single clone of cells.

Monoculture Culture of a single species, *q.v.* Polyculture.

Monosex culture The rearing of a single sex in an attempt to reduce size variability among harvested populations; employed in fish culture to avoid uncontrolled reproduction.

Moulting The act of casting the exoskeleton (Section 2.4) (see Ecdysis).

Mysid A small, swarming crustacean (e.g. *Neomysis integer*) commonly used as fish food in home display aquaria but also a good food for large crustacean larvae such as lobster larvae.

Mysis The stage between zoea and post-larva in the development of penaeid shrimp larvae.

n-3, n-6 Convention for indicating the position of the first double bond in an unsaturated fatty acid, counting from the carboxyl group.

Natural productivity, primary productivity The development of diverse aquatic communities based on phytoplankton growth in culture ponds. These provide food for grazing crustaceans, especially in extensive cultures, and may be enhanced by the addition of fertilizer.

Nauplius(lii) The first stage of crustacean larval development, often feeding only on internal yolk reserves.

Nitrification The aerobic bacterial oxidation of toxic nitrogenous metabolites, ammonia and nitrite to much less toxic nitrate. Occurs in biological filters, also nitrifying bacteria.

Nitrogen load The amount of nitrogenous waste (usually ammonia) that has to be oxidized to nitrate by a biological filter in a given time in order to maintain acceptable levels in the culture water.

NPV Net present value (Section 10.5).

Nursed juveniles See Nursery phase.

Nursery phase The culture of post-larvae from the time of metamorphosis to the time they are stocked in the on-growing ponds or released into the wild.

Ocean ranching See Ranching.

On-growing Growth to market size (see also Growout).

Orange claw male Fast-growing but sub-dominant male *Macrobrachium rosenbergii*, distinguished by large, orange claws (Section 7.3.7).

Organic load The amount of dissolved organic material carried in the water.

Overwintering Adults: Stocks of adults held throughout the cold season (sometimes at elevated temperatures) in order to provide broodstock before wild broodstock are available. Juveniles: Populations of juveniles grown at elevated temperatures through the cold season to provide partly-grown animals for on-growing. Useful in areas where there is only a single or short growing season. May also be called Nursed juveniles.

Ovigerous Carrying eggs (see also Berried).

Oxidative rancidity See Rancid.

Oxygenation Addition of oxygen to water.

Ozonation Addition of ozone to water to break down refactory organic molecules, oxidize waste metabolites and sterilize the water.

P & D See Peeled and deveined.

Paddle-wheel An electrically driven, floating device used to aerate and circulate water in ponds.

Paper-shell Recently moulted crustacean whose shell has started to harden and has turned leathery.

PB Pay-back period (Section 10.5).

Pcs Pieces, term used in the Far East for individual post-larvae or juveniles, e.g. 10 000 pcs means 10 000 juveniles.

Peeled and deveined A processing term describing the removal of shell and gut from headless shrimp.

Peeler A crab that is within one to 14 days of moulting (Section 7.10.9). In *Callinectes sapidus* the new shell can be seen forming beneath the old one.

Pellet Compounded feed rations of graded sizes containing about 10% moisture.

Pereopods The walking legs of crustaceans, commonly spelt pereiopods.

Permeability, coefficient of Measure of soil permeability to water measured in $m\ s^{-1}$ (Section 6.3.3.3).

pH A measure of the hydrogen ion activity (acidity) of water. Strictly the negative logarithm of the hydrogen ion concentration. Seawater has a pH of around pH 8.0–8.2, freshwater pH 6.0–7.5.

Phototactic Swims or moves towards light.

Phytoplankton Microscopic plants, usually single-celled, that grow suspended in the water.

PI Plasticity index (Section 6.3.2).

PL See Post-larva.

Plague Fungal disease of European and Australasian crayfish.

Pleopod The abdominal paddles or swimmerettes of a crustacean, also used to provide attachment for incubation of eggs.

Polychaete Marine worms; many are rich in the specific fatty acids and other nutrients required by broodstock crustaceans.

Polyculture The cultivation of two or more species in the same facility (pond), often but not necessarily at the same time.

Polyhedral inclusion body Resistant, crystalline stage/form of the virus *Baculovirus penaei*. Often appear triangular when viewed under light microscope; also known as occlusion bodies.

Post-larva The stage following the last planktonic larval stage in crustaceans. Frequently the time of transition between planktonic and benthic existence.

ppm. Parts per million.

ppt. Parts per thousand; commonly used to express the salt content or salinity of saline water.

Prawn Natantian decapod crustacean, taxonomically equivalent to shrimp but in use the terms differ between countries. Prawn is increasingly used to describe *Macrobrachium* species while shrimp refers to penaeids.
Prefeasibility study A preliminary assessment of a culture proposal designed to determine the need for or scope of a full feasibility study (Section 9.4).
Premix A mixture of vitamins, minerals and other micro-ingredients which is added to a compounded diet during manufacture.
Pre-operating expenses Costs of feasibility studies, construction and legislative requirements.
Processing The preparation of crustaceans and crustacean products for the market.
Productivity See Natural productivity.
Propeller-aspirator pump An electrically powered, floating device for drawing air into a stream of water discharged at an angle towards the pond bottom. Serves a similar function to a paddle-wheel.
Protanderous hermaphrodite Matures and functions first as a male, later as a female (Section 4.3.2).
Protein skimming See Foam fractionation.
Protozoea The stage between the nauplius and the mysis in the development of penaeid shrimp.
PUD Peeled and un-deveined; see Peeled and deveined.
Pueruli Post-larval stage in the development of spiny lobsters.
PUFA Poly unsaturated fatty acid – two or more double bonds.
Purge The removal of unsightly gut contents by holding the crustaceans without food in clean water for one to two days.

Quarantine Enforced isolation of organisms which are, or may be infected, to prevent transmission of diseases to other organisms or the environment.

Raceway, D ended Rectangular shaped trough with rounded ends, sometimes with a short central barrier to assist creation of water circulation.
Ranching The release of hatchery reared juveniles into the wild, or into enclosed or modified wild habitat.
Rancid The breakdown of lipids and their constituents through poor storage conditions which severely reduces their dietary value.
Recirculation system A culture system incorporating a water treatment unit(s) through which the water continuously passes. A quantity of new water is added periodically or continuously (Section 8.4.4).
Red tides Dense blooms of algae occurring in coastal regions, not always red in colour but often harmful to aquatic life and man.
Refactory organic compound Dissolved organic matter not readily oxidized or broken down by biological filtration (Sections 8.4.3 and 8.6).
Refractometer Optical device for measuring the refractive index of a drop of liquid. Some are specifically calibrated for the measurement of salinity.
Release The act of transferring hatchery reared crustaceans to the wild.
Respiration The acquisition of energy from the oxidation of organic molecules in living cells. During aerobic respiration oxygen and food are consumed and carbon dioxide and ammonia are produced.
Restocking Releasing cultured or wild caught juveniles into the wild or culture environment (Section 5.6.1) (see also Ranching).

Risk capital Equity (Section 10.2.2).
ROI Return on investment (Section 10.5).
Rostrum The forward pointing spine between the eyes of a crustacean.
Rotating biological contactor, biodisc or *biodrum* Slowly rotating discs or drums in the water treatment unit of some recirculation systems. They form effective biological filtration units (see Biofilter).
Rotenone Selective fish poison often applied in the from of dry derris root which contains about 5% active ingredient (Section 8.3.5.2).
Rotifers A group of microscopic aquatic animals whose anterior end bears tufts of cilia used for feeding and locomotion. Important living food resource for larvae, frequently cultured in crustacean hatcheries.
Salinas Shallow ponds built for the production of sea salts by evaporation of seawater.
Salinity A measurement of the total mineral ions content of seawater, often expressed as parts per thousand (ppt.).
Sanitary fish Herbivorous fish introduced into a culture pond to limit the growth of plants.
Saponin see Teaseed cake.
Sashimi Japanese speciality raw seafood.
Secchi A circular plate 20 cm in diameter, the upper surface of which is divided into four quadrants painted alternately black and white. When lowered into a pond on a graduated rope the point at which it disappears provides a measure of turbidity or plankton density (Section 8.3.4.2).
Seed Post-larval or juvenile crustaceans used for on-growing or release.
Seine net, seining Long fishing net which is suspended in the water by floats and is drawn through the pond to encircle the catch.
Semilleros Ecuadorian catcher of post-larval and juvenile shrimp.
Settlement The transition from a planktonic to a benthic existence, usually at or near the first post-larval stage of development.
Shedding trays Shallow trays used to hold adult crustaceans until they moult for the production of soft-shell crustacean delicacies.
Sheepsfoot roller Heavy toothed roller for compacting clay soils during pond construction.
Shellfish Term embracing both aquatic molluscs and crustaceans.
Shell-on Processing term meaning with the outer skin (exoskeleton) left in place, *q.v.* shelled.
Shigueno tanks Circular tanks with a sand-covered, false floor used for the super-intensive culture of kuruma shrimp in Japan (Section 7.2.6.6)
Shrimp Natantian decapod crustacean, commonly *Penaeus* spp. (see prawn).
Skewness Deviations from symmetry. Typically used to describe the distribution of sizes in a population in which the majority of individuals are clustered to the left or right of the mean size. The reason the mean lies outside the bulk of the population is due to the disproportionate influence of individuals that are either extremely large (positive skew) or small (negative skew).
Sluice gate A water control structure governing pond depth at a pond exit, and water flow rate at a pond entrance. Screens prevent escape of stock and entry of predators and competitors (Section 8.2.2.2). Same function as a Monk.
Soft shelled A newly moulted crustacean prior to re-mineralisation of the exo-skeleton. Also a pathological condition in an inter-moult animal.
Sourcing American term for the catching of broodstock from the wild.

Spat Young (seed) bivalves (oysters, clams, mussels) that have settled.

Spawning The natural extrusion of eggs from a female, not to be confused with the hatching of larvae from eggs being incubated by a female.

Specific surface area The surface area of the material in a biofilter available for bacterial colonization, usually expressed as m^2 of surface per m^3 of filter volume (e.g. $m^2\ m^{-3}$).

Spermatophore A packet or capsule of spermatozoa transferred to the female during mating.

Spray-dried algae Larvae food product made by spraying a dense culture of micro-algae into a stream of warm air to evaporate the water.

Standing crop Generally the total weight of animals of all sizes on a farm, or in a pond.

Static water tanks Culture tanks in which the water is changed in batches, *q.v.* continuous flow tanks.

Stock enhancement To improve or increase stocks of fishable crustaceans; in the context of this book by release of hatchery reared juveniles into an existing fishery.

Stocking density The numbers of crustaceans stocked into a body of water per unit volume or bottom area.

Stratification The division of a water body into roughly horizontal layers of different temperature, salinity or oxygen content.

Sub-sand filter Water intake point for a hatchery located beneath a sandy beach in order to obtain a filtered/pretreated water supply (Section 8.4.1). If gravity fed then known as a beach well.

Substrate The material constituting the bottom of the pond, lagoon or bay on or in which the farmed species lives.

Summerlings Juvenile crayfish of one summer old (Section 10.9.2.1).

Surimi Mechanically de-boned, washed and stabilized white fish flesh, which is flavoured and extruded to form analogue products such as 'shrimp' tails and 'crab' sticks (Section 3.3.6).

Sushi Japanese speciality dish of raw seafood on flavoured rice.

Tagging Various methods for the identification of individual or cohorts of crustaceans among conspecifics.

Tambak The Indonesian name for coastal brackish water ponds used in traditional extensive fish culture. Many are being deepened for shrimp culture.

Tangle net Long fine net suspended vertically in the water to trap fish by entangling them; also used to catch spiny lobsters and marine shrimp.

Tax holiday Period of reduced corporate tax liability applied during the early years of a new commercial venture as an incentive to investment.

Teaseed cake Selective fish poison. Contains 10–15% active ingredient, saponin, and is the residue after oil extraction from the seeds of *Camellia* (Section 8.3.5.2)

Telson The central, hindmost segment of the crustacean abdomen which, together with the two leaf-like uropods on either side, make up the tail fan.

Tempura Japanese speciality dish comprising small pieces of seafood or other food deep-fried in batter.

Thelycum The genital region of female shrimp, may be setose (open thelycum) or set within a pouch (closed thelycum).

Tidal impoundments Intertidal areas enclosed by low walls, used in Japan for shrimp culture (Amakusa pens, Section 7.2.6.4) and as nursery areas for juvenile shrimp prior to release into the wild (Artificial tidelands, Section 7.2.9).

Tolerance A measure of the ability of a crustacean to survive or grow in the presence of specified adverse conditions.

Total hardness See Hardness.

TL Total length. Length of a crustacean measured from either the tip of the rostrum or from the posterior margin of the orbit to the tip of the telson or, in some cases, the extremities of the uropods. An imprecise measure, rarely properly defined and subject to errors caused by abdominal flexibility.

Transgenic manipulation Transfer of genetic material from one organism to another.

Transplantation The removal of species from one geographic location to another, often outside the range of natural distribution.

Turnkey operation A complete farm package often including management and investment involvement (Section 9.5.4).

Umbrella net Umbrella-shaped lift-net (see Lift net).

Urea A nitrogenous compound used for the fertilization of ponds.

Uropod See Telson.

UV sterilization Ultra-violet irradiation of water to kill or inhibit bacterial development. Only likely to be effective when the water is free of particulate material which the light cannot penetrate.

Vaccine A preparation of nonvirulent disease organisms or immunogens which still retain the capacity to stimulate the production of antibodies or resistance.

Value-added Processing or presenting crustacean flesh in a more attractive way to increase its market value (Section 3.2.3.1).

Venture capital The portion of capital raised by the sale of equity (Section 10.3).

Venturi An orifice in a pipe which, because of the vacuum created in it by the swift passage of water through the pipe, sucks air into the pipe.

Vivier truck Vehicular transport equipped with facilities for the live transportation of adult crustaceans.

Water exchange rate The partial, exponential replacement of water by a flow of new water into a culture system. Often confused with water flow rate, which only describes the time it would take to fill or empty a vessel.

Water quality A vague but useful term to describe the ability of water to support the cultivated species.

Zeolite Naturally occurring hydrated sodium aluminosilcate mineral. Sometimes applied to intensive shrimp ponds for its ability to bind and remove toxic metabolites (e.g. ammonia) from water. Most effective in freshwater, however.

Zoea The free-swimming larval stages of caridean prawns and lobsters.

Zooplankton Microscopic animal life suspended in the water column.

References

Holmes S. (1979) *Henderson's dictionary of biological terms*. Longman London.

Rosenthal H., Hilge V., Ackefors H., Bucke D., Stewart J.E. & Castell J.D. (1990) *A proposed glossary on biological and technical terms relevant to aquaculture: First draft*. Int. Counc. Explor. Sea CM 1990 F:21 (mimeo).

Index

Tables in bold, *Figures in bold italic*, (Plates in brackets).